Advances in 21st Century Human Settlements

Indexed by SCOPUS

This Series focuses on the entire spectrum of human settlements – from rural to urban, in different regions of the world, with questions such as: What factors cause and guide the process of change in human settlements from rural to urban in character, from hamlets and villages to towns, cities and megacities? Is this process different across time and space, how and why? Is there a future for rural life? Is it possible or not to have industrial development in rural settlements, and how? Why does 'urban shrinkage' occur? Are the rural areas urbanizing or is that urban areas are undergoing 'ruralisation' (in form of underserviced slums)? What are the challenges faced by 'mega urban regions', and how they can be/are being addressed? What drives economic dynamism in human settlements? Is the urban-based economic growth paradigm the only answer to the quest for sustainable development, or is there an urgent need to balance between economic growth on one hand and ecosystem restoration and conservation on the other – for the future sustainability of human habitats? How and what new technology is helping to achieve sustainable development in human settlements? What sort of changes in the current planning, management and governance of human settlements are needed to face the changing environment including the climate and increasing disaster risks? What is the uniqueness of the new 'socio-cultural spaces' that emerge in human settlements, and how they change over time? As rural settlements become urban, are the new 'urban spaces' resulting in the loss of rural life and 'socio-cultural spaces'? What is leading the preservation of rural 'socio-cultural spaces' within the urbanizing world, and how? What is the emerging nature of the rural-urban interface, and what factors influence it? What are the emerging perspectives that help understand the human-environment-culture complex through the study of human settlements and the related ecosystems, and how do they transform our understanding of cultural landscapes and 'waterscapes' in the 21st Century? What else is and/or likely to be new vis-à-vis human settlements – now and in the future? The Series, therefore, welcomes contributions with fresh cognitive perspectives to understand the new and emerging realities of the 21st Century human settlements. Such perspectives will include a multidisciplinary analysis, constituting of the demographic, spatio-economic, environmental, technological, and planning, management and governance lenses.

If you are interested in submitting a proposal for this series, please contact the Series Editor, or the Publishing Editor:

Bharat Dahiya (bharatdahiya@gmail.com) or
Loyola D'Silva (loyola.dsilva@springer.com)

Rosemary Kasimba · Innocent Chirisa
Editors

Disaster Risks and Hazards

Practice for Resilience- Case for Zimbabwe

Springer

Editors
Rosemary Kasimba
Department of Demography Settlement and Development
University of Zimbabwe
Harare, Zimbabwe

Innocent Chirisa
The Vice Chancellor's Office
Zimbabwe Ezekiel Guti University
Bindura, Zimbabwe

ISSN 2198-2546 ISSN 2198-2554 (electronic)
Advances in 21st Century Human Settlements
ISBN 978-981-96-4576-3 ISBN 978-981-96-4577-0 (eBook)
https://doi.org/10.1007/978-981-96-4577-0

This Springer imprint is published by the registered company Springer Nature Singapore Pte Ltd.
The registered company address is: 152 Beach Road, #21-01/04 Gateway East, Singapore 189721, Singapore

Preface

The twenty-first century has experienced a number of disasters that resulted in the loss of lives and property; hence, they are a threat to sustainable development. Disaster risk management efforts and practices have failed to arrest the underlying drivers of increased global risk. No country is immune to disaster risk; thus, resilience has become a critical component of disaster management. It provides an opportunity to confront the social-ecological foundations of risk. However, it has been vaguely conceptualised and lacks a concrete approach to operationalization. Some developing countries are gradually coming to realise that preparedness, hazard mapping, prediction, risk and vulnerability assessment, and planning are critical for increasing resilience and reducing losses following disasters. A paradigm shift is required from reactionary to proactive measures in disaster management as the later focuses more on increasing the resilience of the community. The project seeks to examine how developing countries can employ the concept of resilience in disaster risk reduction with a special emphasis on the priorities of Sendai Framework for Disaster Reduction.

Harare, Zimbabwe — Rosemary Kasimba

Contents

Consilience for Resilience for Disaster Risks and Hazards: An Overview ... 1
Innocent Chirisa and Rosemary Kasimba

Understanding Social Networks and Community Resilience to Pandemics in Zimbabwe ... 9
Rosemary Kasimba and Solomon Muqayi

Community Based Resilience to Climate Change in Chimanimani District, Zimbabwe ... 25
Chipo Mudavanhu, Nelson Chanza, and Rosemary Kasimba

Geo-Spatial Science Use for Disaster Risk Management ... 47
Samuel Kusangaya, Pathias P. Bongo, Beatrice Chikanda, Roy Boy Chiomba, and Albert Maipisi

Linking and Mapping Resilience to Strategies for Management of Disasters and Risks in Zimbabwe ... 73
Innocent Chirisa and Gift Mhlanga

The Ubiquity of Politics and Its Effect on Disaster Resilience in Zimbabwe ... 89
Albert Maipisi, Oripha Chimwara, Pathias Paradzayi Bongo, and Samuel Kusangaya

Gender and Disaster Resilience ... 105
Sandra Bhatasara, Manase Kudzai Chiweshe, and Naume Zorodzai Choguya

Cyclone Damage Assessment Using High Resolution Satellite Imagery, Drone Technology and GIS: Lessons Learnt ... 123
Samuel Kusangaya, Mhosisi Masocha, Isaiah Gwitira, Pianos Gweme, Victor Mukungunurwa, Pathias P. Bongo, Albert Maipisi, and Dominic Mazvimavi

The Resilience of Child Protection Systems During the COVID-19 Pandemic: Perspectives of Social Work Practitioners in Zimbabwe 147
Tanatswa Silvanus Chineka, Edward Muzondo, and Mulwayini Mundau

Zimbabwe's Disaster Preparedness, Readiness and Response: Juxtaposing the COVID-19 Pandemic to the 2019 Cyclone Idai 159
Tapiwa Musasa and Innocent Chirisa

COVID-19 Insights on Geriatric Care Systems' Vulnerabilities and Utility 175
Felistas R. Zimano

Harnessing Local Knowledge for Resilience Building: The Case Study of Lower Muzarabani 187
Vincent Chenzi and Mathew Douglas Chihobo

The Impact of Tropical Cyclones on Human Security: A Case of Mozambique and Zimbabwe (2000–2019) 205
Samuel Banda and Lucky Bassie Bangidza

Natural Resource Governance and Disaster Risk Management in Zimbabwe: Lessons from Past Experiences 235
Sifiso Mpofu, Gay Tapiwa Gweshe, and Barbra Maiwasha

Hurdles to Achievement of Climate Justice in Southern Africa and the Search for Possible Solutions 253
Kudzai Chatiza and Tariro Nyevera

Nuancing Urban Informality in Light of Policy to Contain COVID 19 in Zimbabwe 269
Mervis Chiware and Bernard Nkala

The Future of Managing Disasters and Risks in Zimbabwe and Beyond 285
Gift Mhlanga and Innocent Chirisa

Reviewers and Contributors

About the Reviewers

Alessandra Jerolleman is Associate Professor and Director of the Doctoral Program in Emergency Management at Jacksonville State University's Emergency Management Department. She is a community resilience specialist and applied researcher at the Lowlander Center, as well as a co-founder of Hazard Resilience, the USA-based consultancy providing leadership and expertise in disaster recovery, risk reduction, and hazard policy. In 2019, she published a book titled: *Disaster Recovery Through the Lens of Justice,* and recently released a follow-up edited volume *Justice, Equity, and Emergency Management.* Dr. Jerolleman is one of the founders of the Natural Hazard Mitigation Association (NHMA) and served as its Executive Director for its first seven years. Dr. Jerolleman is a subject matter expert in climate adaptation, hazard mitigation, disaster recovery, and resilience with a long history of working with the public, private, and nonprofit sectors as well as tribal nations. She is involved in various aspects of planning and policy and the national and local levels, including participation in several workshops each year. Her experience includes the following: conducting independent research on disaster risk reduction and hazard mitigation for various organizations such as the National Wildlife Foundation; working as the lead grant writer and emergency planner for the First Peoples' Conservation Council, through her role with the Lowlander Center on coastal community resettlement; community-based resilience planning across the USA; working with Save the Children USA along the Gulf Coast following hurricanes and tornadoes, on a resilience initiative around children's needs in emergencies; hazard mitigation planning at the local, state, and campus levels; community education and outreach regarding mitigation measures and preparedness; development of collaborative networks and information sharing avenues among practitioners; and delivery of training and education to various stakeholders. Dr. Jerolleman speaks on many topics including: Just recovery; hazard mitigation and climate change; campus planning; threat, hazard, and vulnerability assessments; hazard mitigation planning; protecting children in disasters; and public/private partnerships.

Alfredo González-Ruibal, Ph.D., Institute of Heritage Sciences, Spanish National Research Council, alfredo.gonzalez-ruibal@incipit.csic.es. Alfredo González-Ruibal is a senior researcher at the Institute of Heritage Sciences of the Spanish National Research Council. He specializes in the archaeology of the contemporary past and African archaeology, with a particular focus on conflict and resistance. He has conducted fieldwork in Spain, Ethiopia, Somaliland, Sudan, and Djibouti. Recent books include An Archaeology of the Contemporary Era (Routledge, 2019) and The Archaeology of the Spanish Civil War (Routledge, 2020).

Chris Shisanya Kenyatta University, Department of Geography, E-mail: shisanya.christopher@ku.ac.ke; chris.shisanya@gmail.com. Chris Shisanya is currently Full Professor in the Department of Geography at Kenyatta and Fellow of STIAS (South Africa) and AMU (France). He has won over 50 national and international research grant awards in the last twenty years. He has over 100 peer-reviewed journal publications and 10 book titles to his credit. He serves on Editorial Boards of high impact Scientific Journals. His key areas of competence include: research designs, quantitative data analysis using SPSS and R softwares, tracer study designs and implementation of the same, qualitative data analysis using Nvivo software, implementation of field data collection, and supervision of early career researchers.

Dr. Susan Spierre Clark is Assistant Professor and the Director of the Sustainability Leadership Master's Program in the Department of Environment and Sustainability at the University at Buffalo. In these positions, Dr. Clark teaches classes on the topics of sustainability and resilience, mentors sustainability graduate students, and leads externally funded projects on topics related to community resilience and climate change that have been funded by the US Department of Defense, New York State, and the National Science Foundation. Dr. Clark is also heavily involved in community outreach. She currently is Chair of Erie County's Community Climate Change Task Force (C3TF) and takes a leadership role in OUTSTEPS Lower Great Lakes, which is a regional network of scholars, practitioners, and other stakeholders working together to increase the sustainability and resilience across the region in a collaborative, holistic, and inclusive manner.

Professor Kibon is a specialist in Geography and Environmental Management and Risk Analysis. He has about 17 years of work experience in Teaching, Research, and supervision. He teaches database management, application of remote sensing, and GIS in natural resources management and climate change adaptation. Kibon has participated in a number of research activities such as Post-Disaster Need Assessment (PDNA) in 2012 flood scenario in Nigeria, Community Vulnerability Assessment (VCA), Search and Rescue Operation in Lagos International Market Fire. Professor Kibon also attended many workshops on disaster risk management such as regional workshop on harmonization of disaster risk data and information management for risk-informed and sustainable development in the West African Sahel Region in Dakar, Senegal, Ghana, Madagascar, and Niger. He is a member of Association of Nigerian Geographers (ANG), Member of International GIS Focus Group for

Environmental Management, Member of Environmental Management Association of Nigeria (EMAN), Member of International Association for Impact Assessment (IAIA), Member of American Association of Geographers (AAG), and Associate Editor of both internal and international journals.

Contributors

Samuel Banda Zimbabwe National Defence University, Harare, Zimbabwe

Lucky Bassie Bangidza Harare, Zimbabwe

Sandra Bhatasara Faculty of Social And Behavioural Sciences, University of Zimbabwe, Harare, Zimbabwe

Pathias P. Bongo Bindura University of Science Education, Bindura, Zimbabwe;
Department of Geography, Geospatial Sciences and Earth Observation, University of Zimbabwe, Harare, Zimbabwe

Pathias Paradzayi Bongo Women's University in Africa, Harare, Zimbabwe

Nelson Chanza University of Johannesburg, Johannesburg, South Africa

Kudzai Chatiza Development Governance Institute (DGI), Harare, Zimbabwe

Vincent Chenzi Department of Peace, Security and Society (DPSS), University of Zimbabwe, Harare, Zimbabwe

Mathew Douglas Chihobo History Department, University of Zimbabwe, Harare, Zimbabwe

Beatrice Chikanda Department of Geography, Geoinformation Science and Earth Observation, University of Zimbabwe, Harare, Zimbabwe

Oripha Chimwara University of Zimbabwe, Harare, Zimbabwe

Tanatswa Silvanus Chineka School of Health and Social Care, University of Essex, Colchester, UK;
African Center for Migration and Society, University of Witwatersrand, Johannesburg, South Africa

Roy Boy Chiomba Department of Geography, Geoinformation Science and Earth Observation, University of Zimbabwe, Harare, Zimbabwe

Innocent Chirisa Office of the Vice Chancellor, Zimbabwe Ezekiel Guti University, Bindura, Zimbabwe;
Department of Urban and Regional Planning, University of the Free State, Bloemfontein, South Africa

Mervis Chiware University of Zimbabwe, Department of Governance and Public Management, Harare, Zimbabwe

Manase Kudzai Chiweshe Faculty of Social And Behavioural Sciences, University of Zimbabwe, Harare, Zimbabwe

Naume Zorodzai Choguya Faculty of Social And Behavioural Sciences, University of Zimbabwe, Harare, Zimbabwe

Pianos Gweme Department of Geography, Geospatial Sciences and Earth Observation, University of Zimbabwe, Harare, Zimbabwe

Gay Tapiwa Gweshe University of Zimbabwe, Harare, Zimbabwe

Isaiah Gwitira Department of Geography, Geospatial Sciences and Earth Observation, University of Zimbabwe, Harare, Zimbabwe

Rosemary Kasimba Department of Urban and Regional Planning, University of the Free State, Bloemfontein, South Africa;
University of Zimbabwe, Harare, Zimbabwe;
Rhodes University, Grahamstown, South Africa

Samuel Kusangaya Department of Geography, Geoinformation Science and Earth Observation, University of Zimbabwe, Harare, Zimbabwe;
Department of Geography, Geospatial Sciences and Earth Observation, University of Zimbabwe, Harare, Zimbabwe

Albert Maipisi Women's University in Africa, Harare, Zimbabwe

Barbra Maiwasha University of Zimbabwe, Harare, Zimbabwe

Mhosisi Masocha Department of Geography, Geospatial Sciences and Earth Observation, University of Zimbabwe, Harare, Zimbabwe

Dominic Mazvimavi University of the Western Cape, Cape Town, South Africa

Gift Mhlanga Freelance Researcher, Chipinge, Zimbabwe;
University of Zimbabwe, Harare, Zimbabwe

Sifiso Mpofu University of Zimbabwe, Harare, Zimbabwe

Chipo Mudavanhu Higherlife Foundation, Harare, Zimbabwe

Victor Mukungunurwa Zimbabwe National Geospatial and Space Agency, Harare, Zimbabwe

Mulwayini Mundau Department of Social Work, University of Lesotho, Roma, Lesotho

Solomon Muqayi University of Zimbabwe, Harare, Zimbabwe

Tapiwa Musasa Catholic University of Zimbabwe, Harare, Zimbabwe

Edward Muzondo Department of Social Work, Eswatini Medical Christian University, Mbabane, Eswatini

Bernard Nkala Bindura University of Science Education, Department of Economics, Harare, Zimbabwe

Tariro Nyevera Development Governance Institute (DGI), Harare, Zimbabwe

Felistas R. Zimano Great Zimbabwe University, Masvingo, Zimbabwe

Consilience for Resilience for Disaster Risks and Hazards: An Overview

Innocent Chirisa and Rosemary Kasimba

1 Introduction

The twenty-first century has experienced several disasters that resulted in the loss of lives and property worth millions of dollars hence they are a threat to sustainable development. These disasters have "adversely affected humans since the dawn of our existence" [7: 1]. They pose a serious problem across the world and are a threat to the achievement of sustainable development goals. They pose serious impacts that are diverse i.e. loss of life, injury, destruction of property, among other things. Disasters can also cause social and economic disruption and damage to the environment [16]. Furthermore, the emergence of a global economy makes it more and more difficult to contain the consequences of any disaster within one country's borders hence the need to dwell on disaster risk resilience is very critical as it focuses on building communities' resilience to response positively to any disaster [7]. Current disaster risk mgt. efforts and practice have failed to yield desired results. Thus, the underlying drivers of disaster risk have been on the increase worldwide. No country is immune to disaster risk disk thus resilience has become a critical component of disaster risk management. It provides an opportunity to confront the social-ecological foundations of risk. However, it has been vaguely conceptualised, and lacks a concrete approach to operationalization. Some developing countries are gradually coming to realise that preparedness, hazard mapping, prediction, risk and vulnerability assessment and planning are critical for increasing resilience and reducing losses. A paradigm shift is required from reactionary to proactive measures in disaster management as the later focuses more on increasing the resilience of the community. The project

I. Chirisa · R. Kasimba (✉)
Department of Urban and Regional Planning, University of the Free State, Bloemfontein, South Africa
e-mail: rkasimba@gmail.com

I. Chirisa
Office of the Vice Chancellor, Zimbabwe Ezekiel Guti University, Bindura, Zimbabwe

R. Kasimba and I. Chirisa (eds.), *Disaster Risks and Hazards*, Advances in 21st Century Human Settlements, https://doi.org/10.1007/978-981-96-4577-0_1

seeks to examine how developing countries, particularly, Zimbabwe can employ the concept of resilience in disaster risk reduction with a special emphasis on the priorities of Sendai Framework for Disaster risk Reduction.

Zimbabwe is a landlocked country located in Southern Africa. According to Zimbabwe Statistics Agency (ZimStat) [19], it has total population of 13,061,239. Zimbabwe has one rainy season that spans from November to March. It receives relatively low rainfall, with an annual mean of 657 mm and an annual range of 400–1200 mm [10: 590]. Zimbabwe largely has two seasons, rainy and dry seasons. Rainy season commence from November to March while April to October season is dry. Zimbabwe is prone to several disasters. The most common hazards in the country are Hydro-meteorological (droughts and floods), geologic (earthquakes and landslides), biological (gastro-intestinal tract infections and HIV and AIDS) and pandemics such as COVID-19, technological road traffic accidents, warfare and land degradation (deforestation, veldt fires and gold panning) [10: 590]. Zimbabwe was hit hard by several disasters such as environmental, pandemics, outbreaks, accidents, and fires among others. The country, just like any other country has learnt lessons from COVID-19 and Tropical Cyclone Idai regarding ways and methods to enhance the ability of people to bounce back forward should there be any disaster. In addition, The country has also employed several mechanism for disaster risk reduction such as preparedness, mitigation and resettlement scheme.

There are natural and man-made disasters. Some natural disasters are also categorised under man-made for example, flooding. According to Bhaiseni [3: 16] Zimbabwe has experience both natural and human-induced disasters. These include flood, drought, tropical cyclone, pests etc., which are influence largely by climate change. For instance, tropical cyclone that occurred in March, 2019 which ravaged three countries namely: Zimbabwe, Mozambique and Malawi are typical examples.

The cyclone left more than 400 people dead and many others missing. Infrastructure, agricultural produce and schools were affected. In February 2017, Tsholotsho District which is located in the Southern part of the country was hard hit by cyclone Dineo. According to Hills and Nhamire [8], Cyclone Dineo damaged infrastructures, destroyed crops and killed more than 5 people. Beginning from ~~of~~ March 2020, COVID-19 Pandemic came in which has affected the performance of the economy, causing psychological trauma, deaths and impacted seriously on people's social and economic livelihoods. In response, individuals and societies alike have made many attempts to minimize their exposure to the consequences of these disasters, developing measures to minimizes initial impact and post-disaster response and recovery needs [7: 1]. Regardless of its wealth or influence, no county, is advanced enough to be fully immune from disasters' negative effects.

This book focuses on enhancing disaster resilience in Zimbabwe and also to examine how local population and authorities are managing disaster risks across the country. It contains perspectives of scholars from various disciplines such as sociology, Social Work, Public Governance, Political Science, Rural and Urban Planning, Geo-Information Systems, Geography and Environmental Management, Economics and Disaster Management among others. These have examined communitys' resilience in the context of disaster risk reduction with a view to providing models or policies that can be adopted to enhance the resilience of communities in the country in order to response to different types of disasters. Over the past two decades or so, the UN International Decade for Natural Disaster Reduction (IDNDR), the 1994 Yokohama Conference and the 2004 World Conference on Disaster Reduction (WCDR) have contributed to a significant shift in disaster management towards a more comprehensive understanding of the underlying causes of hazards and the development of new strategies for managing disaster risk [15: 14]. Coupled with the increasing number of disasters causing severe damage to property, massive death, and worsening poverty among others, developing resilient communities and organizations is very significant especially if sustainable development Agenda 2063 is to be achieved.

There are many factors that make communities more vulnerable to disasters ranging from political instability, poverty, lack of technology to monitor and provide Early Warning System poor governance, poverty and poor infrastructure among others. With COVID-19 most of the governments were caught unaware and reactionary measures, though saved life, they were not successful, as massive deaths were registered across in many countries. Furthermore, the advent of technology is becoming one of the vehicles to reduce negative impacts of disasters, by detecting and detecting the occurrence of some of the disasters to enable prepare for the disasters. However, there are scenarios where predicted disaster wreak havoc due to poor preparedness, ignorance and lack of resources. These issues have all been discussed in detail in the book and authors suggested measures that can be formulated, adopted and implemented to increase resilience of communities to disasters. The book also brought to the fore, one of the critical aspects of disaster resilience that require local communities, with external help, to be resilient. Thus, authors also pointed out the role of social networks, indigenous knowledge systems and local education in enhancing community resilience to disasters. Disasters are very complex and therefore reducing their impact require a number of stakeholders to work hand and hand. This is the major focus that the book is emphasizing.

2 Defining Disasters and Related Dynamics

Al-Madhari and Keller [1] note that it is not feasible to formulate a universally acceptable definition of disaster that will satisfy all practitioners, but common and agreed upon definitions must be formulated in the various fields and areas concerned with disasters, and where there exist recognizable, common sets of objectives. However, most of the definitions reflect the following characteristics: disruption to normal patterns of life. Such disruption is usually severe and may also be sudden, unexpected, and widespread, with attending effects such as loss of life, injury, hardship, adverse effect on health, social structure such as destruction of or damage to public systems, buildings, communications and essential services. This book is guided by the definition given by Amlôt and Carter [2: XIX] that describes a disaster as an event, natural or human-induced sudden/unexpected, that impacts with such severity that is beyond the coping capacity of local affected community/society. And therefore requires external assistance.

Again, there is no single authoritative definition of community resilience to disasters, a guiding definition, is one that captures most of the aspects provided by several scholars. According to Parker [13: 1] resilience is not a simple re-branding but is a concept that goes well beyond mitigation to embrace adaptation, change and transformation. Parker [13: 1] noted that "some degree of general consensus is emerging that resilience is about attributes (economic, social, informational) or capacities across a number of dimensions (disaster cycle stages)", leading to groups of components of resilience indicators (the use of 'Cs' or capitals, as in the '5Cs' such as physical, natural, economic, social and infrastructural capacities and the '4R's that are attributes, qualities or properties of resilience: rapidity, robustness, redundancy and resourcefulness). According to the Department of International Development (DFID) [14: 6] disaster resilience as 'the ability of countries, communities and households to manage change, by maintaining or transforming living standards in the face of shocks or stresses—such as earthquakes, drought, flood or violent conflict—without compromising their long-term prospects'. Simply put, disaster resilience in this book was conceptualised as the ability of a community, individuals, and the state to bounce back and bounce forward following a disaster and the ability to continue a normal life in the face of a disaster without necessarily being stopped from engaging day to day activities. Resilient entity successfully deals with disasters. Even though, that does not mean that there will be no deaths or negative impacts following a disaster, but an entity or people should be able to at least deal with the disaster to reduce its impacts.

Disaster resilience has become very important especially at a time when there is much emphasis on the achievement of Sustainable Development Goals and agenda 2063. Although, disaster resilience is considered a very critical component in international policy discourse, it is also subjected to criticism. "Some authors contend that disaster resilience does not add anything particularly new to the substance of humanitarian or development assistance Manyena [9, pp. 434–436]. There was also a concern that disaster resilience has been "depoliticised as there is too much emphasis on the individual and wider society rather than on state actors who have the political power to address underlying causes of vulnerability to disasters" Chandler [5, p. 217). More often, scholars differs in terms of their conceptualisation and understanding of disaster resilience. There are efforts made to further clarify disaster resilience.

Dealing and understanding community resilience is always difficult. Community resilience, though still considered as an ambiguous word, has gained momentum in recent years and it is interchanged with sustainability [6]. According to Chirisa et al. Zimbabwe's resilience has not been conceptualised adequately and this is attributed particularly to the nature of capacity building being spearheaded by development partners such as Non-Governmental Organisations. "Recent attempts to develop ways to measure resilience that cross disciplinary boundaries have focused on assessing such elements as technological capacity, skills and education levels, economic status and growth prospects, the quality of environment and natural resource management institutions, livelihood assets, political structures and processes, infrastructure, flows of knowledge and information and the speed and breadth of innovation", Mitchell and Harris [12: 3]. Measuring Resilience is always a problem as different disciplines have different mearing techniques. Despite having the difficulties in measuring disaster resilience efforts to increase community disaster resilience are very significant in contemporary society where several types of disasters are wreaking havoc across the world.

The United Nations International Strategy for Disaster Reduction (UNISDR) [17: p. 3] defined Disaster Risk Reduction as "the systematic development and application of policies, strategies and practices to minimize vulnerabilities, hazards and the unfolding of disaster impacts throughout a society, in the broad context of sustainable development". In fact, there is no single definition of Disaster Reduction but is generally conceptualized as policies and actions put in place to reduce the disaster risk. "In an increasingly integrated world economy built on networks of global supply chains, disasters in one country can easily affect others, and a shock or disruption to one part of the supply chain, such as a production plant or distribution centre, can have a ripple effect throughout the whole chain" [16: 009]. Alternatively: disaster risk reduction could be defined as a systematic approach to identifying, assessing and reducing the risks of disaster. It aims at reducing socio-economic vulnerabilities to disaster as well as dealing with the environment and other hazards that trigger them.

It is also important to distinguish disasters and hazards. A disaster takes place when a society or community is affected by a hazard. The impact of the disaster is heavily influenced by the community's vulnerability to the hazard. Vulnerability is described as characteristics and circumstances of a community, system or asset that make it susceptible to the damaging effects of a disaster [18]. Blaikie et al. [4: 57] defined it as the "characteristics of a person or social group in terms of their capacity to anticipate, cope with, resist and recover (from the impact of a hazard)". Thus, the more people are vulnerable the more they are affected by disasters. It is not very important at this juncture to dwell on vulnerability but to explain how the concept relates to disaster resilience and disaster risk reduction. According to Mercer [11: 249] Successful DRR creates resilient communities, whilst ensuring vulnerability is not increased through development efforts or other initiated activity. The initiated activity can be external or internal. Contemporary disaster management policies give much emphasis on disaster resilience and reduced vulnerability as critical measures for promoting disaster risk reduction.

3 Structure of the Book

Chapter 1 of the book under review focuses on disaster and hazards in the context of consilience. The chapter conceptualises key disaster terms namely disaster, disaster resilience, hazards, disaster risk reduction and disaster management among others.

Chapter 2 investigated the role of social networks in enhancing the capacity of the community to bounce forward following the surge of COVID-19 in 2020. The study explored the nature of social networks that were employed by communities to increase their resilience to COVID-19 disasters and the challenges faced when employed social networks during COVID-19. The study assist in the development of national policy briefs regarding how the government boosts the functioning of social networks in disaster situations.

Chapter 3 looked at Climate change and variability in the context of disaster resilience. A comprehensive analysis of the community-based resilience to climate change in Chimanimani, a district that is increasingly being at risk of meteorological hazards, was conducted. The chapter provides orientation on who, how, when and what aspects of the community-based resilience to climate change. The study examined the nature and the effectiveness of community-based absorptive and adaptive capacity and the challenges facing the community. The findings from the study aid in strengthening community-based disaster risk management and enhance knowledge about disaster resilience.

Chapter 4 examined the utility of Geo-Spatial Science and disaster risk management in Chimanimani District, Zimbabwe, with particular focus on susceptibility to cyclones and associated floods. Chimanimani is part of the country's eastern bloc that experiences periodic episodes of cyclones, the latest being Tropical Cyclone Idai in 2019. The chapter further argued that proper mix of participatory and scientific hazard mapping are critical components to strengthen disaster risk reduction and disaster preparedness. It also posits that use of GIS enhance multi-hazard mapping as part of response would provide a pictorial view of the areas liable or prone to disaster risk for timely preparedness and response.

Chapter 5 analysed Zimbabwean disaster risk management strategies in relationship to the concept of resilience in disaster risk reduction (DRR). The chapter explored and assessed the resilience building strategies employed within the country.

Chapter 6 examined the role of politics on disaster resilience building in Zimbabwe. Chapter 7 utilised an intersectional approach to analyse the gendered dimensions of disaster resilience. The chapter emphasized the need to analyse how gender that is women and girls' response to climate change shocks in view of their background, literacy level and access to resources.

Chapter 8 focused on the importance of increasing availability and accessibility of earth observation data combined with advances in for broader and more innovative applications in damage assessment, monitoring and management.

Chapter 9 analyses the contribution of social workers through the DSD in ensuring the resilience of child protection services under the perturbations instigated by the COVID-19 pandemic, that have unequivocally expanded the vulnerability of children.

Chapter 10 examined how Zimbabwe handled Cyclone Idai, a disaster that occurred in March 2019 and the COVID-19 pandemic that started in December 2019 worldwide but had its first case recorded in Zimbabwe in March 2020, separating the two hazardous events by one year.

Chapter 11 investigated the usefulness of existing systems for looking after the elderly people in different societies. The study digs into literature and follows a comparative approach with the predominantly Western world care home system on one side whilst the home care system widely used in African societies on the other hand. Findings show care home systems as more vulnerable to the effects of contagious diseases like COVID-19. However, the care home systems can also work well if well quarantined as opposed to the home care systems.

Chapter 12 analysed the use of home-grown knowledge towards resilience building in Lower Muzarabani—a natural disaster-prone area from 1980 to 2021.

Chapter 13 analysed the effects of tropical cyclones on human security with regards to the economic security of rural households in Mozambique and Zimbabwe.

Chapter 14 examined resource governance within the context of disaster risk management in Zimbabwe. The chapter emphasised the need to prioritise strengthening natural resource governance policies, frameworks and institutions in order to enable greater community beneficiation from its resource wealth is glaring.

Chapter 15 argues that directly tackling inequalities is critical to climate justice and building resilient communities in Southern Africa. The chapter concluded that there is a need to support community, household and individual measures in fighting climate and economic injustice is also elaborated for purposes of building bottom-up resilience where 'external science' and local technical knowledge are integrated based on the leadership of capable state and regional bodies not captured by local and global capital.

Chapter 16 analysed urban informal sector considering policies aimed at managing the impact of COVID-19 given that Zimbabwe is a low-income country where income is earned and daily for food purchasing. The study revealed that there is lack of integrated approaches in policies employed by the government in building livelihood resilience amongst urban informal businesses in response to COVID-19.

Chapter 17, which is the concluding chapter, explored the available options for the future management of disasters and risks in disaster-prone countries, with particular focus on Zimbabwe.

References

1. Al-Madhari AF, Keller AZ (1997) Review of disaster definitions. Prehosp Disaster Med 12(1):17–21
2. Amlôt R, Carter H (2008) What can emergency planners learn from research on human resilience? Afr Health Sci 8
3. Bhaiseni B (2017) Poverty as a threat to environmental sustainability: implications for social work practice in Zimbabwe. Afr J Soc Work 7(2):25–29
4. Blaikie P, Cannon T, Davis I, Wisner B (2004) At risk: natural hazards, peoples vulnerability and disasters, 2nd edn. Routledge, New York
5. Chandler D (2012) Resilience and human security: the post-interventionist paradigm. Secur Dialogue 43(3):213–229
6. Chirisa I, Mukawri LB, Paradza N (2019) Community resilience under urbanisation and climate change–an overview. In: Community resilience under the impact of urbanisation and climate change: cases and experiences from Zimbabwe, p 1
7. Coppola DP (2006) Introduction to international disaster management. Elsevier
8. Hill M, Nhamire B, Business Live (2017) Tropical cyclone Dineo pounds Mozambique, nears SA. https://www.businesslive.co.za/bd/world/africa/2017-02-17-tropical-cyclonedineo-pounds-mozambigue-nears-sa/
9. Manyena SB (2006) The concept of resilience revisited. Disasters 30(4):434–450

10. Mavhura E (2018) Disaster risk reduction policy and management in Zimbabwe. In: Handbook of disaster risk reduction & management, pp 589–612
11. Mercer J (2010) Disaster risk reduction or climate change adaptation: are we reinventing the wheel? J Int Dev J Dev Stud Assoc 22(2):247–264
12. Mitchell T, Harris K (2012) Resilience: a risk management approach. ODI Background Note 1–7
13. Parker DJ (2020) Disaster resilience–a challenged science
14. Resilience DDD (2011) A DFID approach paper. London, UK, pp 1–20
15. Thomalla F, Downing T, Spanger-Siegfried E, Han G, Rockström J (2006) Reducing hazard vulnerability: towards a common approach between disaster risk reduction and climate adaptation. Disasters 30(1):39–48
16. Twigg J (2015) Disaster risk reduction
17. UNISDR (2004) Hyogo framework for action 2005–2015: building the resilience of nations and communities to disasters. United Nations International Strategy for Disaster Reduction, Washington, DC
18. UNISDR (2012) Making cities resilient. United Nations International Strategy for Disaster Reduction, Washington, DC
19. ZimStat (2012) Zimbabwe population census 2012. National Report, Harare: ZIMSTAT

Rosemary Kasimba holds a Ph.D. in Sociology (Rhodes University) M.Sc. in Sociology and Anthropology, B.Sc. in Sociology (University of Zimbabwe) and Post-Graduate Diploma in Social Work. Her research interests include but not limited to Disaster Management, Peace and Security, International Affairs, Human Resources Management, Industrial Sociology, Migration, Population and Development, Social Policy and Vulnerable Population.

Understanding Social Networks and Community Resilience to Pandemics in Zimbabwe

Rosemary Kasimba and Solomon Muqayi

Abstract Many empirical studies considered social networks as an important tool for disaster risk management. However, in Zimbabwe, few studies provided a comprehensive overview of how social networks enhance community resilience to disasters, particularly pandemics. Thus, there are very few studies, conducted regarding the role of social networks in COVID-19 disaster in the country. This research attempt to investigate the role of social networks in enhancing the capacity of the community to bounce forward following the surge of COVID-19 in 2020. The study was conducted in Luveve, Bulawayo. The research employed qualitative methodology to examine the role of social networks. In-depth interviews and focus group discussions via WhatsApp were used as data collection methods. The study established that social networks helped people to gain more information on how to prevent and manage COVID-19, provide social support to the patients, food and medication. The study also found that some people who fell sick were neglected, discriminated, and did not receive any social support from their friends. The study further revealed that social networks play a pivotal role in disasters in disaster risk management but some people with weak ties were neglected. The study recommends that the government put in place systems that enhance social support in communities in this era where COVID-19 is wreaking havoc.

Keywords COVID-19 · Social networks · Disasters · Disaster risk reduction and community resilience

1 Introduction

Contemporary society is confronted with several crisis and disasters that leaves a trail of destruction and death. Though, resilience is dismissed as an ambiguous term, it entails the ability of a community to bounce forward and back following a disaster Carpenter [4]. Resilience has become a very prominent discourse in the

R. Kasimba (✉) · S. Muqayi
University of Zimbabwe, Harare, Zimbabwe
e-mail: rkasimba@gmail.com

R. Kasimba and I. Chirisa (eds.), *Disaster Risks and Hazards*, Advances in 21st Century Human Settlements, https://doi.org/10.1007/978-981-96-4577-0_2

study of disasters as policy makers are shifting from reactionary to proactive measures regarding disasters management. According to Carpenter [4], "community resilience to disasters promotes an effective recovery." Thus, factors that contribute to disaster resilience are especially important and are of great interest to scholars and practitioners in disaster scholarship. There is an increased attention on the role of social networks in pandemic research. Novel studies revealed that they are ~~a~~ critical pillar of resilience. Nonetheless, the role of social networks in pandemics and disasters remains under explored since there are ~~a~~ few studies on their roles especially ~~on~~ during pandemics in developing countries. This research attempt to create new knowledge to this field of research by looking at how social networks ~~is~~ has fostered resilience in urban areas in Zimbabwe particularly in areas like Luveve Bulawayo. There are very few studies conducted in Zimbabwe on social networks and disaster resilience. Networks depend on norms and trust, and the existence of norms and trust implies networks [16: 20]. According to Islam [16: 26], community networks are critical in disaster resilience because they allow individuals to draw from community resources and increase the likelihoods. Such communities can effectively address their collective concerns. Davison [6], Mayunga [24] and Haines et al. [12] strongly support the idea that social networks are beneficial in promoting disaster resilience and recovery since they allow community members to work together towards a common goal. Where characteristics of a strong community are missing such as, among others, ability to learn new skills and build on past experiences, ability to identify problems and establish priorities, has economic opportunities and has the capacity to respond to disasters proactively, members of that community tend to have less capacity to cope with disasters as they can collect fewer external resources through their networking relationships (Muyunga 2007: 7). This study is an evidence-based policy prescription for disaster management practice in Zimbabwe. It provides information and ~~more~~ new knowledge on how social networks can be cultivated in the formulation and implementation of disaster management policies in the country. Policy makers, development practitioners and other stakeholders develop mechanism that can enhance social network capacities to effectively contribute to disaster resilience.

Furthermore, the study will unveil the salience role of social network in promoting not only resilience to disaster risk but also social welfare, cultural ethics, norms, values, mutual relationships and climate change adaptation in Luveve, Zimbabwe. The researchers will share the key findings of the study with the government and development practitioners. Community members will benefit a lot from the findings as they will learn more about social networks and their role in disaster resilience.

2 Literature Review and Theoretical Framework

The study was informed by the social network analysis that draws from several theoretical foundations that are inclusive of social capital. Social capital has been an influential concept that underpins social network research. Social capital can be understood as 'networks together with shared norms, values and understandings that facilitate co-operation within or among groups' Tselios et al. [31]. Tselios et al. [31] further noted that social capital allows people to trust each other, rely on each other for support, and work together. Although scholars have several views about propounders of the social network analysis, we considered Comte as the father of social network analysis. He showed how components of the social system are interconnected. According to him "Families become tribes and tribes become nations", His view, then, is strikingly contemporary Freeman [9: 14]. He defined society structural terms that are found today in social network analysis. Freeman [9: 3] notes that social network analysis is motivated by a structural intuition based on ties linking social actors, it is grounded in systematic empirical data, it draws heavily on graphic imagery, and it relies on the use of mathematical or computational models. Social networks are a way of looking at a social system on relationships within a system. Actors in social network analysis are most commonly people or organizations, but they can also be countries, teams, species, cities, or businesses [31]. In simple terms, social networks explain interrelationships among people ranging from nuclear families, extended families, tribes, people that share the same totem, friends, relatives, social circles, among others. There are strong and weak networks. Strong networks represent the relations that are close like that of a nuclear family. Weak ties are found in distant relations. However, people can still get adequate assistance from distant relations. Tselios et al. [31] note that actors in a network are characterized by their attributes and these attributes of people may include gender, age, income or for an organization. The basic building blocks of social networks are called dyad and the unit of analysis are the ties between the two actors [31]. There are two main types of social networks namely, egocentric, or personal network, and a whole or full network study. Egocentrics are those networks that individuals have communication with, and they share information with. Whole network encompasses all pre-existing well-defined boundary, such as a workplace, sporting club, or neighborhood, or trade network.

A study by Islam [16], in Coastal Bangladesh showed that bonding and bridging networks provide important support—sharing of food and shelter and providing comfort immediately after a disaster. He further noted that time after the disaster increases, these networks perform less well due to households' limited physical, financial, and human capital, and the uncertainties of their access to natural capital. After a period, bridging networks become less active and sometimes break down due to competition and conflict over access to external relief support. Bonding networks, however, do not break down. Household members continue contributing to the recovery process by reducing food intake, helping with alternative income and livelihood options through temporary migration, among others. Islam [16]. For longer-term recovery, however, disaster victims usually need support through linking social networks—from NGOs (for example local, national, and foreign NGOs), local government and other community-based organisations (CBO). Their links with NGOs and local government provide households with dedicated support in the form of emergency relief, shelter, livelihood assistance, and reconstruction of major community services Islam [16]. Nonetheless, our study focused on community social networks (bonding social networks) networks among community members without an assistance from the government and non-governmental organisations [16]. What necessitated the nature of this study was to explore the role of community social networks in disaster resilience. Contemporary disaster resilience discourse put emphasis on people themselves before external help is offered to the community. Community members have a pivotal role to play in increasing disaster resilience.

For a long time, network theorists have explored the importance of social interactions and network structures Varda et al. [32: 12]. These studies provided a better understanding of the role of networks during a disaster. They also pointed out ~~on~~ how people access resources by increasing their frequency in the use of networks, how community networks help people when dealing with a disaster. Post COVID-19 experience encourage scholars to further explore the role of social networks in disaster risk management. According to Kim and Hastak [20: 86] social media, such as Twitter, WhatsApp and Facebook play a critical role in disaster management by propagating emergency and disseminating information to a disaster-affected community. They further argued that several studies explored social media to understand the networks and extracted critical information to develop pre and post disasters mitigation. A study by Kim and Hastak [20: 86], also applied social network analysis to convert emergency social network data into knowledge and their results showed that social networks consist of individuals, emergency, and organizations. They also established that social networks enabled individuals to actively share information, communicate with each other. A study by Middleton et al. [26], revealed that social media networks was used to develop a GIS-based real time mapping during the Hurricane Sandy in 2012 in New York City. The advancement in technology has made the active role of social network in disaster management timelier and more significant [3: 1]. Lindsay [21] notes that online social networks have been seen as an important vehicle to channel critical resources during emergencies. Thus, Facebook, Telegram, WhatsApp, and other communication platforms are regarded as very vital before, during and after disasters. Chu et al. [5] note that the relationship between online social networks and disaster resilience is a heated topic of debate and according to Lindsay [21] social media was ranked as fourth most important source of disaster information.

A study by Murphy [27] revealed that people with strong networks and relationships are far better withing all phases of disaster cycle from planning to reconstruction than those who do not have access to these types of networks. Furthermore, a study by Zhang et al. [34] found that informal social networks (community people and disaster victims) helped each other and played a prominent role in search and rescue.

Nevertheless, few studies have given a comprehensive overview of how social networks reduce risks during and after disasters. Based on the survey data of Wenchuan earthquakein 2008. Disaster destroy infrastructures, economic capital and the physical resources but social networks are least damaged.

Sadri et al. [29] investigated the role of social capital, personal networks, and emergency responders in post-disaster recovery and resilience: a study of rural communities in Indiana. They conducted a mail survey to collect data on household recovery in four small towns in southern Indiana that were hit by deadly tornadoes in March

2012 [2: 1377]. They specifically focused on how households in these communities are recovering from damage that they experienced and the role of social capital, personal networks, and assistance from emergency responders on the overall recovery experience. Aldrich et al. [2] study found that the density of people's networks is especially important in disaster management. A dense personal network indicates close interpersonal contacts among others and facilitates information and resource sharing among the people. However, our study is different form their study in the sense that they used questionnaires to collect data and ours was qualitative.

Aguirre [1] noted that many types of social networks are being utilised or formed. These include familial, religious, political, economic, medical, educational, scientific, legal, risk management (insurance and first responders), mass media, communications, transportation, energy, food, water, leisure, entertainment, construction, rebuilding, land use, and environmental regulation and protection. Such networks, such as faith-based organizations or citizen volunteer groups, operate in response to a disaster or emergency but are not part of formal disaster planning and management Carpenter [4: 10]. Some social networks develop during a disaster, and these may become a trusted foundation for future resilience to disasters as actors will be using previous success story of these networks and how they may hve utilised them during a disaster.

MacGillivray [22] notes that Theory suggests that social networks should moderate the impacts of climatic and geophysical hazards and shape people's adaptive capacities and recovery trajectories, yet the empirical evidence is more mixed than commonly supposed. This study sought to explore the nature and role of social networks in enhancing community resilience to disasters. Despite having several studies that were conducted on social networks and disaster recovery, the role of social networks in disaster resilience is underexplored. Many scholars conducted studies that linked social capital and disaster resilience were conducted in the context of climate change, floods, and droughts [30], MacGillivary (2018). Kasimba [19] and Islam [16]. A study on the relationship between social networks and rural adaptability to flood and river erosion using a qualitative approach in Bangladesh. Furthermore, a study by Jordan et al. [17] examined the link between social capital and resilience to climate stress using a qualitative approach in two vilages of Bangladesh. In addition, a study by Masud-All-Kamal and Hassan [23] explored the relationship between social capital and disaster recovery based on a qualitative approach in two Aila-affected villages of Shyamnagar Upazila of Satkhira District in Bangladesh. Although, these studies demonstrated that social capital and social networks are important in disaster resilience, a few studies, if not none to the authors' knowledge have conducted a study of that nature in Zimbabwe. In addition to that, a few if not none, were conducted in the context of COVID-19.

Social networks in disaster management have become so important in contemporary society because they encompass an element of community participation. Community can participate in many ways in disaster management. The Hygo Framework for Action 2005–2015 and the new Sendai Framework for Disaster Risk Reduction 2015–2013 emphasized the need to build resilience at all levels, by promoting engagement and networking [10].

Kaniasty and Norris [18: 396] documented the role of social support as a provider of emotional comfort, material goods, self-esteem, or information. According to them there are two modes of social network support namely "the stress buffer and main effects models" Kaniasty and Norris [18: 396]. Stress buffering encompasses how individuals in crisis are protected from resultant stress by the presence of social networks Capernter (2013: 11). Social support inherently benefits physical and psychological health, regardless of the situation. The benefits for community resilience therefore are more apparent when stress is applied; however, the authors believe the inherent benefits also protect the community from physical and mental health vulnerabilities ahead of a disaster [4: 11].

Nakagawa and Shaw [28] conducted a study on the strength of social networks in recovery and rebuilding following a disaster in India and Japan. Capernter (2013) notes that similar neighborhoods in Kobe, Japan (affected by a 1995 earthquake) and Gujarat, India (affected by 2001 earthquake) were studied. Victims in the study area were surveyed using the Integrated Questionnaires for the Measurement of Social Capital that was developed by the United Nations. The findings of the study showed that communities with highest levels of social capital displayed the fastest recovery despite lower income levels [4]. According to Carpenter [4] this scenario reconfigured Japan's disaster management engineering view of disasters to a social and technical issue. However, the studies by Nakagawa and Shaw [28] were quantitative this research was qualitative as researchers sought to understand the role and nature of social networks in promoting disaster resilience in the COVID-19 context. Furthermore, the study by Nakagawa and Shaw [28] used a questionnaire as data collection tool whilst this study used key informant interviews and in-depth interviews as data collection tools. Our study was new in the sense that we sought to understand the nature of social networks in the context of COVID-19 that was not yet conducted in Zimbabwe.

Social networks are considered as significant factor in community resilience to disasters. Nonetheless, there are difficulties in defining and measuring the two concepts namely social networks and resilience. Carpenter [4: 11] notes that resilience is difficult to measure or even recognised in certain situations. Despite these problems, our study attempted to operationalize these concepts to enable an easy understanding of the relations between the two in the disaster management context. This study will fill an existing gap by exploring the role of social networks in the prevention, management, and psychological recovery process of COVID-19. The study also examines the role of social networks in in increasing community awareness about the virus. Thus, the study fills an existing lacuna, the weaknesses of community social capital in disaster resilience. The research will provide evidence-based knowledge to policy makers and development practitioners on the role of community social networks in disaster resilience. The study will, thus contribute to the strengthening of the capacity of the different forms of community social network and design effective disaster resilience models for future pandemics in Zimbabwe and in other countries.

3 Study Area and Methodology

3.1 *Methodology*

The research was basically qualitative approach. This was an exploratory study as researchers sought to understand the role of social networks in increasing community resilience to disasters in the context of COVID-19. The study sought to understand the nature of social networks and its role in enhancing resilience to COVID-19 pandemic in the study area. In-depth interview and key informants techniques were used for data collection.

For in-depth interviews, researchers saved numbers of their respondents and they were called one by one. Those who had WhatsApp recorded audios and send them to the researchers responding to the questions at their own spare time. Researchers would simply send a question to the respondents and they would be given time to respond. Some respondents would take five days before they before they provide feedback. Respondents for in-depth interviews consisted of those who had contacted the virus and those who take care of COVID-19 patients at home. However, there was no 100% response as some people did not reply to their calls. Face to face key informant interviews were used to collect data from respondents who occupied key positions in Luveve, and were the councilor, community health workers, hospital social works and medical doctors. Data was collected at the peak of the second wave of COVID-19, so the researchers saved numbers for all the respondents. In-depth interviews were done over the phone. Researchers were assisted by digital technology which made data collection from the respondents very easy and thematic mode of analysis was used to analyse data.

4 Presentation of Findings

4.1 *The Role of Social Networks in Response to COVID-19 Pandemic*

(a) **Information sharing through social networks**

Findings of the study showed that social networks have helped various communities in sharing information on how to prevent and handle corona virus in the event that one tests positive. The respondents confirmed that they mostly used WhatsApp groups, audios, PDFs, videos and word documents. People who have healed from coronavirus share their subjective experiences on how they managed to save themselves. Some were also getting notes from their friends on Face book concerning the treatment. Respondents highlighted that people are sharing information using different platforms. Some respondents attested said that they did not believe that

COVID-19 was real, but were made to believe through various social media platforms and this helped a lot of communities to manage the virus when their members tested positive. Some of the preventive measures learnt through social networks/ media include steaming, concoction that soften breathing, signs and symptoms, precautionary measures such as distancing, hand sanitising, wearing face marks and avoiding unnecessary movements and public gathering among others.

One of the respondents had this to say:

I never believed that there is COVID-19 pandemic. I thought there were just rumours until I saw videos of people who were severely ill from WhatsApp, Facebook and other platforms. Even though there were some exaggerations on the social platforms about the spread of COVID, but it really help in educating people on the impact of COVID-19 and the need to observe its protocols strictly.

Another respondent added that:

I fell sick and panicked during the first two days as I was having chest pains. My friends who had fallen ill before advised me to seek medical advice and the doctors' instructions.

Another responded said:

I almost died with stress when I had developed breathing difficulties. My friends advised me to cut onion and smear it all over the body as well draining the liquid that I had to put on my chest whilst I was waiting for an oxygen. Had it not been my friends who helped me through WhatsApp with information on how to handle the virus when you are infected, I would have. Suffered.

Respondents did not just share information ~~on~~ via social media platforms, but they also had face to face discussions concerning prevention and handling of the virus when one is infected. They have also highlighted that information sharing helped others to take precautionary measures and motivated others to get vaccinated.

(b) **Medicine and Food Supply Through Social Networks**

Findings from the study showed that social networks enhance relationship between and groups such as friends, family, church and community particularly during and post COVID-19 pandemic. Some respondents who suffered from silent hypoxia and in need of oxygen were being assisted by their friends and relations to raise money for oxygen. Some said that they could not even afford to buy pain reliever, Ivermectin, anti-biotics and other tablets needed for treatment as they were not employed and had to pay rents every month. Some respondents highlighted that they would get food from friends when they were sick as they were not able to cook for themselves.

One of the respondents said:

"……. *COVID-19 has taught me to love no matter the circumstance. The first three days that I tested positive I was able to walk and wash myself. My young children got food assistance from relatives and friends. I was not able to provide for my family. Friends and relatives would bring food until the situation improved.*

One of the respondents also confirmed that:

I had mild flu symptoms and was feeling weak I went for testing, and it was conformed I had contacted the virus. With panic I informed my peers at church on our WhatsApp platform. One of the group members rushed to my house and took my

9-month baby away from me. Two days later I developed breathing problems. I did not have money to go to private hospital. I did not have money to buy oxygen tank. I exited the church group as I thought I was going to die. My church friends raised more than US$600 that was enough to buy an oxygen tank. and calling medical doctor stop treat me at home. Seven days later I was able to breathe on my own. It took a month to fully recover. I thank my church friends together with my family members who saved my life.

Another responded observed that:

As individuals we would not have made it in this community most of us tested positive, our small-scale businesses almost collapsed. Community relations, friendship, ties, and togetherness helped a number of people including myself. I fell sick for more than 32 days but I did not go to the hospital. My neighbours, though we did not have physical contact but they stood with me and my family during such a tough time.

5 Social and Financial Support from Social Networks

Findings from the study also revealed that social networks in the form of family, friends and work mates helped people who recover quickly. Majority of the respondents confessed that they received words of comfort, hope and encouragement when they fell sick, and this helped them to recover quickly as they were not affected much with stress. The study found that some people who were stressed deteriorated quickly after contracting the virus because of stress and lack of social support.

One of the respondents had this to say:

I could not take medication as was prescribed by the medical Doctor because I thought I was going to die so there was no need. I thought everyone who contracts the virus dies because of what I had seen on Aljazeera. Seeing people failing to make it in developed countries, I had no hope of surviving considering that I have another chronic ailment.... With social support, I received from my friends and work mates, I was encouraged to take medication and within few days my breathing becomes normal. If it were not social networks, I would not have survived....

Another respondent has this to say:

I felt sick and as a result I exhausted all my savings as well as the capital for my business for the treatment of Corona virus. This has forced me to go into borrowing from friends and relations. Had it not because of social networks, me and my family would have suffered and become desperate due to lack of communication through which community members will get to know the extent and effects of COVID-19.

The results of the study showed that people in Luveve relied heavily on social networks for communication, enlightenment, medication, financial assistance and social interaction between and among each other during the time when COVID-19 was at its peak.

Findings from the study revealed that community togetherness and a sense of community belonging helped people to manage the spread of COVID-19. People had to educate each other about COVID-19 as some could not believe that the virus existed. Initially, majority refused to wear masks, they were against lockdown, and they considered a lockdown measure as a policy put in place for political reasons. Majority did not believe in the existence of the virus as they thought it would not come to Zimbabwe. They thought it would end in China and other European countries. Majority of people in the community would not practice social distance initially because they did not see the virus killing people. With time, seeing the virus wreaking havoc, people losing their beloved one, some developing some breathing problems. The Zimbabwe Broadcasting Services (ZBC) started sharing with people what was happening in hospitals. Those who believed that the virus was with them, and it was wreaking havoc in the country started conscientizing the masses in the community. People were being encouraged to wear masks and avoid unnecessary movements. Community members had to work together as one family. Community members also started working closely with the police and other health workers reporting those who would run away from quarantine. People became so particular about visitors or returning residents who would not pass-through quarantine. Community members started reporting to the police and those perpetrators would be arrested and taken back to quarantine if they were found guilty of skipping quarantine measures. A sense of community belonging motivated people to be alert all the time. This also reduced the spread of COVID-19.

One of the key informants had this to say:

Togetherness and community togetherness has helped in increasing the community's ability to fight the spread of COVID-19. At first majority did not believe that the virus was real. Many people were very ignorant/. It was difficult for us to tell people to mask up, always sanitise their hands and practice social distance. Some community members would laugh at those whom they would see wearing masks. Those who watched television nd had seen hosiptals in various countries overwhelmed, started educating the masses. This resulted prompted majority to believe that indeed the virus was real, and it was killing.

Another respondent said:

People become alert when the virus started wreaking havoc in this community. Eighty-four cases recorded in one day in Bulawayo raised an alarm. People started dying. The community was forced to put in place measures to make sure that all returning residents had passed though quarantine. Everyone was now alert. People stooped unnecessary movements. There were volunteers who started moving around telling people about the virus.

Another respondent indicated that.

Community members stated sharing messages educating each other when they realised that the virus was real. Many people were now dying, and hospitals were overwhelmed. Community members encouraged each other to always sanitise and watch for the symptoms. Those who contracted COVID-19 would also share their experiences with others and measures they would have taken to survive. This helped several people as they gained a little knowledge on how to manage the virus if they have contracted it.

Another Key informant interview said.

Conquering and control of COVID-19 was made possible because the community was now on high alert Covid needs people that are honest to each other. Community members informed each other that those who exhibit covid-symptoms should stay home, isolate themselves and seek medical attention early. Thus, community interactions are very key in disaster management.

Quotations above explained the importance of the community in disaster resilience. Respondents highlighted that togetherness, sense of belonging and social relations helped people in Luveve deal with COVID-19.

6 Community Networks and Disaster Resilience

When respondents were asked whether community networks enhanced the ability of the community to fight against COVID-19, they had mixed responses. Majority said that the community relied on social networks as it allowed the sharing of information, ideas, prevention, and control and manage the virus. They highlighted that church members, peers, clubs, neighbourhoods, families, and friends among others helped each other in times of crisis. Community supported other people's businesses by given them loans. Some provided funds for hospital care and oxygen. Respondents highlighted that norm of reciprocity worked to increase the resilience of the community to COVID-19 disaster. Families and several social networks groups helped each other in times of need. Some were helping each other by the virtue of sharing a totem.

However, a few of the respondents said that they did not see social networks helped them as they were being segregated by other social network groups. People discriminate each other basing on religious beliefs, class, totem, language, and income.

One of the respondents said:

Social networks lead to discrimination. We were discriminated during COVID-19. My mum fell sick, but no one bothered to help even church members did not come to rescue her. At church were considered as newcomers since we came to say in Bulawayo a year ago when covid-had just started. There is too much nepotism in this community. People do help those that they are close to not newcomers. We got help from out close relatives who would give us food and helped with the money to buy oxygen.

Findings of the study indicated that a few of the community members did not see the usefulness of social networks in disaster management. The study also found that community members were also supporting each other's small businesses in diverse ways. Lockdown came with measures that resulted in the limited flow of money on the market and most of the small businesses suffered financially. Many people lost jobs and many people's income levels dwindled.

7 Discussion of the Findings

Findings of the study showed that residents in the study area belong to an average of four social groups (Family, church, work, social relations, geographic location, and business among others). Many people reported that they had several groups that supported them. People used social networks to improve their risk awareness and preparedness to COVID-19. This substantiated the view that social networks increased the resilience to COVID-19 pandemic even though, there were some challenges highlighted such as poverty, stigmatization lack of access to resources and general understanding of the of the nature of disaster. The study found that a few individuals does misuse social networks for hate speeches, deformation of character etc., but that does not mean social net encourages discrimination. The study further revealed that social networks are incredibly significant in disaster risk management. People in Luveve communicated with their friends, families, and neighbours through social networks in times of emergency. Some also relied on church members who normally settle their hospital bills. These findings of the study are in tandem with Islam's findings (2015) in Bangladesh where he found that social networks help disaster victims to quickly recover from a shock. Another study conducted by Islam and Walkerden [14] also found that linking social networks with NGOs provide strong support through immediate relief (food, water, medicine, household utensils), shelter (building materials, new houses), and livelihood assistance (microcredit, cropping seeds, livestock, fishing boats, and nets). Findings of the study are also in tandem with the study by Saha [30] in Bangladesh where the researcher found that people in flood prone communities depends heavily on social networks for survival and recovery especially after Aila situation that has been characterised by long-term flooding. Community members in Luveve, where able to assist each other by sharing information through social networks prompted people to contribute money for essential humanitarian needs such as oxygen, food items, clothes, drugs.

8 Recommendations

- The government of Zimbabwe and must be encouraged to establish programmes that strengthen community social networks,
- The government of Zimbabwe, people that have wealth, NGOs, external donors such as UN, World bank, UNDP, UNICEP, AU etc., should assist to empower citizens from various communities,
- The government is also encouraged to rope in social workers who can work with communities sensitising them the disadvantages of discrimination,
- In addition, social workers are also needed to inform people about the importance of working as a community,

- The Government should diversify the resources so that more job opportunities will be created. Also, government should embark on enlightenment campaign on causes, nature and impact of disaster,
- Community members must be encouraged to participate in several community activities to increase their visibility to reduce chances of being discriminated under certain circumstances, and
- The government is encouraged to ensure that physical infrastructure, economic structure, and other necessary resources readily available for community members to be able to assist each other in times of a need.
- The government is also encouraged to give much include social networks in its disaster risk reduction policies.

9 Conclusion

Findings of the study revealed that COVID-19 pandemic has no boundaries, it affects the young, old, women, men, the rich, poor, white, and black, among others. It has forced people to share ideas, experiences, and methods of handing the virus. The results further highlighted that the act of solidarity, togetherness, is dependent upon the participation of the household members with a number of formal and informal social networks. In addition, families and individuals who actively participate in several groups and social networks had more social and economic support and thus developed more resilient to the COVID-19 pandemic and related lockdown measures. Community members who had access to information through social networks Significant advantage.

References

1. Aguirre BE (2006) On the concept of resilience
2. Aldrich DP (2017) The importance of social capital in building community resilience. In: Rethinking resilience, adaptation, and transformation in a time of change. Springer, Cham, pp 357–364
3. Annamalai C, Koay SS, Lee SM (2014) Role of social networking in disaster management: an empirical analysis. J Comput Biosci Eng 1(3)
4. Carpenter A (2013) Disaster resilience and the social fabric of space. In: Proceedings of the 9th international space syntax symposium, Seoul, Korea, vol 31
5. Chu H, Liu S, Yang JZ (2021) Together we survive: the role of social messaging networks in building social capital and disaster resilience among minority communities. Nat Hazards 106:2711–2729
6. Davison CB (2008) Leadership characteristics as determinants of business continuity and disaster recovery planning: a correlational research investigation within institutions of higher education (Doctoral dissertation, Capella University)
7. Drabek TE, Key WH (1984) Conquering disaster: family recovery and long-term consequences. Irvington Publishers
8. Dynes RR (2005) Community social capital as the primary basis for resilience

9. Freeman L (2004) The development of social network analysis. Study Sociol Sci 1(687):159–167
10. Gioli G, Hugo G, Costa MM, Scheffran J (2016) Human mobility, climate adaptation, and development. Migr Dev 5(2):165–170
11. Guarnacci U (2016) Joining the dots: social networks and community resilience in post-conflict, post-disaster Indonesia. Int J Disaster Risk Reduction 16:180–191
12. Haines VA, Beggs JJ, Hurlbert JS (2011) Neighborhood disadvantage, network social capital, and depressive symptoms. J Health Soc Behav 52(1):58–73
13. Islam R, Walkerden G (2014) How bonding and bridging networks contribute to disaster resilience and recovery on the Bangladeshi coast. Int J Disaster Risk Reduction 10:281–291
14. Islam R, Walkerden G (2015) How do links between households and NGOs promote disaster resilience and recovery? A case study of linking social networks on the Bangladeshi coast. Nat Hazards 78(3):1707–1727
15. Islam R, Walkerden G (2017) Social networks and challenges in government disaster policies: a case study from Bangladesh. Int J Disaster Risk Reduction 22:325–334
16. Islam M (2015) Social capital and cyclones: how households' social networks contribute to disaster resilience and recovery in Bangladesh
17. Jordan T, Preining W, Curry GD, Leflar JJ, Glasser M, Loyear R, Grey B, Ong L, Sobron JM (2015) How social media is transforming crisis management and business continuity. J Phys Secur 8(2)
18. Kaniasty K, Norris FH (1995) In search of altruistic community: patterns of social support mobilization following Hurricane Hugo. Am J Commun Psychol 23(4):447–477
19. Kasimba R (2018) Understanding the role of social capital in enhancing community resilience to natural disasters: a case study of Muzarabani District, Zimbabwe
20. Kim J, Hastak M (2018) Social network analysis: characteristics of online social networks after a disaster. Int J Inf Manage 38(1):86–96
21. Lindsay BR (2011) Social media and disasters: current uses, future options, and policy considerations
22. MacGillivray BH (2018) Beyond social capital: the norms, belief systems, and agency embedded in social networks shape resilience to climatic and geophysical hazards. Environ Sci Policy 89:116–125
23. Masud-All-Kamal M, Hassan SM (2018) The link between social capital and disaster recovery: evidence from coastal communities in Bangladesh. Nat Hazards 93(3):1547–1564
24. Mayunga JS (2007) Understanding and applying the concept of community disaster resilience: a capital-based approach. Summer Acad Soc Vulnerability Resilience Build 1(1):1–16
25. Mickoleit A (2014) Social media use by governments: a policy primer to discuss trends, identify policy opportunities and guide decision makers
26. Middleton SE, Middleton L, Modafferi S (2013) Real-time crisis mapping of natural disasters using social media. IEEE Intell Syst 29(2):9–17
27. Murphy DM (2007) The role of information and communication in disaster response: an overview
28. Nakagawa Y, Shaw R (2004) Social capital: a missing link to disaster recovery. Int J Mass Emerg Disasters 22(1):5–34
29. Sadri AM, Ukkusuri SV, Lee S, Clawson R, Aldrich D, Nelson MS, Seipel J, Kelly D (2018) The role of social capital, personal networks, and emergency responders in post-disaster recovery and resilience: a study of rural communities in Indiana. Nat Hazards 90(3):1377–1406
30. Saha SK (2020) Social capital and post-disaster response and recovery: cyclone Aila in Bangladesh, 2009
31. Tselios V, Noback I, van Dijk J, McCann P (2015) Integration of immigrants, bridging social capital, ethnicity, and locality. J Reg Sci 55(3):416–441
32. Varda DM, Forgette R, Banks D, Contractor N (2009) Social network methodology in the study of disasters: issues and insights prompted by post-Katrina research. Popul Res Policy Rev 28(1):11–29

33. Yandong Z (2013) Social networks and reduction of risk in disasters: an example of Wenchuan earthquake. Econ Stress Hum Capital Families Asia Res Policy Challenges 4:171
34. Zhang X, Yi L, Zhao D (2013) Community-based disaster management: a review of progress in China. Nat Hazards 65:2215–2239

Rosemary Kasimba holds a Ph.D. in Sociology (Rhodes University) M.Sc. in Sociology and Anthropology, B.Sc. in Sociology (University of Zimbabwe) and Post-Graduate Diploma in Social Work. Her research interests include but not limited to disaster management, Human Resources Management, Industrial Sociology, Migration, Population and Development, Social Policy and Vulnerable Population.

Solomon Muqayi (Ph.D.) is a Senior Lecturer in the Department of Governance and Public Management at the University of Zimbabwe. He specialises in international relations, political science, foreign policy and diplomacy, border management and regional integration. He holds a Ph.D. in International Relations (UZ), Masters in International Relations (UZ) and Honors Degree in Public Administration (UZ). He has published many peer reviewed journals, books and book chapters.

Community Based Resilience to Climate Change in Chimanimani District, Zimbabwe

Chipo Mudavanhu, Nelson Chanza, and Rosemary Kasimba

Abstract Climate change today is the most significant threat of the 21st century-endangering the sustainability of the world's environment, its agriculture and subsequently the health and well-being of its people. Focus on resilience to climate change has received considerable attention but most research considers resilience at either national or household scale whereas little is known about community-scale interventions. Questions on how community-based resilience is conceptualized and operationalized have not been adequately interrogated yet these are useful for understanding communities' response to climate change while maintaining and developing critical community functions. The assumption is that knowledge and capacity to respond to risks must be built from outside, a situation that may not be desirable for enhancing community resilience. Several studies demonstrated that community awareness and preparedness determine the outcome of the impact of any hazards. (This study therefore offers a comprehensive analysis of the community-based resilience to climate change in Chimanimani, a district that is increasingly being at risk of climatic hazards.) Using a mixed approach, the study examines the nature and the effectiveness of community-based absorptive and adaptive capacity and the challenges facing the community. The findings will assist in strengthening community-based disaster risk management and enhance knowledge about disaster resilience.

Keywords Community based resilience · Absorptive capacity · Adaptive capacity · Transformative capacity and meteorological hazards

C. Mudavanhu (✉)
Higherlife Foundation, Harare, Zimbabwe
e-mail: chipomuzenda@gmail.com

N. Chanza
University of Johannesburg, Johannesburg, South Africa

R. Kasimba
Rhodes University, Grahamstown, South Africa

R. Kasimba and I. Chirisa (eds.), *Disaster Risks and Hazards*, Advances in 21st Century Human Settlements, https://doi.org/10.1007/978-981-96-4577-0_3

1 Introduction

Human-induced climate change could be the most disturbing global environmental challenges facing humanity today. Climate change is a disaster on its own as it affects people's livelihoods in different ways, contributing to food insecurity and causing poverty and diseases, among others. Scientifically, the challenge is being exacerbated by urbanisation, poor methods of farming, industrialisation, burning of fossil fuels and deforestation. Poor communities are further impoverished. Furthermore, in the era where pandemics like COVID-19 are wreaking havoc, climate change makes it very difficult for agro-based economies to recover quickly. Thus, this study sought to establish community-based resilience to climate change with a view to document the strategies and assess how best they can be improved to increase local people's resilience to climate change calamities. Contemporary climate change discourse requires the participation of local people if resilience is to be strengthened. Community participation is a "critical element of responding to the climate change challenge and this translates to new scientific insights into innovative policy and practice on the ground and enable community to move forward and deal with climate change" (Chang et al. 2009). Chang et al. (2009) noted that climate change adaptation strategies require new forms of engagement between scientists, policy-makers, and wider communities who can make a valuable contribution to more informed climate policy and practice. Thy further argue that climate change agenda has moved from one of scientific problem framing (climate science and identification of climate impacts) to one more concerned with the implementation of societal responses (mitigation and adaptation), a detailed political, ethical, social and normative analysis has become increasingly important. Mitigation and adaptation result in the increase of the community's resilience to climate change.

Climate change, according to the Intergovernmental Panel on Climate Change (IPCC), is understood as a change in the state of the atmosphere that can be measured by changes in elements of weather or climate system that can persist for an extended period. Climate variability on the other hand refers to variations in the mean state if climate at spatial and temporal scales beyond that of individual weather events. Both change and variability can be triggered by natural internal processes or external causes, that can be natural or anthropogenic [13]. Most scholars treat them jointly as 'climate change and variability,' a position adopted by this chapter. This study therefore offers a comprehensive analysis of the community-based resilience to climate change in Chimanimani, a district that is increasingly being at risk of climatic hazards.

First, the chapter examines the concept of community-based resilience in the context of growing risks that are induced by climate change. A scholarly treatment is given to show existing thoughts and gaps about community-based. Second, a description of the study area is presented. Third, the chapter describes the case-based study methodology used to apply the concept of community-based resilience to climate change. Fourth, the chapter presents the research findings. Fifth, the implication of the findings in relation to community-based resilience are discussed. Based on these insights, the chapter concludes by highlighting major aspects on what constitutes community-based resilience.

2 Community-based resilience to climate change

The concept of resilience is increasingly gaining some traction in climate change research, policy and practice. The IPCC [13] and United Nations International Strategy for Disaster Risk Reduction [33] attribute this surge to the growing interest in challenging existing climate change interventions at national and global levels. Yang et al. ([36]: 1) looks at resilience as "a process of bouncing back to the previous condition after facing any adverse effect". He further notes that disaster resilience also focus on the "integrated function of the adaptive, absorptive and transformative capacity of a social unit such as individual, community or state for facing any natural disaster". However, this definition has its own short falls as the new trend in disaster resilience has moved on to focus on bouncing forward. For example, some scholars like Parker [28] critiqued the notion of bouncing back following a disaster. The most comprehensive definition of resilience can be drawn from the Intergovernmental Panel on Climate Change (IPCC) [13: 1772], that defines it as: "the ability of a system, community or society exposed to hazards to resist, absorb, accommodate to and recover from the impacts of a hazard in a timely and efficient manner; including through the preservation and restoration of its essential basic structures and functions". Accordingly, community-based resilience entails the sustained ability of community to use its available resources to response and recover from adverse disturbances. This suggests that communities are said to be resilient to climate change when they can harness their locally available capital to respond to climatic events and bounce back, or even improve their livelihoods after learning from such disturbances. Thus, the meaning of resilience inferred in this conceptualisation indicates that communities may find it very difficult to be resilient in the milieu of worsening climatic disturbances. This is particularly challenging given the confluence of severe climatic shocks and underlying non-climatic vulnerabilities that were highlighted by Defe and Matsa [9].

Noting the increased frequency and unpredictability of extreme weather and climate events in developing countries, Dapilah et al. [10] pointed out the need to support communities towards building stronger resilience. It is a worrying situation that the economies of developing countries face a great deal of a challenge [14]. The overdependency on agriculture and ~~a~~ low income have been cited as the major sources of vulnerability in these communities [31]. Unless risk reduction measures are put in place, drought is feared to these communities to serious disasters. Some scholars argue that living in disaster prone areas or in places susceptible to natural hazards does not necessarily mean damage and vulnerability, but it is the lack of resilience combined with limited knowledge and the perceptions of a population about the risk of potential damage that causes a problem [15, 16, 18].

The research conducted by Dai [8] revealed that an increase in aridity in most parts of the world has been associated with the changing global climate since 1950. Climate change rendered people vulnerable to food insecurity. The vulnerability of the food production systems implies that the increasing global population will face food insecurity challenges [22]. Mensah et al. [22] further argued that about 70% more food will be needed to feed 9.1 billion people by 2050. The ability of the land to supply food to the growing population will thus be undermined by climate change and variability. Given that the developing world is expected to experience significant impacts of climate change, understanding the extent of applied adaptation research specifically taking place in the global South is critically important [35]. The inability to manage climate change leads to abandonment of farms and loss of interest in agriculture, that also contribute to poverty and food insecurity. This indicates a low level of resilience to the crisis. Thus, community-based resilience to climate change is very critical in reducing the impact of climate change. Communities are therefore required to employ skills, develop adaptive and absorptive measures and increasing their transformative capacity when facing any disaster [17, 29, 37]. Several countries in Sub-Saharan Africa developed policies that help them address community vulnerability to climate change and variability (Ahadzie 2020). For instance, Kenya, Liberia, Uganda, Ghana and Tanzania prepared National Adaptation Plan (NAP) in line with the United Nations Framework Convention on Climate change (UNFCCC) to develop climate-resilient agriculture. This points to the importance of community-based climate change adaptation measures in ensuring the ability of communities to bounce forward following climate change disasters.

There are several studies on community resilience to disasters that demonstrate the need for communities to develop permanent measures that can help in reducing the impact of climate change. Ahadzie et al. (2020) conducted a study on climate change resilience in Ghana where they examined Climate Smart Agricultural (CSA) practices. They outlined key CSA practices that farmers have prioritised, their benefits, challenges, and possible remedial actions. Their study found that local strategies and CSA measures improved food security. Mapfumo et al. [21] conducted a study on small holder farmers in Makoni and Hwedza, Zimbabwe to understand local perceptions of the impacts of climate change, and how indigenous knowledge could be utilized together with scientific approach to help farmers develop appropriate response to any emerging meteorological disaster. Their study further revealed evidence of increased climate variability and heightening vulnerability in farming systems. They also found that rainfall seasons have shortened, thereby impacting on food sources and ecosystem services, which in turn increasing pressure on women and children as traditional roles change. This clearly indicates that climate change is a cause for concern among many agrarian-based communities in Africa. Accordingly, there is need for robust adaptation pathways to enable communities to be resilient to climate change and variability.

Local measures need to be documented to assess and implement mechanism that can further increase resilient adaptation pathways. Chikowo et al. (2016) found out that communities in Hwedza and Makoni depended primarily on indigenous knowledge and local biological and geographical indicators of seasonal forecasts in making major decisions on crop production, including management of food stocks and social safety nets. However, a major concern is the continued reliance by farmers on indigenous knowledge against the continued loss of the resource base upon that the knowledge is built (Chikowo et al. 2016). Chadhary et al. (2012) investigated community-based climate change adaptation for building local resilience in the Himalayas. Their findings established that community-based biodiversity management approach has contributed to conserving and utilising the genetic resources and associated knowledge, increasing access to genetic resources through exchange and diversifying income sources. The approach has "promoted community seed banks, value addition of farm products, the identification of crops that thrive best in harsh climate, the establishment of group fund and the strengthening of local social institutions" (Chadhary et al. 2012). Thus, community-based interventions are very critical if the communities are to deal with climate related problems. External driven adaptation interventions, particularly those involved non-governmental organisations have been criticised for failing to promote sustainable development trajectories [5].

Effah et al. (2014) developed a matrix for community-based resilience to climate change. This matrix has three types of resilience namely ecological, engineering, and socio-economic. They used the matrix to assess the resilience of the community to climate change in northern Ghana. Community maintains diverse resources (including livestock, poultry, or through fishing), shares knowledge of climate and ecological risks (floods and droughts) and engages in diversified sources of livelihood or income. For example, non-farming jobs like trading or food processing or as blacksmiths are considered as socio-economic resilience to climate change. Flood or drought protection measures other than early warning systems, reliable early warning systems and access to irrigation are considered as engineering resilience. Crop diversification by community members is considered as an aspect of ecological resilience. It can be learnt from this model that community-based strategies can improve agricultural productivity and reduce the negative impacts of climate change in the community, aspects that can promote climate resilient communities.

From this discussion therefore, it is critical to evaluate community-based interventions in dealing with climate change and establish the challenges facing communities in the face of growing climatic risks. This assessment can guide policy directions towards climate resilient pathways that address the unique needs of communities experiencing climate change in their settings. The next section sets to show how Chimanimani has been used as a particular case to illustrate community-based resilience practices because of increasing frequency and severity of cyclones and floods being experienced in the area.

3 The study area

Chimanimani District is located in Manicaland province, eastern side of Zimbabwe. The district has 32,578 households with a total population of 134,940 people [38]. It covers a total of 3,548,050 km^2 distributed over five natural regions found in

Zimbabwe [6]. There are 23 wards in the district: 18 communal, 1 resettlement area, 3 commercial and 1 urban and peri-urban. The climate in Chimanimani is quite distinct because of a giant mountain that runs north-south for about 50 km, dividing the district into east and west Chimanimani. Thus, the eastern side has a cool and wet climate, whereas the western part is hot and dry. The altitude is very irregular, ranging from 142 m in the generally flat Save and Odzi valleys in the west to 2436 m in the mostly rugged and mountainous eastern side. Much of the rainfall in the district is received in summer, that stretches from October to February or up to April in the eastern side especially if the season is prolonged. Temperature averages 16 °C in the eastern side and 25 °C in the western side. Precipitation follows this irregularity with the wet eastern side receiving up to 2000 mm, whereas the dry western side gets a mean annual rainfall of just 500 mm [25].

The effects of poor watershed management have been identified as leading to biodiversity threats, the alarmingly low river volumes, and the threats to forest cover, rangelands and the increasing vulnerability of rural inhabitants. Declining rainfall, late onset of rainfall season, increasing mid-season dry spells, extreme temperatures, and increasing frequency and severity of cyclones are major sources of climatic vulnerabilities in the district [5]. The vulnerability situation is worsened by other non-climatic stressors as poverty, high population density and human pressure on the environment. The district has a mixed economy, with agriculture being the main economic activity. The livelihoods of communities are largely dependent on rain-fed agriculture. However, livestock production and crop yields have been declining as a result of climate change [26].

4 Methodology

This study used a mixed method approach in analysing community resilience to climate change and variability. A questionnaire-based survey was used to collect household quantitative data. Ten key informants from government institutions, eight focus group discussions (FGDs) and field observations were used to collect qualitative data. To identify the participants for this study, researchers used multi-stage sampling techniques where samples were selected by using combinations of different sampling methods to ensure that all interest group interests are represented. A 10% sample of the eight wards with 10 338 was considered and 1034 households were selected through simple random sampling. Using a case study approach, this study analyzed the factors that contribute to resilience that helped to determine the level of community resilience to climate change. Quantitative data was collected using Tablets running on KoboCollect. Using. This approach helped in data quality control checks. The questionnaire used for household data collection focuses on quantitative information. The household questionnaire collected quantitative information related to the transformative, absorptive and adaptive capacities based on 22 indicators (Table 1).

Table 1 Capacity indicators and their ranking based on contribution to resilience

Resilience capacities	Factor	Sub factor
Absorptive capacity	Active disaster management action plan	None, inadequate, well established but not updated, updated and functional
	Community preparedness for shocks	None, inadequate, well prepared
	Community's access to resources	Poor distribution of resources by community leadership during crisis, sharing resources
	Geographical location	Spatial distribution pattern of physical features Steep slopes, gentle slopes, flat land
	Social net works	Large number of people are not socially connected
	Experienced a shock in the last 12 months	No, yes
	Functional early warning systems	None, inadequate, functional EWS
	Soil fertility level	Poor soil fertility, fertility declining, good soil fertility
Adaptive capacity	Household size	Below 3, 3–6, 7+
	HHH occupation	Farmer, informal employment, formally employed, student, trader
	Dependency level	None, 1–2, 3+
	Status of livelihoods	Less than 3 livelihood options, 3+ options
	Age HHH	24–64, 65+
	HHH educational attainment	Primary level, secondary education, graduate, post-graduate, non-formal education advanced level, skills training and employed, tertiary
	HH Income level per month	Below $50, 50–100, 101–150, 151–200, 200+
Transformative capacity	Level of technology see educational attainment	None, inadequate, well established
	Access to social facilities	Access to road in > 5 km or > 1 h walking distance from home Access to road in < 5 km or < 1 h walking distance from home
	Access to health care facilities	Access to clinics, dispensaries, primary health care, secondary health care, national hospitals etc. road in > 5 km or > 1 h from home, access to road in < 5 km or < 1 h from home
	Access to education facilities SSS	Access in > 5 km or > 1 h from home, access to road in < 5 km or < 1 h from home
	Access to market	Access in > 5 km or > 1 h from home, access in < 5 km or < 1 h from home

(continued)

Table 1 (continued)

Resilience capacities	Factor	Sub factor
	Community relationship with authorities	Community not working with the authorities, community working?
	Conflicts management	Community does not participate in conflict resolution to sustain peace and security, community participates in conflict resolution

The data was analysed using SPSS Version 21 and Microsoft Excel. At each household, eight absorptive capacity indicators, seven adaptive and seven transformative capacity indicators were rated and combined using weighted average to produce a composite resilience index. The factors were selected since several studies cited them as major contributors to resilience in an area [2, 4]. A model for calculating resilience to climate change was developed based on procedure described in Adab et al. [1] and Mudavanhu et al. [24]. Adab et al. [1] idea of rating vulnerability factors using weighted average scale was adopted by this study to calculate the composite resilience index. The factors were were assigned a ranking value and the higher the score, the higher the contribution of the factor to the overall resilience index. The use of different weights based on relative importance of factors results in subjectivity of the resilience index. In order to eliminate this subjectivity, on the relative importance of the factors considered in this study, equal weights of 1 was used for all vulnerability factors. Resilience was further reclassified into 5 categories: very low resilience (0–0.2), low resilience (0.2–0.4), moderate resilience (0.4–0.6), high resilience (0.6–0.8) and very high resilience (0.8–1).

5 Access to Socio-economic Facilities Takes Care of 1. Access to Education 2. Access to Market

5.1 *Findings*

5.1.1 Community's Perceptions on the Signs Factors and Causes of Affecting Climate Change and Variability

The finding of the study revealed that local population in the area perceived that climate change is real as evidence by gradual changes in most of the elements of weather. When asked to recall to mention any observable changes in overall weather conditions that they had observed all the participants (100%) were of the opinion that climate is changing as evidenced by fluctuation in rainfall, temperature changes. Participants (Respondents) stated that excessive heat and unreliable rainfall were the main indicators of climate change that had been observed. Figure 1 shows that

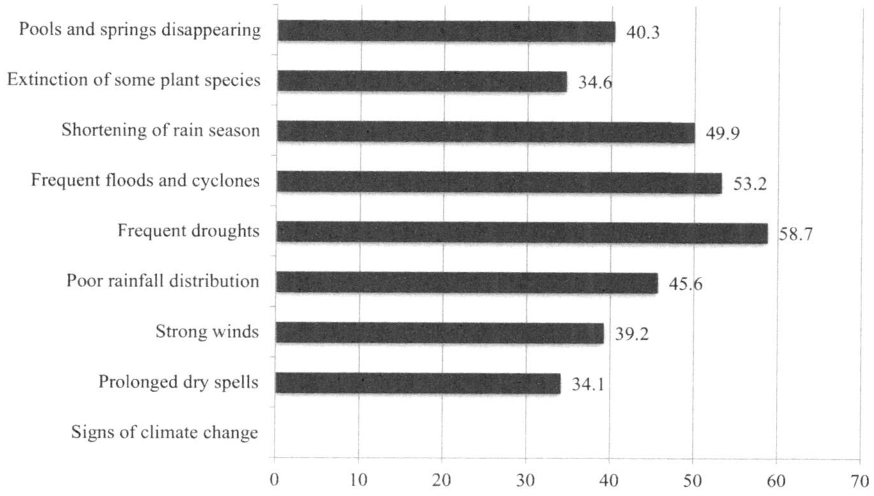

Fig. 1 Signs of climate change and variability in Chimanimani

more than half of the respondents, 58.7% and 53.2% indicated that they have been experiencing frequent droughts and cyclones respectively. The least common signs of climate change were prolonged dry spells (34.1%) and extinction of some plant species (34.6%). About (45.6%) of the respondents who perceived rainfall fluctuations as evidence of climate change added that the length of rainy season has drastically reduced with increased in dry season. The results indicates that communities are aware and very conscious about the on-going climate change that had occurred in the past. The community's perceptions and observations were also consistent with experts' observations. Interviews with one of the officials from the Meteorological Services Department indicated that climate is changing as evidenced by extreme weather events with temperature and rainfall fluctuations.

To ascertain if community's perceptions of changes in rainfall and temperature patterns correspond to actual long-term climatic records, climate data of Chimanimani District was assessed. Results indicated that declining rainfalls, late onset of seasons, increasing mid-season dry spells, extreme temperatures and decreasing crop yields have been reported in the district. Summer temperature increases have been projected to range between 1.75 and 2.5 °C and annual average rainfall has been projected to decrease by 10–15% [7].

The community's perception on the causes of climate change and variability were also investigated. Results showed that 100% of the respondents perceived that the main causes of climate change were the anthropogenic factors with a slightly higher proportion of participants (41%) claiming deforestation as the main factor. Further investigations on the causes of deforestation revealed that toe-excavations of slopes in the name of development were the major culprits for deforestation and land degradation. One of the participants during FGD supported this saying: '*We have seen people clearing the forests for infrastructural development, mining activities and farming that are the major culprits of land degradation.*' Although some participants indicated that poor farming methods (22%), population growth (26%) and natural causes (11%) were also contributing to climate change and variability, it was then agreed during the FGDs that all these were contributing to deforestation and land degradation.

5.1.2 Impacts of Climate Change and Variability in Chimanimani

The study results revealed the occurrence of extreme weather events as the major impacts of climate change and variability in Chimanimani. Participants had witnessed several natural hazards, including floods and cyclones, extreme temperatures, including the delay in the farming season. The frequency of occurrence of these hazards is presented in Table 2. A significant number of respondents indicated that extreme temperatures (22.9%) are becoming more frequent followed by drought and windstorms. Although more than 50% indicated that tropical cyclones, windstorms and droughts are occasional they highlighted that their frequency have increased in the past 20 years. Apart from frequency of occurrence of these climate related hazards, this study noted that the hazards are becoming more intense at a scale previously unimagined as indicated by the key informants saying:

> Rainfall and temperature patterns have changed. We are experiencing different climate related hazards these hazards have become frequent and more intense than before. We experienced Cyclone Elline in 2000, Japhet in 2003, Flavio in 2007, Dineo in 2017, Kenneth and Idai in 2019, Chalaine and Eloise in 2020. This shows that cyclonic activities are increasing in the region. Key Informant 1.

> Climate related hazards are now a cause for concern in our district. The frequency and intensity is now worrying us. Climate change is real and is affecting us. We are experiencing loss of lives, damage to property and loss of livelihoods because of these extreme weather events. Key Informant 5.

The results presented in Table 3 revealed a decline in crop yield due to extreme weather events, new pests and diseases, loss of arable land due to degradation, loss of livestock to drought and flooding, conflicts over resources and extinction of animal and tree species as the major impacts of climate change and variability in Chimanimani District. These findings suggest that climate change and variability impacts the

Table 2 Frequency of identified climate related hazards in Chimanimani

	Cyclones	Extreme temperatures	Hail	Windstorms	Drought	Landslides
Frequently	2.9	22.9	12.3	15.3	18.0	1.9
Occasionally	59.1	58.5	46.2	55.8	52.7	23.6
Seldom	35.8	16.8	37.3	26.3	25.6	65.2
Never	2.2	1.7	4.2	2.6	3.7	9.3

Table 3 Impacts of climate change and variability in Chimanimani District

Impact	Frequency (%)
General decline in crop yields per ha	80.3
Loss of livelihoods due to flooding	63.3
Low yields due to extreme weather events (drought and floods)	70.6
Loss of arable land	58.1
Conflicts over land resources	61.1
Loss of livestock to outbreak of diseases and disasters	20.9
Emergency of new pests and diseases	23.7
Extinction of some plant and animal species	34.6
Pools and springs disappearing	20.8

Chimanimani community negatively. Climate change causes households with few survival options to be more vulnerable. They disrupt social functioning of people by destroying the common forms of survival such as food security, health, education and other related aspects of humanity.

5.1.3 Community-Based Interventions in Dealing with Climate Change and Variability

To understand the level of community resilience to climate change and variability, this study assessed how communities have adjusted to the changes in weather conditions and the environment. Results indicated the community's 10 common interventions to reduce the impacts of climate change and variability. Most respondents (93.7%) have resorted to livelihood diversification as a way of cushioning the main livelihood activity. As shown in Table 4, the most common adaptation strategies included diversification of off-farm and on-farm activities, intercropping in case one fails and also to maximise the available land, early planting, and construction of contour ridges to reduce soil erosion during periods of heavy down pour and community-based by-laws to manage the environmental resources. All the identified strategies were agro-based. None of the strategies was in response to property that could be damaged by extreme weather events such as flooding or heavy storms. This is because the community mainly depends on subsistence agriculture for food and income. Therefore agro-based adaptation strategies were critical given that all the household respondents revealed that they depend on agriculture as their main source of livelihood.

5.1.4 Level of Community Resilience to Climate Change and Variability

This study quantified indicators of resilience to climate change based on the three capacities that are absorptive, adaptive and transformative. The eight absorptive

Table 4 Community's climate adaptation strategies

Adaptation strategy	Frequency (%)
Conservation farming *(Pfumvudza)*	88.7
Intercropping	33.5
Livelihood diversification	93.7
Crop rotation	78.2
Investing in irrigation schemes	16.8
Early planting	37.6
Growing short seasoned variety crops	56.7
Contour ridges	78.3
Afforestation in steep slopes	53.6
Community based by-laws on use of land	73.8

capacity indicators assessed were the existence of functional early warning systems, community preparedness, and household experienced a shock in the last 12 months, fertility of the soil, location of the homesteads, awareness of the impacts of climate change, community sharing resources and the existence of an updated disaster risk management plan. The adaptive capacities considered the level of income, socio demographic status (gender of household head, household size, and occupation, level of education, dependency and age of household head), livelihood options, and use of technology in livelihood activities such as irrigation. Lastly, the transformative capacities included access to health and education facilities, access to road and market, ability to resolve conflicts, powered dynamics and the community relationship with local authorities. As shown in Table 5, the weighted mean values of the 22 indicators of resilience ranged from 0.15 to 1. Table 5 indicates that the absorptive capacities ranged from 0.395 (Ngorima) to 0.531 (Nyahode) indicating very low to low capacities in all the eight selected wards. Figure 2 shows the major factors contributing to very low to low absorptive capacities such as the non-existence of functional and updated DRM plans and early warning systems, poor soil fertility as a result of poor farming methods and land degradation as well community locations. Analysis of how the location is contributing to low resilience, this study revealed that most people are settled in fragile and dangerous locations such as steep slopes and along the flood plains.

Even though the communities have low absorptive capacities, community sharing and networking is commendable (with average of 1). Generally, people had the habit of sharing the available resources and work together whenever they experience a climate related shock (see Fig. 2). One of the participants supported this saying:

> Yes, we are having challenges, but we are united. In case of an emergency people work together to an extent of helping the disadvantaged with shelter, food, clothing and also providing psychosocial support services.

This was also confirmed by one of the local authorities during KIIs who said that:

Table 5 Determinants of community resilience to climate change

WARD	Chikukwa	Ngorima	Nyahode	Ngangu	Gwindingwi	Martin	Biriiri	Tilbury	RS
Absorptive capacity									
Functional disaster management plan	0.09639	0.14815	0.13131	0.05833	0.31081	0.25688	0.19608	0.04167	0.154953
Community preparedness for shocks	0.54217	0.40741	0.75758	0.65	0.82432	0.51376	0.72549	0.58333	0.625508
Community access to resources	1	1	1	1	1	1	1	1	1
Location	0.075	0.02174	0.23899	0.14375	0.00763	0.06897	0.30435	0.12651	0.123368
Social net works	0.97541	0.95652	1	0.99379	0.90152	1	1	0.95808	0.973165
Experienced a shock in the last 12 months	0.49398	0	0.65657	0.44167	0.16216	0.23853	0.33333	0.57639	0.362829
Functional early warning systems	0.2623	0.36957	0.24224	0.4472	0.34091	0.36571	0.62319	0.34132	0.374055
Soil fertility	0.19672	0.26087	0.2236	0.28571	0.17424	0.29714	0.21739	0.38323	0.254863
Absorptive capacity score	0.4552463	0.3955325	0.5312863	0.5025563	0.4651988	0.4676238	0.5499788	0.5013163	0.483592
Adaptive capacity									
Household size	0.775	0.80587	0.72242	0.71876	0.73659	0.716	0.76029	0.70467	0.74245
HH occupation	0.81967	0.97826	0.68323	0.63975	0.63636	0.93143	0.97101	0.75449	0.801775
Dependency	0.60656	0.71739	0.57143	0.58385	0.50758	0.52	0.44928	0.61078	0.570859
Status of livelihoods	0.81967	0.8587	0.88043	0.79814	0.82576	0.81714	0.72464	0.79641	0.815111
Age HHH	0.59836	0.80435	0.80745	0.83851	0.59848	0.72571	0.66667	0.88024	0.739971
HHH education	0.62705	0.65217	0.67702	0.74534	0.58523	0.62714	0.61232	0.68263	0.651113
HH income	0.46721	0.32609	0.15528	0.42857	0.08333	0.22286	0.17391	0.17964	0.254611
Adaptive capacity	0.67336	0.73469	0.6424657	0.6789886	0.5676186	0.6514686	0.6225886	0.6584086	0.653699
Transformative capacity									
Use of technology	0.78333	0.43478	0.67925	0.7375	0.29771	0.57471	0.37681	0.57831	0.5578
Access to social amenities	0.51807	0.55556	0.39394	0.61667	0.40541	0.24771	0.19608	0.74306	0.459563

(continued)

Table 5 (continued)

WARD	Chikukwa	Ngorima	Nyahode	Ngangu	Gwindingwi	Martin	Biriiri	Tilbury	RS
Access to Health facilities	0.51639	0.19565	0.62112	0.26087	0.70455	0.61143	0.78261	0.42515	0.514721
Access to education facilities	0.37705	0.13043	0.79503	0.3354	0.34848	0.46286	0.71014	0.41916	0.447319
Access to market	0.6747	0.40741	0.33333	0.46667	0.55405	0.47706	0.86275	0.39583	0.521475
Community relationship with authorities	0.79189	0.75556	0.81856	0.79664	0.82895	0.7717	0.87692	0.87536	0.814448
Conflicts management	0.49167	0.69565	0.49057	0.5875	0.61832	0.57471	0.89855	0.75301	0.638748
Transformative capacity score	0.5933	0.4535771	0.5902571	0.5430357	0.5367814	0.5314543	0.67198	0.5985543	0.564868
Composite resilience score	0.5739688	0.5279332	0.588003	0.5748602	0.5231996	0.5501822	0.6148491	0.586093	0.567386

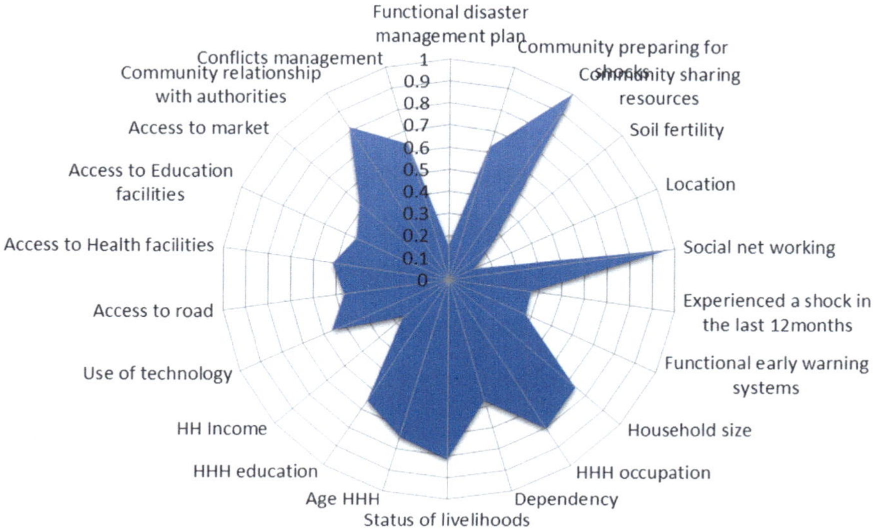

Fig. 2 Factors determining level of resilience in Chimanimani District

> Yes, we may say that communities are vulnerable, but they have certain capacities, knowledge, skills and resources that they have to assist each other in times of shocks and stresses (that is resilience). For example, during Tropical Cyclone Idai, people managed to rescue some of the survivors, provided food and accommodation. They were also sharing information on what was happening in different communities.

Ngorima had the lowest absorptive capacities, the community had the highest adaptive capacity score (0.735). As shown in Table 5, Gwindingwi had the lowest adaptive capacities (0.567). The transformative capacities ranged between 0.453 and 0.598, contributing to low resilience in the study area. Although the communities are in fragile environments, this study also revealed that the communities had generally moderate to high adaptive capacities. As presented in the previous section on the strategies adopted to respond to climate change, it was noted that all household had adopted different livelihoods strategies (livelihood diversification) to address the impacts of climate change. As such the livelihood status ranged from 0.7 to 0.9. Although all the other factors were generally high, the level of income among the communities was worrisome (0.08–0.4). Further investigations on the level of income revealed that, although the communities have growing and diversified livelihoods the income was generally low due to limited access to the market and inaccessibility of other communities by road especially during the rainy season. This has caused some of the produce not to reach the market on time. These factors also contributed to low transformative capacities. It was also noted that the low income was due to lack of value addition of the products as 100% of the produce were agro-based where price is determined at the market.

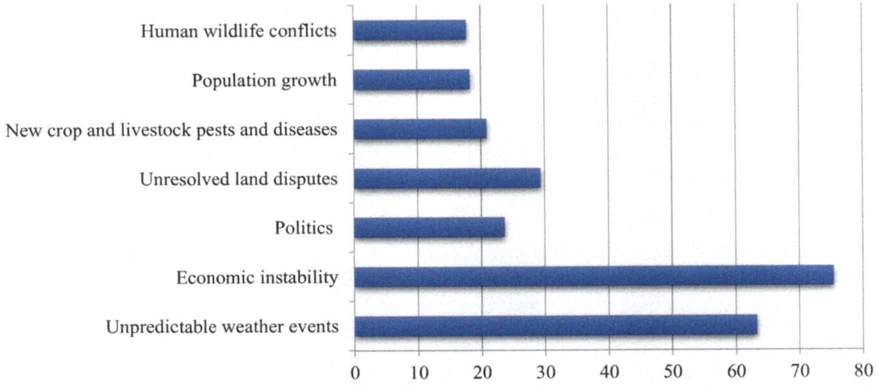

Fig. 3 Challenges facing community members in responding to climate change

Further analysis revealed that although the communities are managing to adapt to climate change with a certain degree of resilience, this study noted that most of the adaptive strategies are not very effective. The challenges faced by farmers ranges from unpredictable weather events to human wildlife conflicts as presented in Fig. 3. More than half of the respondents indicated that the economic instability (75.6%) and unpredictable events (63.3%) were the major challenges in climate change adaptation. The extreme weather events are threatening agro-based livelihood strategies where 100% of the respondents rely on.

> As human beings, we come up with various solutions to address our problems but sometimes it is difficult to plan due to the sudden and frequency of occurence of weather hazards. In 2018 we had serious drought and our livelihhods were affected. In 2019 we had Tropical Cyclone Idai and we lost hectares of arable land, another year of food insecurity. In 2020 we were threatened by Cyclone Eloise and Chalaine. We do not know what is coming in 2021. (A statement made by community leader during the FGD).

Figure 4 illustrates the number of households at each level of community resilience in each of the eight sampled wards. Generally, the community resilience is low implying that most of the households are vulnerable to climate change and climate variability as derived from the analysis of resilience components where on average 63.6% of the households fall under the low to very low resilience level. Only 12% were classified as high and very high resilient. None of the households in Ngorima were classified as highly resilient to climate change with 76.1%. Ngorima, Nyahode and Tilbury had more than 70% of the people who are highly vulnerable to climate change proving that resilience is low (Fig. 4).

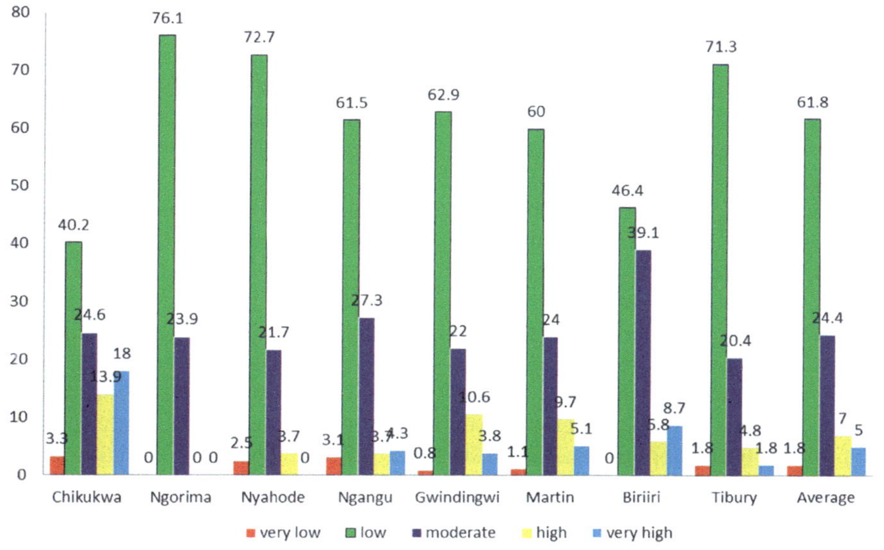

Fig. 4 Level of community resilience

6 Discussion

The findings of this study have shown that adopting an in-situ participatory assessment involving people experiencing climate change phenomena can be a sound way of understanding the resilience situation of communities. The scope of resilience analysis has been traced from the people's perception of climate change, their understanding of drivers and causes of climatic disturbances to assessment of response practices that the community adopt when they face these disturbances. Communities' construct to climate change as a phenomenon driven by human activities and their capability to observe climatic impacts can be understood as a necessary foundation towards resilience building. Knowledge of climate change has been found to be very strong. Local perceptions of climate change corroborated with recorded data and was consistent with expert observations. The indicators identified by the community were consistent with those existing from past studies, that have been summarised as interruptions in rainfall and temperature regimes, increase in frequency and severity of extreme weather and climatic events such as drought and floods, irregular seasons, short rainy season being interrupted by frequent and long dry spells, and ecosystem and hydrological changes. Similarly, Savo et al. [32] revealed that observations of increased temperature by local communities in many parts of the world showed patterns consistent with projected trends from global climate models. On a different note, a study conducted in Gweru, Zimbabwe by Makuvaro et al. [19] showed that although temperature increases reported by local people corroborated with established evidence of a warming environment, the farmers' perceptions on rainfall deviated from historical climatic trends. In Chimanimani District, the observed climatic events have disruptive effects on communities and are the main threats to their resilience. However, this high level of knowledge has not translated into community resilience.

Although, the communities studied are managing to adapt to climate change with a certain degree of resilience, this study noted that most of the adaptive strategies are not effective enough. The resilience situation is quite heterogenous in line with the variations in the resilience components studied. While exploring the relationship between climate change perceptions and adaptation practices in Chimanimani, Mutandwa et al. [26] observed that indigenous knowledge systems and non-governmental organizations increased the likelihood farmers' acknowledgement of climate change that in turn influenced uptake of climate services. Similar to what this study reveals, the probability of adopting adaptation strategies was influenced by education level, land tenure and access to extension services. There is now growing recognition of the inclusion of local and indigenous knowledge into community-based adaptation projects that most scholars understand as critical to successful climate adaptation [11, 27, 34]. Resilience and community development practitioners would need to make use of this information to make ward-specific adaptation interventions.

This study has shown that the resilience terrain consists of numerous factors that are both natural and anthropogenic, and endogenous and exogenous to the communities experiencing climate change disturbances. Within endogenous forces, social capital that existing in the form of resource sharing and social networking, tend to strengthen the communities' absorptive capacity that positively influence community resilience. Characterisation of these forces is important in order to identify strategic resilience pathways in the face of growing climatic and non-climatic forces in the Zimbabwean environment. There is agreement amongst adaptation scholars that climatic and non-climatic shocks influence community resilience. Defe and Matsa [9] indicated that shocks emanating from natural hazards reinforce with socio-economic stressors such as epidemics and poverty to undermine resilience building initiatives in many places at risk of climate change. In a similar study with agriculture-dependent communities in Zimbabwe, Gwenzi et al. [12] identified extreme climatic hazards such as drought, dry spells and high temperatures as factors impinging on community resilience.

With focus on community-based resilience, this study has broadened the scope of adaptation assessment beyond the common focus on vulnerability assessments (e.g., by Rurinda et al. [30]) and adaptive capacity assessment (e.g., by Chanza et al. 2018). It has comprehensively assessed the resilience notion by isolating the resilience elements (namely absorptive capacity, adaptive capacity and transformative capacity). Thus, all these elements have been found to depict the resilience status of communities at risk of climate change in Chimanimani. Community-based resilience can be comprehensively understood by joint treatment of these components. For example, there were situations where some wards scored highly in one resilience component but their cumulative resilience indices were found to be very low. Resilience interventions would need to deal with threats on the communities' agrarian system (crop and livestock production), address drivers of poverty (education, income and markets), give user-friendly climate information services through early warning systems that promote early action against climatic hazards and emergencies, while taking measures to reverse environmental degradation in order to maintain ecosystem services. This observation resonates with insights given by Defe and Matsa [9]. The scholars argued that resilience building interventions need to be comprehensive to counter shocks to be able to protect livelihoods and realise recovery gains. They recommended strengthening interventions in crop production, livestock production and human well-being.

7 Conclusion

This chapter has shown that the concept of resilience can best be understood through participatory engagement with communities affected by climate change. Our framing of community-based resilience in a context such as Chimanimani has been informed by the disruptive effects of climate change that the communities were aptly able to construct. The existing adaptation practices have been found to be relatively weak to withstand the observed and anticipated climate change and variability. It is therefore critical to understand that community-based resilience to climate change can be assessed from the broad spectrum of climate change indicator identification, understanding of drivers of vulnerability, impact identification and assessment to the joint evaluation of responses that the communities and other adaptation stakeholders implement. In order to orient communities towards a 'bouncing forward trajectory', demanded by the current definition of resilience, there is need to focus on this whole spectrum, with communities being seen at the centre of the investigation. To do this, the resilience research, practice and policy community would need to target all the three components of resilience as absorptive capacity, adaptive capacity and transformative capacity. Our study has contributed towards addressing the intriguing question of 'whose resilience', a subject that continues to characterise resilience discourse owing to failure by vulnerability and adaptation studies to put communities at the centre of research. It achieved this by focusing on the who, how, when and what aspects of the community-based resilience to climate change.

References

1. Adab H, Kanniah KD, Solaimani K (2011) Gis-based probability assessment of fire risk in grassland and forested landscapes of Golestan Province, Iran. In: 2011 international conference on environmental and computer science Ipcbee. Iacsit, Singapore
2. Alam MS, Chakraborty T, Islam MD (2019) Community resilience of urban slums to climate change induced events: a case study of five major slum in Khulna City, Bangladesh. Conference Paper March 2019. https://www.researchgate.net/publication/332632886

3. Antwi EK, Otsuki K, Saito O, Obeng FK, Gyekye KA, Boakye-Danquah J, Boafo YA, Kusakari Y, Yiran GA, Owusu AB, Asubonteng KO (2014) Developing a community-based resilience assessment model with reference to Northern Ghana
4. Asmamaw M, Mereta ST, Ambelu A (2019) Exploring households' resilience to climate change-induced shocks using climate resilience index in Dinki watershed, central highlands of Ethiopia. PLoS ONE 14(7):e0219393. https://doi.org/10.1371/journal.pone.0219393
5. Chanza N, Siyongwana PQ, Williams-Bruinders L, Gundu-Jakarasi V, Mudavanhu C, Sithole VB, Manyani A (2020) Closing the gaps in disaster management and response: drawing on local experiences with tropical cyclone Idai in Chimanimani, Zimbabwe. Int J Disaster Risk Sci. https://doi.org/10.1007/s13753-020-00290-x
6. Chimanimani Rural District Council (2019) District profile http://www.chimanimanirdc.org.zw/index.php/16-about-us/2-district-profile. Accessed 7 June 2021
7. Chimanimani District Climate Change Response & Watershed Management Strategy (2017) https://www.birdlife.org/sites/default/files/attachments/final_chimanimani_district_ccwsm_strategy.pdf. Accessed 8 July 2021
8. Dai A (2013) Increasing drought under global warming in observations and models. Nat Clim Change 3:52–58
9. Defe R, Matsa M (2021) Resilience building initiatives to counter shocks and stressors affecting rural communities in Chiredzi District, Zimbabwe. In: Behnassi M, Gupta H, El Haiba M, Ramachandran G (eds) Social-ecological systems (SES). Springer, Cham. https://doi.org/10.1007/978-3-030-76247-6_8
10. Dapilah F, Nielsen JØ, Friis C (2020) The role of social networks in building adaptive capacity and resilience to climate change: a case study from northern Ghana. Clim Dev 12(1):42–56
11. Gadzirayi TC, Manatsa D, Mutandwa E (2020) Tailoring conservation farming to climate change in the smallholder farming sector: case of lowveld areas in Zimbabwe. Afr J Sci Technol Innov Dev 12(5):581–590. https://doi.org/10.1080/20421338.2020.1800242
12. Gwenzi J, Mashonjowa E, Mafongoya PL (2020) Coping with extreme weather in arid areas, a case study of Uzumba Maramba Pfungwe District, Zimbabwe. Clim Change Manage 701–716. https://doi.org/10.1007/978-3-030-37425-9_35
13. IPCC (2014) Climate change 2014: impacts, adaptation, and vulnerability. In: Barros VR, Field CB, Dokken DJ, Mastrandrea MD, Mach KJ, Bilir TE, Chatterjee M, Ebi KL, Estrada YO, Genova RC, Girma B, Kissel ES, Levy AN, MacCracken S, Mastrandrea PR, White LL (eds) Part B: regional aspects. Contribution of working group II to the fifth assessment report of the intergovernmental panel on climate change. Cambridge University Press, Cambridge, United Kingdom and New York, NY, USA, p 688
14. Javadinejad S, Dara R, Jafary F (2021) Analysis and prioritization the effective factors on increasing farmers resilience under climate change and drought. Agric Res 10:497–513
15. Javadinejad S, Eslamian S, Ostad-Ali-Askari K (2019) Investigation of monthly and seasonal changes of methane gas with respect to climate change using satellite data. Appl Water Sci 9(8):180
16. Javadinejad S, Ostad-Ali-Askari K, Eslamian S (2019b) Application of multi-index decision analysis to management scenarios considering climate change prediction in the Zayandeh Rud River Basin. Water Conserv Sci Eng 4(1):53–70 23
17. Kates RW, Travis WR, Wilbanks TJ (2012) Transformational adaptation when incremental adaptations to climate change are insufficient. Proc Natl Acad Sci USA 109(19):7156–7161. https://doi.org/10.1073/pnas.1115521109
18. Kelman I (2019) Axioms and actions for preventing disasters. Prog Disaster Sci 2:100008
19. Makuvaro V, Murewi CTF, Dimes J, Chagonda I (2018) Are smallholder farmers' perceptions of climate variability and change supported by climate records? A case study of Lower Gweru in Semiarid Central Zimbabwe. Weather Clim Soc 10(1):35–49. https://doi.org/10.1175/WCAS-D-16-0029.1
20. Mango N, Makate C, Tamene L et al (2018) Adoption of small-scale irrigation farming as a climate-smart agriculture practice and its influence on household income in the Chinyanja Triangle. Land, Southern Africa

21. Mapfumo P, Mtambanengwe F, Chikowo R (2015) Building on indigenous knowledge to strengthen the capacity of smallholder farming communities to adapt to climate change and variability in southern Africa. Clim Dev 8(1):72–82. https://doi.org/10.1080/17565529.2014.998604
22. Mensah H, Ahadzie DK, Takyi SA, Amponsah O (2020) Climate change resilience: lessons from local climate-smart agricultural practices in Ghana. Energy Ecol Environ 5(5):353–364. https://doi.org/10.1007/s40974-020-00172-4
23. Ministry of Lands, Agriculture, Water, Climate and Rural Settlement (2017) Rainfall characteristics in the five natural regions of Zimbabwe. http://www.moa.gov.zw/index.php/zim/. Accessed 5 June 2021
24. Mudavanhu C, Manyangadze T, Mavhura E, Pedzisai E, Manatsa D (2020) Rural households' vulnerability and risk of flooding in Mbire District, Zimbabwe. J Nat Hazards. https://doi.org/10.1007/s11069-020-04145-y
25. Mugandani R, Wuta M, Makarau A, Chipindu B (2012) Re-classification of agro-ecological regions of Zimbabwe in conformity with climate variability and change. Afr Crop Sci J 20(s2):361–369
26. Mutandwa E, Hanyani-Mlambo B, Manzvera J (2019) Exploring the link between climate change perceptions and adaptation strategies among smallholder farmers in Chimanimani district of Zimbabwe. Int J Soc Econ 46(7):850–860. https://doi.org/10.1108/IJSE-12-2018-0654
27. Nalau J, Becken S, Schliephack J, Parsons M, Brown C, Mackey B (2018) The role of indigenous and traditional knowledge in ecosystem-based adaptation: a review of the literature and case studies from the Pacific islands. Weather Clim Soc 10(4):851–865. https://doi.org/10.1175/WCAS-D-18-0032.1
28. Parker DJ (2020) Disaster resilience – a challenged science. Environ Hazards 19(1):1–9. https://doi.org/10.1080/17477891.2019.169485
29. Pelling M (2010) Adapting to climate change: from resilience to transformation. Taylor & Francis, Routledge
30. Rurinda J, Mapfumo P, van Wijk MT, Mtambanengwe F, Rufino MC, Chikowo R, Giller KE (2014) Sources of vulnerability to a variable and changing climate among smallholder households in Zimbabwe: a participatory analysis. Clim Risk Manage 3:65–78
31. Sarker A, Chakraborty S, Roy M (2020) Dark red kidney bean (Phaseolus vulgaris L.) protein hydrolysates inhibit the growth of ox- idizing substances in plain yogurt. J Agric Food Res 2:100062. https://doi.org/10.1016/j.jafr.2020.100062
32. Savo V, Lepofsky D, Benner JP, Kohfeld KE, Bailey J, Lertzman K (2016) Observations of climate change among subsistence-oriented communities around the world. Nat Clim Change 6:462–473. https://doi.org/10.1038/NCLIMATE2958
33. United Nations Office for Disaster Risk Reduction (2015) The Sendai framework for disaster risk reduction 2015–2030. United Nations
34. Whitney CK, Frid A, Edgar BK, Walkus J, Siwallace P, Siwallace IL, Ban NC (2020) "Like the plains people losing the buffalo": perceptions of climate change impacts, fisheries management, and adaptation actions by Indigenous peoples in coastal British Columbia, Canada. Ecol Soc 25(4):33. https://doi.org/10.5751/ES-12027-250433
35. Xu C, Kohler TA, Lenton TM, Svenning J-C, Scheffer M (2020) Future of the human climate niche. Proc Natl Acad Sci 117(21):11350–11355
36. Yang Y, Liu B, Wang P, Chen WQ, Smith TM (2020) Toward sustainable climate change adaptation. J Ind Ecol 24(2):318–30. https://doi.org/10.1111/jiec.12984
37. Ziervogel G, Enqvist J, Metelerkamp L, van Breda J (2021) Supporting transformative climate adaptation: community-level capacity building and knowledge co-creation in South Africa. Clim Policy. https://doi.org/10.1080/14693062.2020.1863180
38. ZimStat (2013) Census 2012: Manicaland Province. Zimbabwe National Statistical Agency (ZimSTAT), Harare, Zimbabwe

Chipo Mudavanhu holds a PhD in Disaster Risk Management, MSc in Natural Resource management and Environmental Sustainability. Currently heading the Disaster Relief and Preparedness at Higherlife Foundation and worked as a Senior Lecturer in the Department of Disaster Risk Reduction at Bindura University of Science Education. Her research interests are in community resilience building, climate change adaptation, disaster risk reduction, public health emergencies, gender and development.

Nelson Chanza holds a Ph.D. in Environmental Geography. He is currently a Research Fellow in the Department of Urban and Regional Planning, University of Johannesburg. Before this, he worked as a Senior Lecturer in the Department of Geography at the Bindura University of Science Education, and Lecturer in the Department of Rural and Urban Planning at the University of Zimbabwe. His research contributions are largely in the fields of local knowledge applications in disaster management and climate science, in particular, the use of community-based knowledge in climate change impact assessment, adaptation and mitigation strategies that are understood and used by local communities. His research interests are in exploring the sustainability science discourse through examining the interface between climate change, disasters and society.

Rosemary Kasimba holds a Ph.D. in Sociology (Rhodes University) M.Sc. in Sociology and Anthropology, B.Sc. in Sociology (University of Zimbabwe) and Post-Graduate Diploma in Social Work. Her research interests include but not limited to disaster management, Human Resources Management, Industrial Sociology, Migration, Population and Development, Social Policy and Vulnerable Population.

Geo-Spatial Science Use for Disaster Risk Management

Samuel Kusangaya, Pathias P. Bongo, Beatrice Chikanda, Roy Boy Chiomba, and Albert Maipisi

Abstract In southern Africa, changes in the intensity, frequency, and duration of weather extremes have resulted in increasing incidences of tropical cyclones, resulting in recurrent droughts and floods. Projected increases in the frequencies and magnitudes of drought and flood events through to 2050, coupled with increasing population in the region, implies increased vulnerability and people at risk. This chapter posits that the use of geo-spatial science to enhance multi-hazard mapping as part of preparedness, mitigation, response and rehabilitation would have contributed to a different narrative with regard to the outcomes of tropical cyclone Idai in 2019. Consequently, geospatial techniques can be utilized effectively in the entire spectrum of the disaster cycle, henceforth enhancing disaster risk management. This chapter demonstrated that biophysical; socioeconomic and geo-spatial data sourced from remote sensing, census, and other spatial and non spatial databases can be employed to evaluate tropical cyclone-related risk. Furthermore, it was demonstrated that a variety of geo-spatial methods; comprising of inductive, deductive, and multi-criteria evaluation such as AHP can be used to determine elements at risk. Through the use of geo-spatial science, meteorological hazards such as cyclones, can be better mapped, understood and ultimately anticipated. In conclusion, it was noted that risk, vulnerability, and hazard have a spatial component, that which can best be handled using geo-spatial science. Even though this chapter largely focused on application of geo-spatial science it argues for a blend of such with community-level participatory mapping, which can be a powerful method for disaster risk management. It was therefore concluded that a proper mix of participatory and scientific hazard mapping are critical components to strengthen hazard risk reduction and disaster preparedness.

S. Kusangaya (✉) · B. Chikanda · R. B. Chiomba
Department of Geography, Geoinformation Science and Earth Observation, University of Zimbabwe, Harare, Zimbabwe
e-mail: kusangayas@yahoo.com

P. P. Bongo
Bindura University of Science Education, Bindura, Zimbabwe

A. Maipisi
Women's University in Africa, Harare, Zimbabwe

R. Kasimba and I. Chirisa (eds.), *Disaster Risks and Hazards*, Advances in 21st Century Human Settlements, https://doi.org/10.1007/978-981-96-4577-0_4

Keywords Disaster risk · Geo-spatial · Multi-hazard mapping · Cyclone · Preparedness · Response

1 Introduction

In southern Africa, changes in the intensity, frequency, and duration of weather extremes have resulted in increasing incidences of tropical cyclones resulting in recurrent droughts and floods. It has been noted that the El Niño-Southern Oscillation effect has intensified in the past four decades (1900–2010) resulting in increased frequency of floods and droughts in Southern Africa [49]. Kabat et al. [40] reported on the projected increase in the frequency of droughts which is likely to reduce flows and water storage. On the other hand, Mazvimavi [57] reported the increase in the frequency and magnitude of extreme precipitation events associated with flooding. Countries like Mozambique, South Africa, Malawi, and Zimbabwe are already experiencing recurrent floods and droughts [48]. Projected increases in the frequencies and magnitudes of such events through to 2050, coupled with increasing population in the region, suggest that the vulnerability of weather-sensitive sectors, such as water resources and agricultural production [49, 85] will also increase.

Several studies from southern Africa have shown that extreme hydrological events are increasing in frequency and magnitude [15, 34, 63, 86]. This has largely been driven by the El Niño-Southern Oscillations which are becoming more intense because of climate change (e.g., [16, 52, 59]). For example, the frequency of droughts is projected to increase and will most likely increase the frequency of extremely low flows and low storage episodes [24]. These will inevitably result in drought and flood related disasters, affecting aquatic ecosystems, water supply, irrigation, leisure, and hydropower generation. To note is that tropical cyclone activity in southern Africa has been recorded with increased frequency in recent years causing massive damage and loss of life mainly due to flooding for example in Mozambique, Zimbabwe, Zambia, Swaziland, South Africa, Madagascar, and Mauritius [68]. These observations ultimately signify a need for more preparedness to cyclone related disasters.

Data from the Emergency Events Database has shown that (a) Africa has the second highest number of extreme events after Asia, and, (b) Africa has the highest number of hydrological extreme events [30]. A significant number of these hydrological (e.g., floods) and climatological phenomena (e.g., droughts) are occurring in southern Africa. In addition, it was observed that (a) the actualisation of natural hazards into disasters in the last two decades (1991–2000 and 2001–2010) is by far higher than for the previous decades, with the 2001–2010 decade having the highest occurrences (Fig. 1) and (b) the occurrence of water-related disasters has been increasing at an exponential rate (Fig. 2). The number of such disasters in the last decade is in most cases equivalent to the total number of disasters in the previous three decades (Fig. 2). Predominant among these is the occurrence of cyclone-induced flooding. For better response as well as preparedness to flooding disasters, the use of

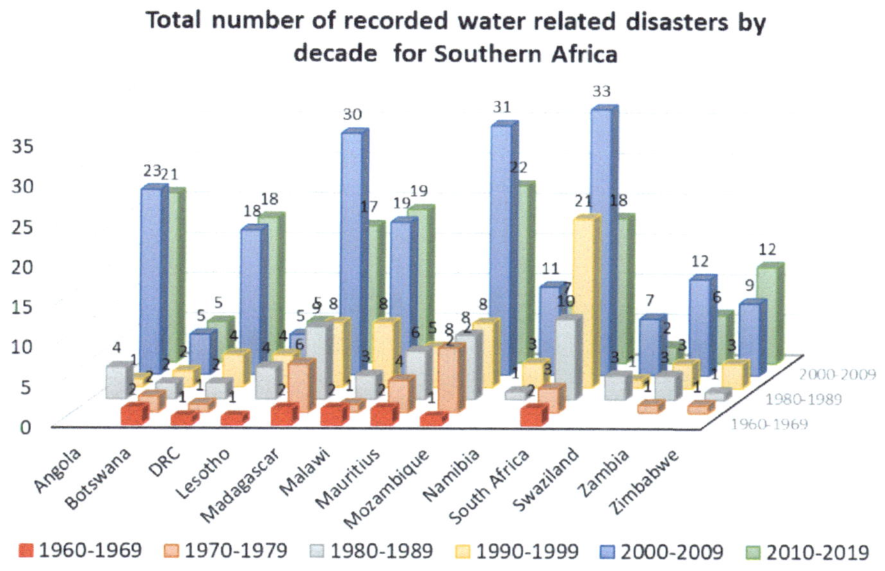

Fig. 1 Trends in the occurrence of water-related disasters in southern Africa by decade. *Source Data*: Source: EM-DAT: The Emergency Events Database—Universite catholique de Louvain (UCL)—CRED, D. Guha-Sapir—www.emdat.be, Brussels, Belgium (accessed July 4 2019)

geo-spatial technologies presents and effective and efficient near real time method that can be used in effectively dealing with the related risk.

Figure 2 shows the trends in the occurrence of water-related disasters for selected southern African countries. What is evident is that the occurrence of droughts has been increasing exponentially in the following countries: Lesotho, Malawi, Mozambique, South Africa, and Madagascar. Flood occurrence also exhibits the same pattern as droughts as can be seen in South Africa, Zambia, Mozambique, Malawi, Botswana, and Angola. Floods also results in indirect adverse effects such as epidemics [43, 45]. The increasing cases of recorded epidemics are also closely related to the increasing occurrence of floods due to increased cyclone activity as in the case of Mozambique, Zimbabwe, Lesotho, and Malawi. Recently, the region has also been experiencing heat waves and extensive wildfire phenomena. The upsurge in heat waves and wildfires is attributable to increasing temperatures.

1.1 *Flooding Disasters*

Flooding is one of the most frequent devastating disasters worldwide [44]. Flooding is referred to as a hydrological event that is characterized by high discharges or water levels that can lead to inundation of land adjacent to streams, rivers, lakes, wetlands, and other water bodies [5]. Floods have a profound effect on people and

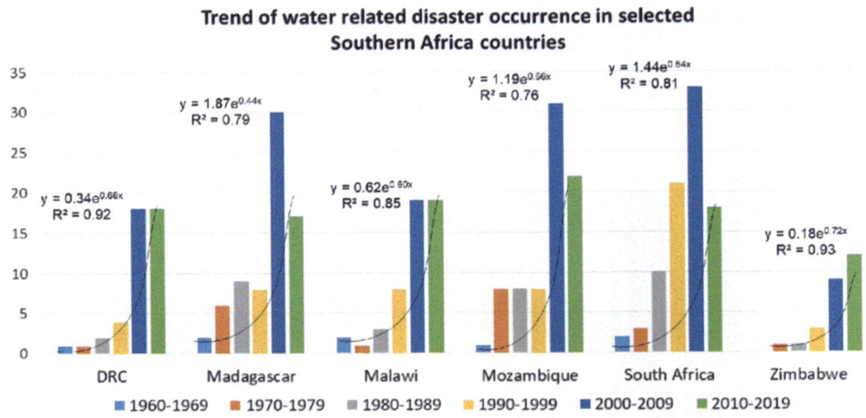

Fig. 2 Trend of water-related disaster occurrence in selected southern African countries

the economy and on an annual basis, they leave over 3 million people homeless [39]. Since the dawn of civilization, destructive floods have jeopardized settlements located in river valleys and floodplains causing the largest amount of deaths and property damage [75]. For example, floods that occurred between 1971 and 1995 affected more than 1.5 billion people worldwide [64], including 318,000 persons who were killed and over 81 million left homeless [72]. In 2003, floods accounted for 3723 fatalities around the world [84]. A ten-year comparative study of the world's great flood disasters from 1950 to 1998 showed that the number of flood events increased nearly threefold, while the economic losses of the 1990s are ten times as high as in the 1950s amounting to $250 billion globally [83].

Injuries that occur after floods not only result from the direct physical damage from the event but are also determined by the prevailing socio-economic and health conditions of the community. Without worldwide investment in flood defences, the flooded land area will increase by 48% [73]. The population impacted will increase by 52% (about 287 million people) and the infrastructure affected will increase by 46%, which will comprise about 20% of the global GDP [73]. Consequently, it has been noted that flooding is the leading cause of losses from natural disasters particularly cyclones and is responsible for damaging large areas more than any other type of natural event [26]. It is reported that flood disasters account for about a third of all-natural hazard-induced disasters by number and economic losses [83]. In addition, floods are responsible for over half of the deaths associated with all such disasters [78]. The continuous heavy rains worsen the situation as it can lead to mudflows and landslides, resulting in more loss of life and property, and a reduction in the asset base of households and communities through the inundation of agricultural lands that exacerbates food insecurity, and the destruction of infrastructure and machinery [26, 83]. This chapter posits that the use of geo-spatial technology to enhance multi-hazard mapping as part of preparedness, mitigation, response and rehabilitation would have contributed to a different narrative with regard to the outcomes of tropical cyclone Idai.

Overall, recurring floods [4, 65] have resulted in food insecurity [77], exacerbated health conditions, such as increased incidences of cholera [37, 38] and malaria [8, 51], loss of shelter and even loss of life as happened in the 2000–2001 Cyclone Eline induced floods which affected Mozambique, Zimbabwe and South Africa [41, 68]. In Zimbabwe, Manicaland Province was also heavily hit by Tropical Cyclone Idai on 15–17 March 2019 resulting in disrupted livelihoods, destruction of houses, infrastructure damage and loss of agricultural land [58].

2 Disaster Risk Reduction

Social and economic under-development, disease epidemics and the impacts of HIV and AIDS exacerbate the impacts of disasters, posing significant threat to the SADC region [48] and the ability to achieve the United Nations Millennium Development Goals and sustainable development [9]. As a response measure, governments need to implement disaster risk management measures that ensure the safety and protection of communities and economic assets. Disaster risk management measures include preparedness, mitigation, response, rehabilitation and recovery. Disaster risk management involves the application of disaster risk reduction policies and strategies, so as to prevent new disaster risks, reduce existing disaster risks, and manage residual risks, contributing to the strengthening of resilience and reduction of losses [79]. According to Le Cozannet et al. [20], disaster risk management actions can be divided into the different steps of the disaster management cycle, which include: prevention, preparedness and crisis management, often focused on saving lives, and post-crisis management aimed at re-establishing services which support human activities. According to Carreño et al. [10] disaster risk is defined as the potential economic, social, and environmental consequences of hazardous events that may occur in a given period of time. In order to evaluate risk, the assessment should consider not only the expected physical damage, the number and type of potential casualties or the economic losses, but also the conditions related to social fragility and lack of resilience that favour the second-order effects (indirect effects) that amplify impacts when a hazardous event occurs in an area. Disaster risk cannot be eliminated completely, but it can be assessed and managed in order to mitigate the impact of disasters [74].

Consequently, geospatial techniques can be utilized effectively in the entire spectrum of the disaster cycle, hence enhancing disaster risk management. For example, biophysical and socioeconomic data sourced from remote sensing, census, and other spatial databases can be employed to evaluate tropical cyclone-related risk. A variety of geo-spatial methods; comprising of inductive, deductive, and multi-criteria evaluation can be used to determine elements at risk. This chapter therefore was aimed at showing the utility of geo-spatial technologies in disaster risk management using Chimanimani district in Zimbabwe as a case study. According to Manna [53], geo-spatial technologies opened a new era for disaster risk management, whereby meteorological hazards such as cyclones, can be better monitored, understood and ultimately anticipated. This chapter describes a geo-spatial based methodological approach which enables more rigorous evidence-based decision-making in disaster risk management, and notably for assessing the degree to which it increases preparedness for the adverse effects of hazards in vulnerable areas.

3 Theoretical Framework

In dealing with extreme events, many of the critical problems that arise are inherently spatial. Whether an analyst is assessing the potential impact of a hazard, or an emergency manager is identifying the best evacuation routes during a disaster, or a civil engineer is planning a rebuilding effort following a disaster, all these individuals face tasks with a strong spatial component. For this reason, geographical space is a valuable framework for reasoning about many problems that arise in the context of emergency management [19]. Geo-spatial science has emerged as one the key tools used for disaster risk management worldwide. While maps have facilitated much of humanitarian history, the geospatial renaissance in which shared spatial data is the fulcrum for enhanced coordination is still largely driven by voluntary organizations [35]. Voluntary organizations advocate for open platforms and shared data across sectors and stakeholders. Geo-spatial science facilitates the integration of spatial and non-spatial data critical for in-depth analysis and management of disasters [70]. In this context, geo-spatial data particularly remote sensing data, can provide effective and timely information for monitoring floods and for damage assessment. In addition, the potential of new air- and space-borne imaging technologies for improving hazard evaluation and risk reduction is continually being explored [1, 61].

Geo-spatial technologies are relatively inexpensive and can provide information on several parameters that are crucial in disaster management such as saving lives, limiting damage, and reducing the costs associated with disasters [62]. In a nutshell, the growth of geospatial technology positively interacts with, and influences all aspects of disaster risk management—such as mitigation (modelling hazards and vulnerability), preparedness (formulating emergency response and evacuation plans), response (executing such plans), and recovery (assessing damages, rebuilding, preventing recurrence, and educating the public) [19, 36]. As a visualization tool, GIS can assist in locating, identifying, and understanding relationships between areas of social vulnerability and potential hazard exposure [36]. Greenough and Nelson [35] posit that thematic maps have long been used to inform humanitarian actors regarding the location of affected populations, access routes, and contextual topography. The visual representation of data transcends linguistics and creates an efficient and effective mechanism supporting preparedness, response, and coordination. Geo-spatial sciences have evolved from basic cartography into a field of geo-statistics and modelling. While operational mapping has day-to-day utility for humanitarian work, the analytic power of geospatial information can be transformative for disaster risk management.

When examining the role of GIS in emergency management identifying a conceptual framework can assist in understanding how to operationalize GIS in phases of the disaster management cycle. This chapter adopts the comprehensive emergency management framework, after Drabek and Hoetmer [27]. This relies on the temporal dimension of disasters to organise the emergency management process into a cycle of four, often overlapping, phases: mitigation, preparedness, response, and recovery [19] (Fig. 3).

Mitigation involves actions that are taken to eliminate or reduce the degree of long-term risk to human life and property from hazards. Preparedness is concerned with actions that are taken in advance of an emergency to develop operational capabilities and facilitate an effective response to an emergency. The response phase involves actions that are taken immediately before, during, or directly after an emergency occurs, to save lives, minimise damage to property, and enhance the effectiveness of recovery. The recovery phase is characterised by activity to return life to normal or improved levels. In examining the GIS literature, perhaps it is more appropriate to reduce the four phases of comprehensive emergency management into three phases: (1) mitigation, (2) preparedness and response, then (3) recovery. This is simply because GIS systems developed for the preparedness phase can also be utilised in the response phase. In other words, systems designed to help emergency managers

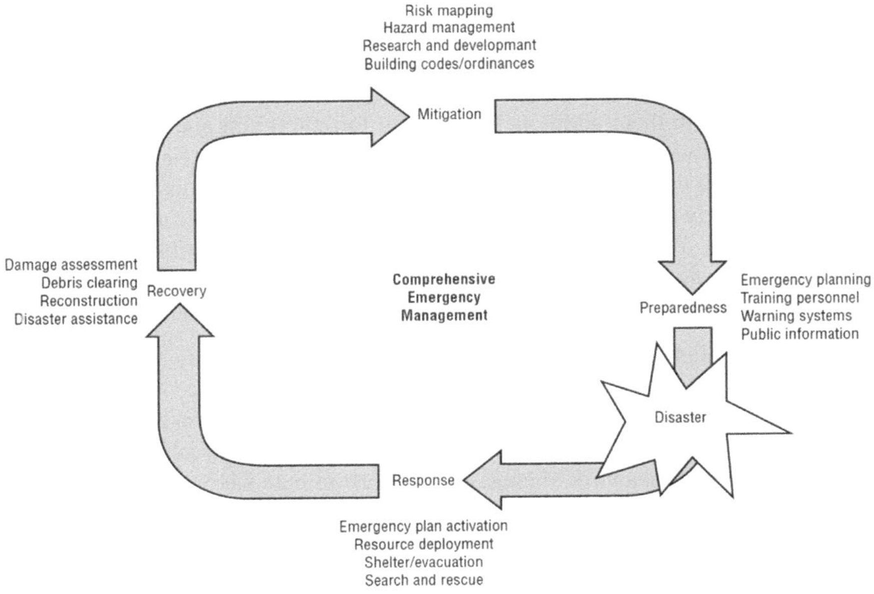

Fig. 3 Comprehensive emergency management framework. *Source* Godschalk [33]

respond to an actual disaster are frequently utilised to train emergency personnel and develop preparedness plans [19].

4 Materials and Methods

The Sendai Framework for Disaster Risk Reduction 2015–2030 is the roadmap for disaster risk reduction, and it identifies the centrality of geo-spatial technologies for disaster risk management [81]. Disaster risk management is the set of pre- and post-disaster actions toward facing negative consequences of hazards [31]. Pre-disaster actions include risk transfer (i.e., insurance), risk assessment (monitoring, mapping, etc.), mitigation works (i.e., infrastructure strengthening, etc.), and preparedness [71]. Post-disaster actions include emergency response (aid, clean-up, etc.) and loss sharing for rehabilitation and reconstruction (i.e., private-public and national-local resources [71]. This methodology will showcase geo-spatial pre-disaster risk assessment mapping and preparedness and use of geo-spatial science for pre- and post-disaster actions.

5 Tropical Cyclones That Have Affected Zimbabwe

Tropical cyclones are among the most powerful and destructive weather systems on earth. At peak intensity, surface winds around the inner core have been observed to exceed 90 m s^{-1} (333 km h^{-1}) [66]. In addition to their violent surface winds, tropical cyclones bring with them additional hazards of torrential rain and powerful storm surges, which individually may pose even greater threats than the wind, especially to those living along the coast [67]. The following cyclone descriptions were obtained from https://reliefweb.int/ and https://en.wikipedia.org/wiki/Tropical_cyclones_in_Southern_Africa. The respective paths of all the cyclones described below were digitised in GIS, as well as the areas affected by each and every cyclone. Data for cyclone tracks were collected from http://www.bom.gov.au/cyclone/history/tracks/.

Cyclone Eline (February 2000) travelled over 11,000 km during its 29-day trek through the Indian Ocean, in February 2000. The cyclone formed on February 1, 2000, in the Australian basin as Tropical Cyclone Leon, and was renamed Eline after crossing 90 E into the South-West Indian Ocean. Late on February 17, Eline made landfall near Mahanoro, Madagascar, with 10-minute winds of 165 km/h. The storm rapidly weakened over land, but restrengthened in the Mozambique Channel to reach peak 10-minute winds of 185 km/h, making it an intense tropical cyclone. On February 22, Eline made landfall about 80 km (50 mi) south of the city of Beira, Mozambique, near peak intensity. The cyclone tracked almost 2000 km across southern Africa toward the cool southeast Atlantic and led to substantial rainfall over arid to semiarid southern Namibia. Eline quickly weakened over land as it moved across Southern Africa, finally dissipating over eastern Namibia on February 29. Cyclone Eline caused 91 deaths, 357 injuries, destroyed 59,187 houses and resulted in about 2.7 million people being affected [11]. Transport routes were destroyed, trees uprooted, houses destroyed and agricultural land flooded.

Cyclone Japhet developed on February 25, 2003 near the southwest coast of Madagascar, and was "very strong" characterised by strong maximum winds of 110 km/h, with gusts up of up to 140 km/h. Cyclone Japhet quickly intensified in the Mozambique Channel, reaching maximum winds of 175 km/h, sustained over 10 min. After stalling briefly, the cyclone turned to the northwest, weakening slightly before striking Mozambique just south of Vilankulo on March 2. Japhet slowly weakened while progressing inland, dissipating over Zambia on March 6, 2003. Along its path, Japhet resulted in heavy rainfall causing widespread flooding. The excessive precipitation caused heavy crop damage. In two provinces in Mozambique, the cyclone damaged or destroyed 25,000 houses, leaving at least 23,000 people homeless. There were 17 deaths in Mozambique. Further inland, remnant rainfall destroyed a bridge and several houses in Zimbabwe, killing eight people. In Zimbabwe the cyclone affected the eastern Highlands of Zimbabwe and also left destruction mostly in the Zambezi basin area of Muzarabani, going as far as Gutu in Masvingo in the south and south-eastern part of the country (https://reliefweb.int/).

Cyclone Favio (February 2007) was an intense tropical cyclone which was formed in on February 11, 2007 east of Madagascar, and moved southwest. On February 18, the storm was upgraded into a severe tropical storm. Then, it turned west towards Mozambique and was upgraded to a tropical cyclone early on February 19, which then made a landfall on February 22 in the Inhambane Province in Mozambique. As Favio crossed the Mozambique Channel it reached a peak intensity of 232 kilometres per hour on the early morning of February 22, making it a Category 4 storm. The cyclone then weakened slightly before slamming into southern Mozambique with sustained winds estimated at 204 km/h. Maximum sustained winds were still estimated to be 167 km/h. The associated heavy rain triggered floods along rivers in Central Mozambique and southern Madagascar. Tropical Cyclone Favio brought widespread damage to Vilanculos in Mozambique, where four people were killed and 70 injured, destroying about 80% of the town. Overall, a total of ten people were killed by the storm and nearly 100 others were injured, and combined with an earlier flood, the storm caused $71million (2007 USD) in damage. Around 130,000 homes suffered damage and 130 schools were levelled by the tropical cyclone. Across the nation, a total of 33,000 people were left homeless. Cyclone Favio destroyed 277,000 ha of crops. In Zimbabwe and 400 electricity poles were downed.

Cyclone Idai (March 2019) was a tropical cyclone that originated from a tropical depression formed over the Mozambican Channel on 4 March 2019 (http://www.meteofrance.re/cyclone/trajectoire). This storm moved initially over land in Mozambique and Malawi during the 4–9 March 2019 period, and then returned to the Mozambican Channel where it made a rapid intensification to become a tropical cyclone. On 11 March 2019 at 06.00 UTC the system had become an intense tropical cyclone with wind speed up to 165 km/h and heading towards the mainland Mozambique. The cyclone which was now a category 4 made landfall immediately north of Beira during the early hours of 15 March 2019. This system weakened on the same day into at tropical depression as it moved overland. By the evening of the same day this tropical depression had moved some 230 km from the coast into eastern Zimbabwe affecting the Chimanimani and Chipinge districts. The system further moved as far as the central part of Zimbabwe and started dissipating on 16 March 2019 having moved overland for 470 km from Beira. From being a tropical depression to dissipation, Cyclone Idai lasted 12 days (4–16 March 2017).

Cyclone Kenneth (April 2019) occurred towards the end of the cyclone season during late April 2019, and was formed over the northern limits of cyclone formation at 9.5° S latitude and 53.3° E[1]. This system started moving westward and by 23 April 2019 and was classified as a tropical depression, and then a tropical cyclone on 24 April 2019 that was affecting the Comoros Island. Cyclone Kenneth had wind speed of about 205 km/h made landfall on 25 April 2019 over Pemba a north coastal town in the Cape Delgado Province of Mozambique. Once the cyclone moved overland, it started to weaken and by 00.00 UTC on 26 April 2019. The system dissipated on 28 April 2019 after moving 540 km overland. Thunderstorm activity developed off the coast of Mozambique on 27 April as the system began drifting northward. During the Cyclone Kenneth period, Pemba received 570 mm of rainfall by 29 April 2019 and other parts of Cape Delgado and Nampula Provinces received 100–200 mm of rainfall.

Cyclone Ana (Jan 2022) was a moderate tropical storm that affected the African nations of Madagascar, Malawi and Mozambique, and brought heavy rain to Zimbabwe. Cyclone Ana developed from an area of convection that was designated as Invest 93S northeast of Madagascar. The system rapidly developed into a tropical storm coded Ana which then made land fall on Mozambique on 24 January. By the 25 January, it had reached Zimbabwe as an overland depression. Moderate to heavy rains with daily maximums of 80 mm accompanied by strong winds of up to 80 km/h affected several communities in the northern and North-eastern boarder of Zimbabwe. In Zimbabwe, Tropical Storm Ana caused destruction and damage in at least six provinces, with Manicaland Province hardest hit. At least 3000 people were affected by heavy rains and flooding, with schools, bridges and roads being damaged. A total of 12 affected districts over the 5 provinces: Mashonaland West, East and Central, Masvingo and in Manicaland. The associated rainstorms and winds caused mild to extensive damages to public infrastructure in Zimbabwe (classrooms, roads, electricity lines and bridges).

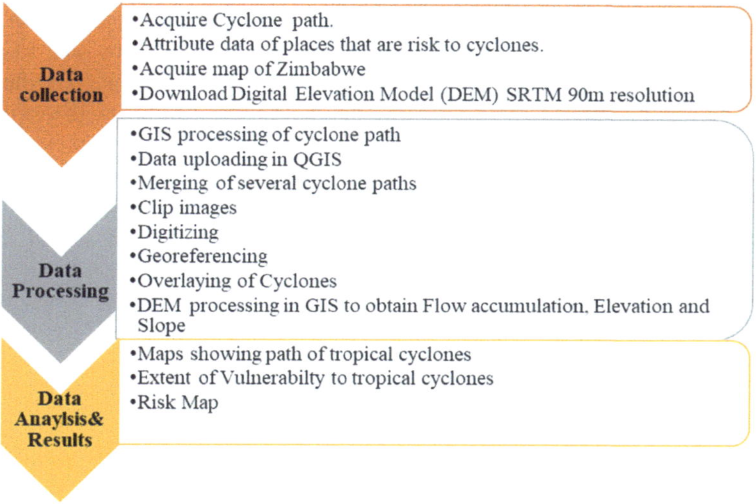

Fig. 4 Pre disaster mapping and geo-spatial analysis actions for disaster preparedness

6 Pre-disaster Risk Assessment Mapping

For the pre-disaster risk assessment mapping the following steps were taken: geo-spatial data collection, processing, analysis and visualisation. Administrative boundary maps (provinces, districts and wards) of Zimbabwe were then used to subset the data. In addition, a digital elevation model (DEM) was downloaded from https://earthexplorer.usgs.gov/ and used to derive slopes, flow accumulation, and as input to risk modelling. The cyclone paths were digitized and stored as shapefiles for the different cyclones. The cyclone paths were then classified according to intensity with the highest intensity values assigned to the centre of the cyclone path. The flow diagram below illustrates the geo-spatial steps taken for pre-disaster mapping actions (Fig. 4).

7 Disaster Risk Modelling for Floods

The risk map was obtained from the digital elevation model using the risk assessment model framework (Risk = Hazard + Vulnerability) given by Wisner et al. [82] and as explained by Maskrey [56]. Hazard potential was calculated as a factor of elevation, slope and flow accumulation. Vulnerability was derived as a function of population for the respective districts and wards. Each hazard potential factor was weighted according to weights calibrated using the Analytical Hierarchical Process (AHP) process. AHP is a mathematical multi evaluation decision-making method

[60]. It utilises a system of relationships in a hierarchic structure arising from inter-relationships of the system and analyses these relationships step by step to aid in the systemic and numerical decision-making process [69]. Using the AHP the most impacting environmental factor is given a higher ranking than the one with less impact and the thematic rankings were based on literature evaluations and field experience. As such, AHP is a semi-qualitative, semi-quantitative analysis method for weighting in which the construction process of the judgment matrix is completely dependent on the user. For this study, the model utilised is as follows:

$$\text{Hazard} = (\text{Flow Accumulation} * 0.1187) + (\text{Slope} * 0.0414) + (\text{Elevation} * 0.071 \quad (1)$$

$$\text{Vulnerability} = (\text{Number of Population} * 0.0369) \quad (2)$$

$$\text{Risk} = \text{Hazard} + \text{Vulnerability} \quad (3)$$

The GIS raster calculator was used in the determination of the areas at risk of cyclones activity as well as areas that can be affected by secondary effects of cyclones such as flooding. The severity of cyclone was determined as a function of distance from the cyclone centre. The process undertaken is shown in Fig. 5 and Table 1.

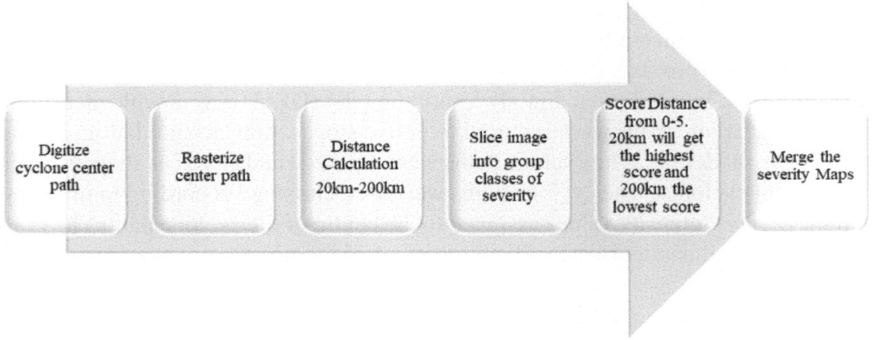

Fig. 5 Cyclone severity mapping procedure

Table 1 Ranking of cyclone severity based on distance from the centre path of the cyclone

Distance from centre path (km)	Severity rank	Category
0–20	5	Extremely severe
21–60	4	Very severe
61–100	3	Severe
101–140	2	Low severity
141–200	1	Very low severity

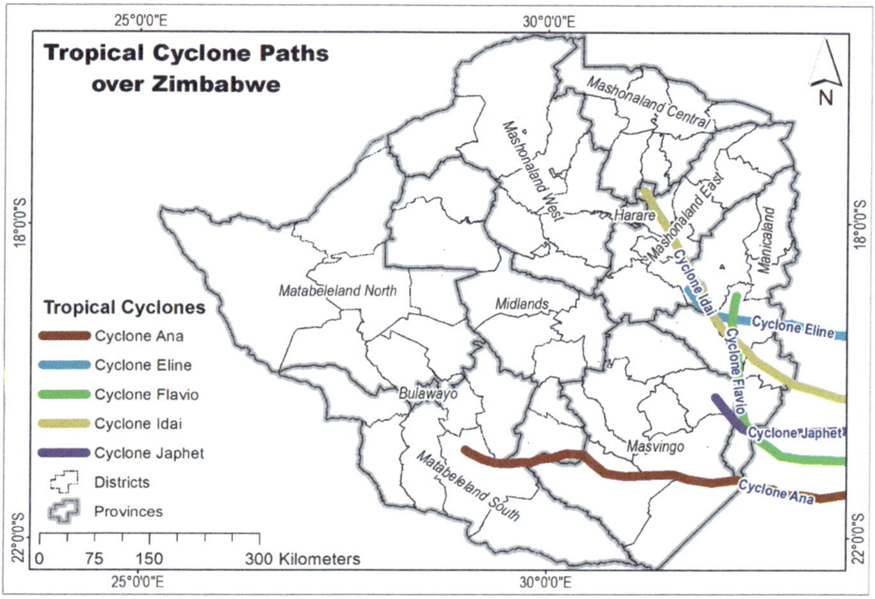

Fig. 6 Digitised tracks of the different tropical cyclones which have affected Zimbabwe

8 Results

8.1 Pre-disaster Risk Assessment Mapping

Geo-spatial science was used in the mapping of tropical cyclones path affecting Zimbabwe. Figure 6 shows the digitised tracks of the different tropical cyclones under consideration and the districts and provinces which were affected and Fig. 7 shows the extent of all the areas affected by cyclones, up to 20 km radius from the centre path of the cyclone

8.2 Cyclone Severity over Zimbabwe

As stated previously, distance from the centre path of the cyclone was used to establish cyclone severity, with scores ranging from 1 to 5 for each cyclone path. One implied a very low risk (further away from the cyclone centre path) and 5 implied extremely severe risk (at the centre path of cyclone) (Fig. 8).

Using the raster calculator, the severity risk of all cyclones was added together to obtain a compound severity map for all the cyclones under consideration (Fig. 9).

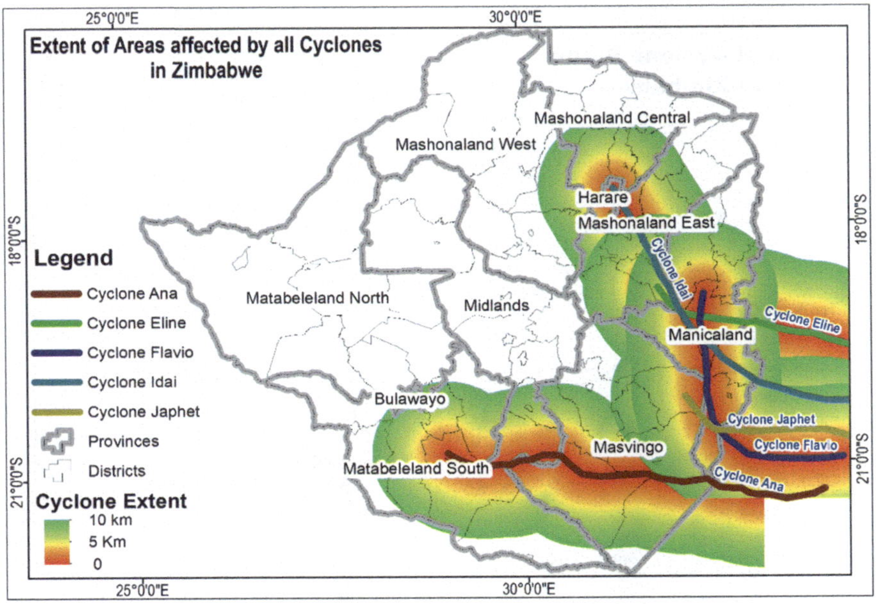

Fig. 7 Extent of all the areas affected by cyclones, up to 20 km radius from the centre path of each cyclone

8.3 *Risk Map of Affected Areas*

The risk map was created by using Eq. 3 above given as: Risk = Hazard Potential + Vulnerability. The risk map (Fig. 10) showed that areas in the path of merged cyclones as well as areas with low elevation have a very high risk of being affected by tropical cyclones and areas with high elevation have a low risk of being affected by cyclones. Figure 9 shows that districts such as Chikomba, Marondera, and Chirimhanzu are at low risk, while Chiredzi, Mwenezi, and Beitbridge districts are at high risk. It was noted that the higher the elevation the lower the risk and the lower the elevation the higher the risk of an area to flooding after a cyclone.

8.4 *Managing Disaster Risk Using Geo-spatial Science*

The management of disaster risk can best be exemplified through the analysis of land cover changes in Kopa, Chimanimani from February to March 2020. Chimanimani experienced a huge number of changes due to floods which affected Kopa area between February and March 2019. Sentinel images were classified for analysing the land cover changes dure to the flooding experienced. The images were classified into four major land uses namely water, built-up (roads, pavements, and buildings),

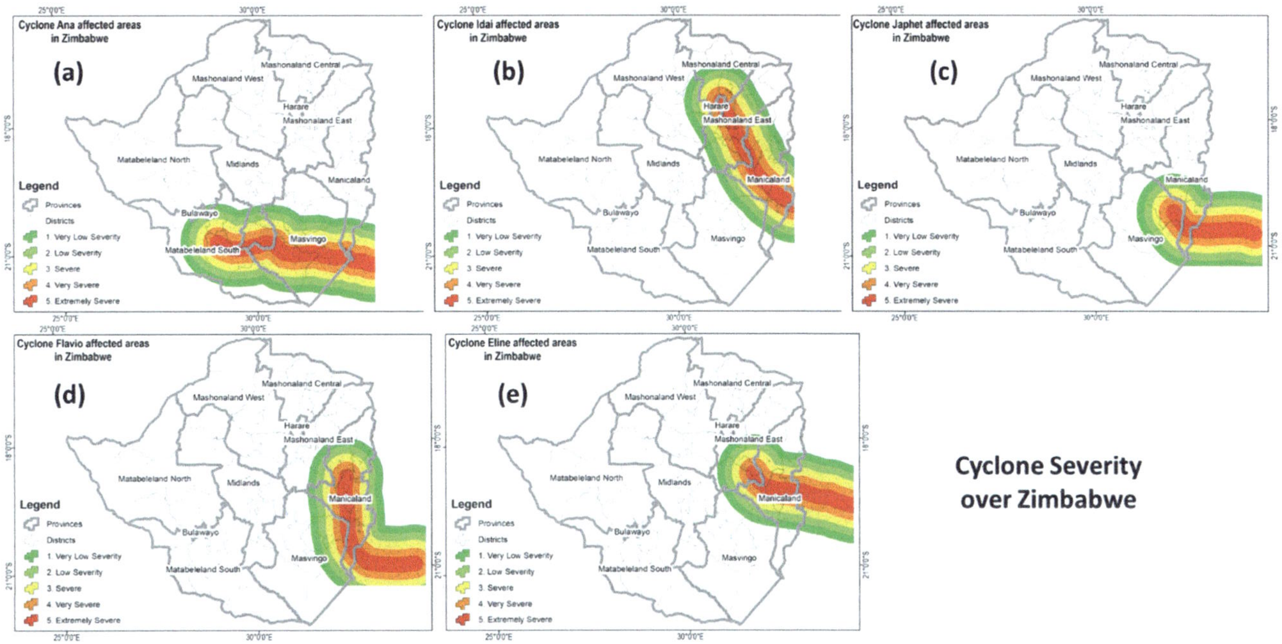

Fig. 8 Cyclone Severity Maps for cyclones: **a** Ana **b** Idai **c** Japhet **d** Flavio **e** Eline

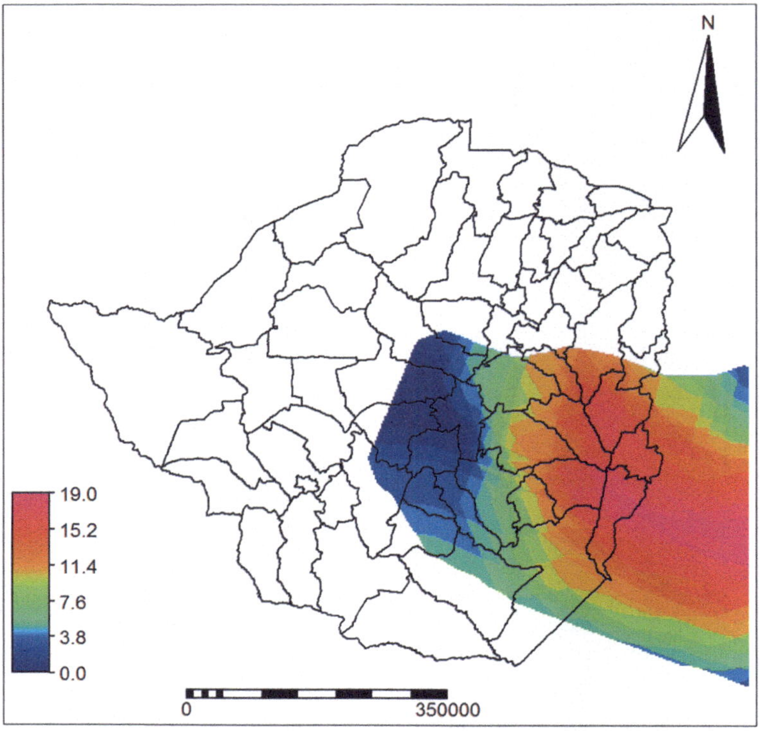

Fig. 9 Severity map of combined cyclones path

vegetation (trees, croplands, and grasslands), and bare lands. Figures 11 and 12 show the classified land uses before and after the flood occurrence respectively.

Results from the classification showed that in February 2019 the area was dominated by dense vegetation and built-up spaces. Vegetation covered 15020.54 (34%) whereas built-ups covered 14030.47 h (33%). Water covered 319.3 h 10 (9 %) whilst bare land covered 13802.25 h (32 %) of the total area. By March 2019, vegetation covered of 10291.73 h (23.8%) whilst built-ups occupied 9276.8 h (21.5%). Water covered 1320.68 h (3.1%) and bare land had a composition of 22283.56 h (51.6%) of the total area. The overall classification accuracy of all images was above 80% as recommended by Thakkar [76].

From the classified images, in February 2019, only 0.9% of the area was covered with water, but after the cyclone, 3.1% of the area was under water. This showed that floods affected all the land uses. Flooding washed away 33% of the built-up area making victims homeless. In addition, an extensive area covered by vegetation was flooded as evidenced by the percentage decline from 34.8 to 24.8% of vegetated areas. On the other hand, from February to March 2019, the built-up areas decreased by approximately − 4753.67 ha while water increased by + 1001.37 ha. Vegetation

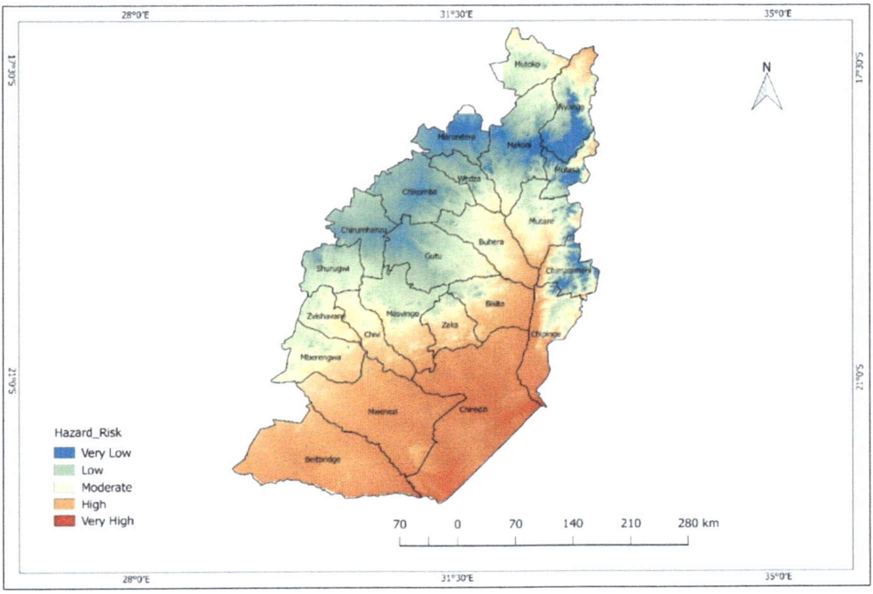

Fig. 10 Risk map of tropical cyclones in Manicaland and Masvingo Provinces, Zimbabwe

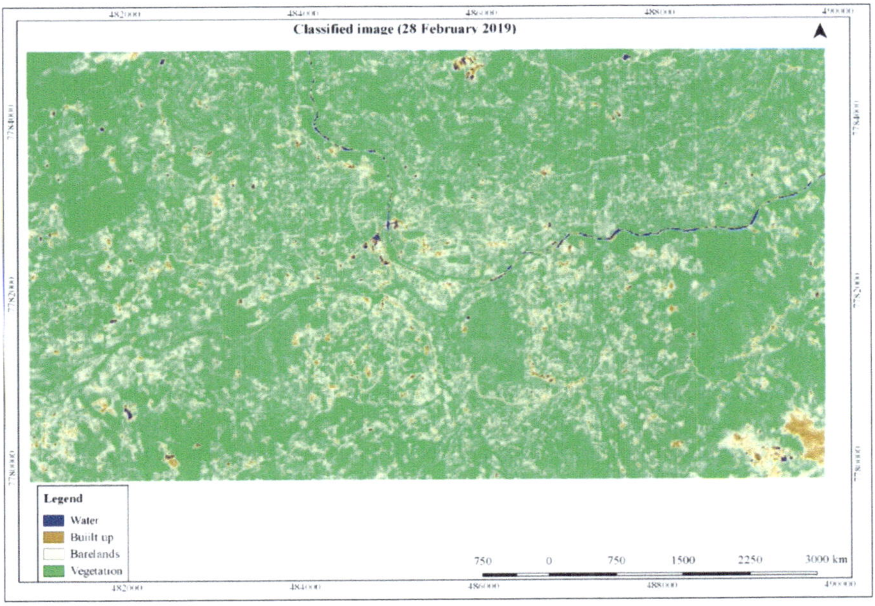

Fig. 11 February 2019 classified image of Kopa area, Chimanimani

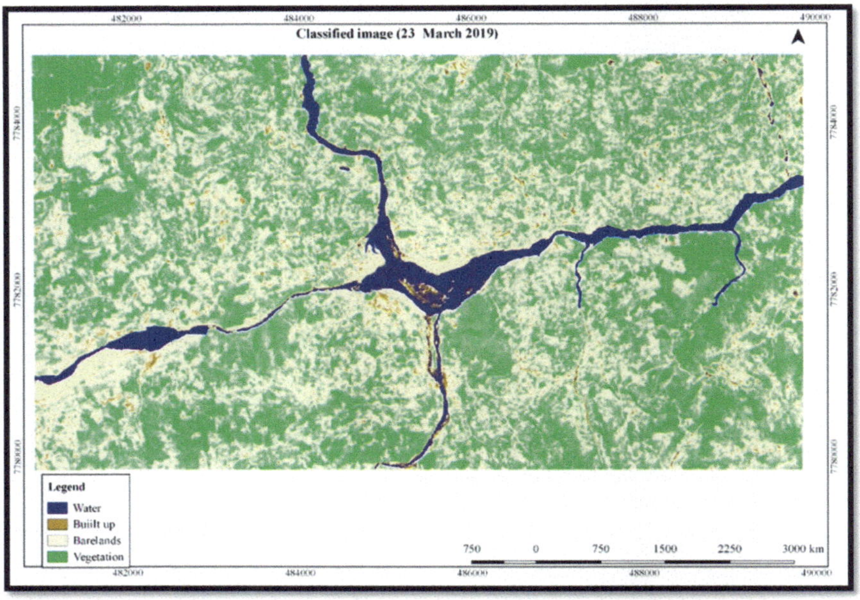

Fig. 12 March 2019 classified image of Kopa area, Chimanimani

decreased by − 4728.81 ha whereas open surface increased by + 8481.31 ha. Generally, it was evident that due to flooding most vegetation and agricultural land were under water, resulting in crops being washed away. This means that communities were negatively impacted as they were left without crops, homes, and infrastructure.

8.5 *Discussion and Conclusion*

The purpose of the research was to argue for the utility of geospatial science for hazard mapping in disaster risk management by demonstrating the use of GIS to enhance multi-hazard mapping as part of response and preparedness. In view of the devastating impact of Tropical Cyclone Idai on Chimanimani and other districts, the use of GIS to inform preparedness, response and recovery could have contributed to a different narrative about the damage of the cyclone. It has already been revealed that there was a time lag of 5 days from the day the cyclone hit Chimanimani to the first day of organized response from authorities due to a myriad of reasons [58]. The cyclone paths presented showed that the Eastern Highlands is prone to be affected by cyclones, hence better preparedness is required in this area.

It is therefore concluded that using geo-spatial science to identify and map the paths of tropical cyclones is imperative in determining areas at risk as well as vulnerable communities. Cyclones influence or trigger flooding depending on the magnitude and intensity of the tropical cyclones. However, the cyclone path alone is not sufficient to explain vulnerability and risk, since it is possible to be exposed but not vulnerable [2]. Risk is a function of hazard and vulnerability, however, vulnerability is determined by the population of the areas that the cyclone path follows. All these (risk, vulnerability, and hazard) have a spatial component, that which can be handled using geo-spatial science. The risk map is a factor of vulnerability henceforth, where there is a high risk of being affected by cyclones there are high chances of vulnerability, if the area is settled by people of houses critical infrastructure. Risk is determined by environmental factors and the population of the areas likely to be affected by cyclones. Liu et al. [50] also utilised geo-spatial science in mapping risk in China, and demonstrated the centrality of geo-spatial science in disaster risk management. This is also supported by Cova [19] who pointed out that geo-spatial science is designed to support geographical inquiry and, ultimately, spatial decision-making. The value of GIS in emergency management arises directly from the benefits of integrating a technology designed to support spatial decision-making into a field with a strong need to address numerous critical spatial decisions [19].

The empirical results from this research can serve to support the argument that incorporating GIS systems in decision-making for disaster management is very important, reducing considerably the risk of wrongfully selecting facilities, and using suboptimal routes for humanitarian supplies and overall logistics to support people affected by disasters. Thus, there is scope for use of GIS in humanitarian logistics for disaster response and recovery, as most of its use has been largely limited to preparedness. Concerning logistics activities, the assessment of infrastructure can be beneficial, e.g., for route planning, road damage assessment, and the identification of best locations for facilities supported by mapping activities.

Study findings will be useful to inform local authorities, policymakers, the Department of Civil Protection, and key players in the disaster risk management sector on the use of geospatial technology in optimizing emergency response and recovery. In relation to humanitarian logistics, this optimization relates to a reduction in time taken in the collection of data on post-disaster infrastructural damage, identification and mapping of disaster hotspots, the best location for facilities, and optimal routes for transporting humanitarian aid to affected communities, and other elements at risk, among others.

It is also noted that despite the proliferation of geo spatial technologies, such as earth observation, these are still not fully exploited for disaster and risk management [7]. Their successful implementation requires a solid base of political support, laws, and regulations, institutional responsibility, and trained people. Knowledge should be transferred from geoscience specialists and international bodies to professionals and decision-makers working on disaster risk management with different technological backgrounds [3]. It is commendable that the Department of Civil Protection, through support from donors like the World Bank, is taking the first step of embarking on capacity building of its staff in GIS and Remote Sensing.

As observed by Chatiza [14], disaster risk in Zimbabwe is not yet fully mainstreamed across sectors. As such, no holistic picture exists of common risks, disaster risk assessments, plans, and programmes, making responses difficult. This is because of resource constraints and also low institutional coherence. Disaster risk assessments, early warning systems, and planning in Zimbabwe remain inadequately resourced. This is the gap that geo-spatial science can close when it comes to disaster risk management in the country. Key institutions lack access to robust data about risks. This makes it difficult for the different agencies to fully inform policymakers, implementers, and citizens about risks and the necessary mitigation measures. This is worsened by a lack of near real-time and geospatial data that can better inform policymakers and decision-makers and enrich their decisions. In the context of disaster risk management, community members should feel good enough, and safe enough and the environment should be open enough to identify and map hazards, express their vulnerability and capacity, and deal with uncertainties and difficulties in order to construct appropriate responses to the realities obtaining in their community [6]. Hazard mapping is one of the avenues to achieve this.

9 Conclusion

Even though this chapter largely focused on application of geo-spatial science it argues for a blend of such with community-level participatory mapping, which can be a powerful method for disaster risk mitigation and preparedness. To date this has been argued for by Kienberger and Steinbruch [42] who advocated for this approach in Mozambique. Participatory mapping is driven by input from the beneficiary participants, this benefits the plan output with a broader overview of the area while allowing the communities to be involved. Local people can, using simple maps that they have created, quickly see, and analyse important patterns in the risks they face. This is because, at their core, maps are visual expressions of measurements. You can compile sketches, data from assessments, and notes into representations of the region you are looking at using simple tools like pen and paper. In a situation with no time or resources, a map drawn on paper can be enough to help to identify the most at-risk areas of a settlement and to mark the location of valuable services such as water resources [54].

Cova [19] submits that the level of GIS activity in emergency management may be increasing, but in the eyes of the emergency management community, GIS has a long way to go to meet their needs. Numerous challenges still exist in the application of geo-spatial science in this area, and many of its related fields like hazard mitigation. Coppock [18] as well as Cova [19] lists five major limitations regarding the application of geo-spatial science for disaster risk management:

- lack of data and the weakness of existing data;
- difficulty in developing, and understanding the error in, GIS models;
- deficiencies of available software, particularly commercial GIS;

- failure to consider the needs of end-users adequately;
- lack of lead organisations and necessary infrastructure.

For Zimbabwe, the above limitations also still need to be addressed in part or in full for the country to adequately use geo-information science for disaster risk management. This has, to date, been pointed out by Chatiza [14], Chapungu [12], Chanza et al. [11] and Mazvimavi [58].

References

1. Al-Tahir R, Saeed I, Mahabir R (2014) Application of remote sensing and GIS technologies in flood risk management. Flooding and climate change: sectorial impacts and adaptation strategies for the Caribbean region, pp 137–150
2. Alam E, Collins AE (2010) Cyclone disaster vulnerability and response experiences in coastal Bangladesh. Disasters 34(4):931–954
3. Altan O, Backhaus R, Boccardo P, Tonolo FG, Trinder J, Van Manen N, Zlatanova S (2013) The value of geoinformation for disaster and risk management: benefit analysis and stakeholder assessment. Joint Board of Geospatial Information Societies. Joint Board of Geospatial Information Societies, Copenhagen
4. Artur L, Hilhorst D (2012) Everyday realities of climate change adaptation in Mozambique. Glob Environ Change 22(2):529–536 (Elsevier Ltd). https://doi.org/10.1016/j.gloenvcha.2011.11.013
5. Bayley PB (1991) The flood pulse advantage and the restoration of river-floodplain systems. Regul Rivers Res Manage 6:75–86. https://doi.org/10.1002/rrr.3450060203
6. Bongo PP, Manyena SB (2015) 'From "government" to "governance": tensions in disaster-resilience leadership in Zimbabwe. J Disaster Risk Stud 7(1):10 Art. #188 (Jàmbá). https://doi.org/10.4102/jamba.v7i1.188
7. Brivio PA, Colombo R, Maggi M, Tomasoni R (2002) Integration of remote sensing data and GIS for accurate mapping of flooded areas. Int J Remote Sens 23(3):429–441. https://doi.org/10.1080/01431160010014729
8. Bunce M, Brown K, Rosendo S (2010) Policy misfits, climate change and cross-scale vulnerability in coastal Africa: how development projects undermine resilience. Environ Sci Policy 13(6):485–497
9. Campbell DA (2017) An update on the United Nations millennium development goals. J Obstet Gynecol Neonatal Nurs 46(3):e48–e55
10. Carreño ML, Cardona OD, Barbat AH, Suarez DC, Perez MDP, Narvaez L (2017) Holistic disaster risk evaluation for the urban risk management plan of Manizales, Colombia. Int J Disaster Risk Sci 8(3):258–269. https://doi.org/10.1007/s13753-017-0136-7
11. Chanza N, Siyongwana PQ, Williams-Bruinders L, Gundu-Jakarasi V, Mudavanhu C, Sithole VB, Manyani A (2020) Closing the gaps in disaster management and response: drawing on local experiences with cyclone Idai in Chimanimani, Zimbabwe. Int J Disaster Risk Sci 11(5):655–666. https://doi.org/10.1007/s13753-020-00290-x
12. Chapungu L (2020) Mitigating the impact of cyclone disasters: lessons from Cyclone Idai. South African Institute of International Affairs, Johannesburg. https://media.africaportal.org/documents/Chapungu_-_Final.pdf. Accessed 22 Mar 2022
13. Charvériat C (2000) Natural disasters in Latin America and the Caribbean: an overview of risk (SSRN scholarly paper no. ID 1817233). Social Science Research Network, Rochester, NY. https://doi.org/10.2139/ssrn.1817233
14. Chatiza K (2019) Tropical cyclone Idai in Zimbabwe: an analysis of policy implications for post-disaster institutional development to strengthen disaster risk management. Oxfam Briefing Paper, November, Oxfam in Zimbabwe

15. Christensen JH, Hewitson B, Busuioc A, Chen A, Gao X, Held R, Jones R, Kolli RK, Kwon W, Laprise R (2007) Regional climate projections, climate change, (2007). In: The physical science basis. Contribution of working group I to the fourth assessment report of the intergovernmental panel on climate change. Cambridge University Press, Cambridge, UK, pp 847–940
16. Collins M, An SI, Cai W, Ganachaud A, Guilyardi E, Jin FF, Jochum M, Lengaigne M, Power S, Timmermann A, Vecchi G (2010) The impact of global warming on the tropical Pacific Ocean and El Niño. Nat Geosci 3(6):391–397
17. Collins JM (2011) Temperature variability over Africa. J Clim 24(14):3649–3666
18. Coppock JT (1995) GIS and natural hazards: an overview from a GIS perspective. In: Carrara A, Guzzetti F (eds) Geographical information systems in assessing natural hazards. Kluwer, Dordrecht, pp 21–34
19. Cova TJ (1999) GIS in emergency management, geographical information systems, vol 2, issue 12. Wiley, New York, pp 845–858
20. Le Cozannet G, Kervyn M, Russo S (2020) Space-based earth observations for disaster risk management. Surv Geophys 41:1209–1235. https://doi.org/10.1007/s10712-020-09586-5
21. Crowley J, Chan J (2011) Disaster 2.0: the future of information sharing in humanitarian emergencies. Harvard Humanitarian Initiative. https://hhi.harvard.edu/publications/disaster-relief-20-future-information-sharinghumanitarian-emergencies. Accessed 19 June 2019
22. de Moel H, Aerts JCJH (2011) Effect of uncertainty in land use, damage models and inundation depth on flood damage estimates. Nat Hazards 58(1):407–425. https://doi.org/10.1007/s11069-010-9675-6
23. Desanker P (2002) The impact of climate change of life in Africa: climate change and vulnerability in Africa. Washington, DC, USA
24. Desanker P, Magadza C (2001) Africa. In: McCarthy JJ, Canziani OF, Leary NA, Doken DJ, White KS (eds) Climate change 2001: impacts, adaptation and vulnerability. IPCC working group II. Cambridge University Press, UK
25. Disaster (CRED) (2015) The human cost of natural disasters: a global perspective. Report for Centre for Research on the Epidemiology of Disasters (CRED)
26. Doocy S, Daniels A, Murray S, Kirsch TD (2013) The human impact of floods: a historical review of events 1980–2009 and systematic literature review. PLoS Currents 5
27. Drabek TE, Hoetmer GJ (eds) (1991) Emergency management: principles and practice for local government. International City Management Association, Washington DC
28. EM-DAT (2013) The international disaster data base. Centre for research on the epidemiology of disasters—CRED. Université Catolique de Louvain, Brussels. www.emdat.be/database. Accessed 30 Oct 2013
29. Etinay N, Egbu C, Murray V (2018) Building urban resilience for disaster risk management and disaster risk reduction. Procedia Eng 212:575–582
30. Field CB, Barros V, Stocker T, Qin D, Dokken D, Ebi K, Mastrandrea M, Mach K, Plattner GK, Allen S (2012) IPCC, 2012: managing the risks of extreme events and disasters to advance climate change adaptation. A special report of working groups I and II of the intergovernmental panel on climate change, vol 30, no 11. Cambridge University Press, Cambridge, UK, and New York, NY, USA. pp 7575–7613
31. Freeman P, Mechler R (2001) Public sector risk management in Mexico for natural disaster losses. In: Innovations in managing catastrophic risks: how can they help the poor? Proceedings of Wharton-World Bank Conference, Washington, DC, pp 12–18
32. GoZ (2019) Tropical cyclone Idai international humanitarian assistance appeal, April 2019. Government of Zimbabwe
33. Godschalk DR (1991) Disaster mitigation and hazard management. In: Drabek TE, Hoetmer GJ (eds) Emergency management: principles and practice for local government. International City Management Association, Washington DC, pp 131–160
34. Goodess CM (2013) How is the frequency, location and severity of extreme events likely to change up to 2060? Environ Sci Policy 27:S4–S14
35. Greenough PG, Nelson EL (2019) Beyond mapping: a case for geospatial analytics in humanitarian health. Confl Heal 13(1):1–14

36. Holser M (2016) The importance of GIS in emergency management. Harvard Environmental Law Review, November. https://harvardelr.com/2016/11/24/the-importance-of-gis-in-emergency-management/. Accessed 20 June 2021
37. Hulme M, Arntzen JW (1996) Climate change and Southern Africa: an exploration of some potential impacts and implications for the SADC region. Climatic Research Unit, University of East Anglia Norwich, UK
38. Hulme D, Scott J (2013) Governing development: power, poverty and policy. In: Governing the World? Routledge, pp 28–45
39. Huq S, Kovats S, Reid H, Satterthwaite D (2007) Reducing risks to cities from disasters and climate change. Environ Urban 19(1):3–15. https://doi.org/10.1177/0956247807078058
40. Kabat P, Schulze RE, Hellmuth ME, Veraart JA (2002) Coping with impacts of climate variability and climate change in water management: a scoping paper. International Secretariat of the Dialogue on Water and Climate Wageningen, The Netherlands
41. Kadomura H (2005) Climate anomalies and extreme events in Africa in 2003, including heavy rains and floods that occurred during northern hemisphere summer. Afr Study Monogr Supplementary Issue 30:165–181
42. Kienberger S, Steinbruch F (2005) P-GIS and disaster risk management: assessing vulnerability with P-GIS methods–experiences from Búzi, Mozambique. International Conference on Participatory Spatial Information Management and Communication, Nairobi, Kenya
43. Knox JW, Rodríguez Díaz J, Nixon D, Mkhwanazi M (2010) A preliminary assessment of climate change impacts on sugarcane in Swaziland. Agric Syst 103(2):63–72
44. Kundzewicz ZW (2002) Non-structural flood protection and sustainability. Water Int 27:3–13. https://doi.org/10.1080/02508060208686972
45. Kundzewicz Z, Mata L, Arnell NW, Döll P, Jimenez B, Miller K, Oki T, Şen Z, Shiklomanov I (2008) The implications of projected climate change for freshwater resources and their management. Water Resour Res 2:32–45
46. Kundzewicz ZW, Budhakooncharoen S, Bronstert A, Hoff H, Lettenmaier D, Menzel L, Schulze R (2002) Coping with variability and change: floods and droughts. In: Natural resources forum, vol 26, no 4. Blackwell Publishing Ltd., Oxford, UK and Boston, USA, pp 263–274. https://doi.org/10.1111/1477-8947.00029. Accessed 31 Mar 2021
47. Kusangaya S, Toucher MLW, van Garderen EA (2018) Evaluation of uncertainty in capturing the spatial variability and magnitudes of extreme hydrological events for the uMngeni catchment, South Africa. J Hydrol 557:931–946
48. Kusangaya S, Warburton ML, Van Garderen EA, Jewitt GP (2014) Impacts of climate change on water resources in southern Africa: a review. Phys Chem Earth Parts a/b/c 67:47–54
49. Kusangaya S, Mazvimavi D, Manatsa D, Shekede MD (2021) Chapter 5: climate change impact on hydrological regimes and extreme events in Southern Africa. In: Diop S et al. (eds) Climate change and water resources in Africa. Springer Publishing. https://doi.org/10.1007/978-3-030-61225-2_5
50. Liu B, Fan J, Siu YL, Mitchell G (2019) A multi-hazard map of China. Environ Plan A: Econ Space 51(1):8–10
51. Mabaso M, Kleinschmidt I, Sharp B, Smith T (2007) El Niño Southern oscillation (ENSO) and annual malaria incidence in Southern Africa. Trans Roy Soc Trop Med Hyg 101(4):326–330
52. Manatsa D, Behera SK (2013) On the epochal strengthening in the relationship between rainfall of East Africa and IOD. J Clim 26(15):5655–5673
53. Manna AJ (1985) 25 years of Tiros satellites. Am Meteorol Soc Bull 66:421–423
54. MapAction (2011) Field guide to humanitarian mapping (2nd ed). https://mapaction.org/wpcontent/uploads/2016/12/mapaction_field_guide_to_humanitarian_mapping.pdf
55. Maskrey SA, Mount NJ, Thorne CR, Dryden I (2016) Participatory modelling for stakeholder involvement in the development of flood risk management intervention options. Environ Model Softw 82:275–294
56. Maskrey SA (2017) Supporting local flood risk decision-making using participatory modelling (PhD thesis). University of Nottingham. Retrieved from http://eprints.nottingham.ac.uk/39963/

57. Mazvimavi D (2010) Climate change, water availability and supply. SARUA Leadersh Dialogue Ser 2(4):81–97
58. Mazvimavi D (2019) Cyclones Idai and Kenneth in South African region in 2019. In: Hiroki K (ed) HELP global report on water and disasters 2019, secretariat of high-level experts and leaders panel on water and disasters (HELP), https://www.wateranddisaster.org/cms310261/wp-content/uploads/2019/07/HELP-Global-Report-on-Water-and-Disasters-D9-20190607_s.pdf
59. McMichael AJ, Woodruff RE, Hales S (2006) Climate change and human health: present and future risks. Lancet 367(9513):859–869
60. Melese T, Belay T (2022) Groundwater potential zone mapping using analytical hierarchy process and GIS in Muga Watershed, Abay Basin, Ethiopia. Glob Chall 6(1):2100068
61. Metternicht G, Hurni L, Gogu R (2005) Remote sensing of landslides: an analysis of the potential contribution to geo-spatial systems for hazard assessment in mountainous environments. Remote Sens Environ 98(2–3):284–303
62. Molinari D, Ballio F, Menoni S (2013) Modelling the benefits of flood emergency management measures in reducing damages: a case study on Sondrio, Italy. Nat Hazards Earth Syst Sci 13:1913–1927. https://doi.org/10.5194/nhess-13-1913-2013
63. Osbahr H, Twyman C, Adger WN, Thomas DS (2008) Effective livelihood adaptation to climate change disturbance: scale dimensions of practice in Mozambique. Geoforum 39(6):1951–1964
64. Owolabi TOS, Ekechi CO (2014) Communication as critical factor in disaster management and sustainable development in Nigeria. Int J Dev Econ Sustain 2(3), 58–72
65. Patt AG, Schröter D (2008) Perceptions of climate risk in Mozambique: implications for the success of adaptation strategies. Glob Environ Change 18(3):458–467
66. Ramsay H (2017) The global climatology of tropical cyclones. In Oxford research encyclopedia of natural hazard science. https://doi.org/10.1093/acrefore/9780199389407.013.79
67. Rappaport EN, Blanchard BW (2016) Fatalities in the United States indirectly associated with Atlantic tropical cyclones. Bull Am Meteorol Soc 97(7):1139–1148
68. Reason C, Keibel A (2004) Tropical cyclone Eline and its unusual penetration and impacts over the southern African mainland. J Weather Forecast 19(5):789–805
69. Saaty TL (2001) Deriving the AHP 1-9 scale from first principles. In: Proceedings 6th ISAHP. Berna, Suiza, pp 397–402
70. Saifuzzaman M (2015) Land use and land cover dynamics in the Ganges delta region, Bangladesh. Doctoral Dissertation, Queen's University
71. Saldana-Zorrilla, Sergio O (2015) Assessment of disaster risk management in Mexico. Disaster Prev Manage ISSN: 0965-3562
72. Seaton JM (2013) Flood control and poverty. Phillips County, AR and Guyana
73. Shimi AC, Parvin GA, Biswas C, Shaw R (2010) Impact and adaptation to flood: a focus on water supply, sanitation and health problems of rural community in Bangladesh. Disaster Prev Manage Int J 19(3):298–313. https://doi.org/10.1108/09653561011052484
74. Smith K, Petley DN (2009) Environmental hazards: assessing risk and reducing disaster. Routledge, London, p 306
75. Tali MPA (2012) Land use/land cover change and its impact on flood occurrence: a case study of upper Jhelum floodplain. Master of philosophy thesis submitted to the University Kashmir. University of Kashmir
76. Thakkar AK, Desai VR, Patel A, Potdar MB (2017) Impact assessment of watershed management programmes on land use/land cover dynamics using remote sensing and GIS. Remote Sens Appl Soc Environ 5:1–15
77. Tirado MC, Clarke R, Jaykus LA, McQuatters-Gollop A, Frank JM (2010) Climate change and food safety: a review. Food Res Int 43(7):1745–1765
78. Turner BA (2020) The organizational and interorganizational development of disasters. In: Risk management. Routledge, pp 145–164
79. UNDRR (United Nations Office for Disaster Risk Reduction) (2022) Disaster risk management. https://www.undrr.org/terminology/disaster-risk-management. Accessed 10 Nov 2022

80. UNISDR (United Nations International Strategy for Disaster Reduction) (2015) Sendai framework for disaster risk reduction 2015–2030. UNISDR, Geneva
81. UNISDR (United Nations International Strategy for Disaster Reduction) (2017) UNISDR annual report 2017. UNISDR, Geneva
82. Wisner B, Blaikie P, Cannon T, Davis I (2014) At risk: natural hazards, people's vulnerability and disasters. Routledge
83. World Bank (2016) Natural disasters force 26 million people into poverty and cost $520bn in losses every year, new World Bank analysis finds. World Bank, Washington, DC. https://www.worldbank.org/en/news/press-release/2016/11/14/natural-disasters-force-26-million-people-into-poverty-and-cost-520bn-in-losses-every-year-new-world-bank-analysis-finds. Accessed 19 June 2020
84. Wuebbles DJ, Hayhoe K (2004) Climate change projections for the United States Midwest. Mitig Adapt Strat Glob Change 9(4):335–363. https://doi.org/10.1023/B:MITI.0000038843.73424.de
85. Yamba FD, Walimwipi H, Jain S, Zhou P, Cuamba B, Mzezewa C (2011) Climate change/variability implications on hydroelectricity generation in the Zambezi River Basin. Mitig Adapt Strat Glob Change 16(6):617–628
86. Yanda PZ (2010) Climate change impacts, vulnerability and adaptation in Southern Africa. In: Kotecha P (ed) Climate change, adaptation and higher education: securing our future. SARUA Leadership Dialogue Series, Cape Town, South Africa, pp 11–30

Samuel Kusangaya is a physical geographer with a Ph.D. in Hydrology (UKZN) MPhil in GIS and Remote Sensing (Cambridge), M.Sc. in Land and GIS (UZ), and B.Sc. in Geography and Mathematics (UZ). He is currently employed as a senior lecturer at the University of Zimbabwe lecturing in applied hydrology, water resources management, GIS, remote sensing, and climate change and climate modelling. His research interests include but not limited to geo-science applications in hydrology, water resources management and disaster management.

Albert Maipisi holds a Ph.D. in Disaster Management from the University of Free State in South Africa, Ph.D. taught courses in Public Administration from the National University of Public Service in Hungary, M.sc. In Disaster Management from the National University of Science and Technology (NUST) in Zimbabwe as well as a B.Sc. Hons. Degree in Administration from the University of Zimbabwe (UZ). He is currently a Senior Lecturer with the Women's University in Africa domiciled in Zimbabwe and the national Institutional Development Component Lead for the New York State University Rockefeller Foundation's Parliament Support Programme in Zimbabwe. Immediately before joining the University, he held the position of Committee Clerk/Coordinator for the Parliamentary Portfolio Committee on Mines and Mining Development under the Parliament of Zimbabwe and also have introduced and currently coordinates the Master of Science Degree in Disaster Risk and Livelihoods Management Programme at the Women's University in Africa. Research interests are disaster risk reduction, poverty reduction, parliamentarism, nationalism and public policy.

Linking and Mapping Resilience to Strategies for Management of Disasters and Risks in Zimbabwe

Innocent Chirisa and Gift Mhlanga

Abstract The chapter examines the Zimbabwean disaster risk management strategies in relationship to the concept of resilience in disaster risk reduction (DRR). The study is buttressed by the fact that the country is a signatory to the SENDAI Disaster Risk Reduction Framework, whose main thrust is resilience in disaster risk management. As such, it is expected that the strategies promulgated towards disaster risk reduction in the country enhance resilience in disaster-prone environments and situations. It is therefore the aim of this chapter to explore and assess the resilience building strategies employed within the country. The chapter is based on a desktop study involving interrogation of literature and documents mined from Google Scholar, Ebsco and websites with news and policies, statutes, and related materials. For data analysis, the study mainly engaged thematic analysis. The responsibility for the administration and implementation of the Civil Protection Act, the regulatory foundation of the disaster risk management strategies in the country, lies with the Department of Civil Protection (DCP), that falls under the Ministry of Local Government and Public Works. It is further observed that the implementation of the Disaster Risk Reduction (DRR) policy is derailed by a limited capacity to support the mooted practices on the ground. As such, disaster risk strategies in the country have remained reactive, and implemented on an ad hoc basis at local levels. It is concluded that resilience planning is a denominator in disaster risk management strategies in the country. Therefore, the chapter recommends that, since the government has the mandate to offer resilient service delivery in disaster risk management, there is need to strengthen governance to manage disaster risk by implementing effective institutions, particularly a comprehensive disaster risk management policy. Furthermore, local authorities, with the aid of other stakeholders, such as

---

I. Chirisa (✉)
Office of the Vice Chancellor, Zimbabwe Ezekiel Guti University, Bindura, Zimbabwe
e-mail: chirisa.innocent@gmail.com

Department of Urban and Regional Planning, University of the Free State, Bloemfontein, South Africa

G. Mhlanga
University of Zimbabwe, Harare, Zimbabwe

R. Kasimba and I. Chirisa (eds.), *Disaster Risks and Hazards*, Advances in 21st Century Human Settlements, https://doi.org/10.1007/978-981-96-4577-0_5

non-governmental organizations (NGOs), need to take centre-stage in implementing resilience strategies.

Keywords Disaster risk management · Resilience building · Theory of change

1 Introduction

Zimbabwe is a disaster-prone country with both man-made and natural disasters suppressing development efforts within the country. Noteworthy is the fact that the impact of disasters in the country is heterogeneous, with rural areas having lower capacity than urban areas to cope with the disasters. This resilience divide between rural areas and urban areas in Zimbabwe is explained by various factors and some of them include the following: limited livelihood options in rural areas, lower disposable income, and infrastructure development is more pronounced in urban areas than in rural areas [1]. These factors confirm that disaster vulnerability and exposure are functions of socio-economic development. As such, poverty-stricken areas are vulnerable to hazards that subsume to disaster due to increased exposure and lower capacities to deal with the challenges.

At national level, some provinces are inherently exposed to hazards due to factors such as geography and political economy, and these lead to a disparity in the impact of disasters. Nonetheless, a national, multi-hazard disaster risk management framework attempts to consolidate all localised disaster risk management strategies so much that there is a resultant limited disparity in disaster risk management at geo-spatial and sectoral levels. However, Zimbabwe does not have a comprehensive and multi-sectoral regulatory framework to ensure a coordinated disaster risk management system at national level—the Civil Protection Act of 1989 (Chapter 10: 06) has become archaic in dealing with the evolutionary disaster landscape [2, 3]. As such, the country's policies aim to achieve resilience through a haphazard approach whereby sectoral disaster risk management strategies are implemented. It is therefore, the aim of this chapter to explore and assess the resilience building strategies which are being employed at local levels in Zimbabwe to realise resilience against disasters.

2 Background and Overview

According to Mavhura et al. [4: 01], "Zimbabwe is currently on the drive to build resilience to disasters so as to ensure continued progress towards the improvement of the wellbeing of its citizens." This corresponds with the observation that Zimbabwe does not have a holistic and multi-hazard regulatory framework for disaster risk management. Of concern is the fact that Zimbabwe disaster risk management strategies are guided by the Civil Protection Act that does not recognise such concepts as resilience, vulnerability, mitigation, contingency plans and DRR [3]. The legislative arm of the government is yet to pass the Disaster Risk Management Bill of 2011 into law, taking into consideration that the consultations for this Bill had been underway as early as 2004 [5].

Therefore, although the country is a signatory for both regional and global resilience frameworks (notably the Sendai Framework) that advocate for the alignment of disaster risk management with national institutional frameworks, these provisions are yet to be fulfilled. This misalignment of the necessary legislative structure and the disaster risk reduction efforts contribute to the ineffectiveness of Zimbabwean strategies towards resilience building. Nonetheless, it is noted that the country is inclined towards climate change adaptation as evidenced by the following frameworks: the Zimbabwe Climate Change Response Strategy of 2014; National Adaptation Plans (NAPS); National Adaptation Programme of Actions (NAPAs); the Nationally Appropriate Mitigation Actions (NAMAS); and the Climate Change Policy of 2016. Moreover, the country is a signatory to the following climate change frameworks: United Nations Framework on Climate Change Convention (UNFCCC), the Kyoto Protocol, and the Paris Agreement. These frameworks and initiatives, have a significant impact to disaster risk management if they are implemented and coordinated accordingly. However, Olowu [6] argues that most African governments hardly move beyond the scope of formal commitment to declarations and high-level meetings.

Inherent resilience of areas is determined by both natural and human factors. Zimbabwe is divided into five agro-ecological zones depending on the prevailing climatic conditions and the resultant vegetation type. As such, areas in regions IV and V for example, are susceptible to droughts, and Matabeleland South is a suitable case on this. On the other hand, areas in region 1 and 2 coincide with the natural paths of cyclones and this is exemplified by the Eastern Highlands. Whereas the natural factors signify that the natural hazards are exogenous occurrences that cannot be controlled, the human factor turns the hazards into disasters if the coping capacity of the community systems is overwhelmed, leading to a disruption of the status quo. In Zimbabwe, owing to Tropical Cyclone Idai, "At least 344 people have been recorded dead and at least 257 people are still missing, and about 60,000 people have been displaced by the Cyclone" [7]. On the other hand, the ReliefWeb [8] points out that "The impact of Tropical Cyclone Idai is reflected in the fact that 270,000 people across nine districts were affected with more than 50,000 people displaced." This portrays a failed disaster resilience system in the country since the disasters are recurrent, and their intensity is increasing [1].

Besides cyclones, other natural hazards which affect Zimbabwe include droughts and floods. The occurrence and duration of droughts are determined by the El Nino Southern Oscillation phenomenon [9]. As such, the agricultural year 2015/2016 was declared a national disaster as the drought effects were nationwide. On the other hand, UNOCHA [10: 05] submits that "The impact of 2018/2019 drought combined with continuing macroeconomic challenges have resulted in high-level of food insecurity…5.5 million people in the rural areas are food insecure." Food insecurity in the country is compounded by floods. In this regard, Oxfam [5: 15] points out that "The cyclone [Idai] also caused flood damage to arable land in other districts in Manicaland, Mashonaland East and Masvingo, alongside loss of livestock and stored cereal/grain, destruction of farm infrastructure, equipment, forestry resources and orchards." Moreover, floods rip away livelihoods of people in lowlands such as Muzarabani that are susceptible to the hazards due to the proximity to the Zambezi catchment area and the Mavhuradonha mountain ranges [11]. These disasters are fostered by poor disaster risk management strategies at national level, with the local communities circumventing the worst impacts of the disasters through implementing indigenous knowledge systems and other coping strategies. However, effective disaster risk management needs an inclusive approach whereby all stakeholders play their part. Such an inclusive approach leads to resilience building as the localized strategies are complemented by the input of other players, and this may include technical advice, financial backup, institutional development, and capacity building.

By and large, the most common disasters that affect Zimbabwe include droughts, floods, and livestock diseases [12]. These are all linked to climate change as Nyahunda and Tirivangasi [13] submits that the climate change risks in Zimbabwe include water stress, drought, increased incidences of pests and diseases, and death of livestock. Of concern is the fact that these disasters are more acute in rural areas where the people of low capacities to counter the disasters. ZIMVAC [14, 15] portray that the majority of people in rural areas are not able to cope with the disasters. This is reflected in Fig. 1.

According to ZIMVAC [15], most communities rely on rain-fed agronomy-based livelihoods, indicating that only 27% of communities across the country have irrigation schemes. As such, reliance on rainfall increases the exposure to dry spells and droughts. By and large, Beitbridge (90%), Nkayi and Buhera (75%) and Rushinga

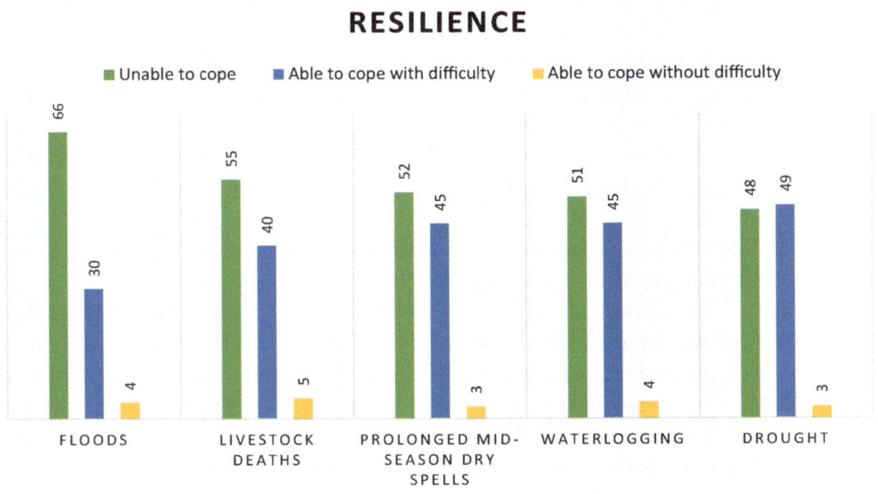

Fig. 1 Households' perception of their ability to cope with future shocks. *Source* ZIMVAC [14, 15]

(70%) had the highest proportion of households that reported drought as a shock. As for livestock deaths and diseases as a shock, the highest proportions of households were recorded in Mt Darwin (54%), Buhera (51%) and Beitbridge (46%). This disaster mapping is also confirmed by ZIMVAC [12] that reveals that provinces impacted most by drought were Matabeleland South (74%), Matabeleland North (58%), Midlands (56%), Masvingo (54%) and parts of Manicaland (40%). Beitbridge is in Matabeleland South, and Buhera is in Manicaland. Therefore, this confirms that these areas have inherently lower resilience capacities.

3 Conceptual Framework

Resilience building flourishes when all stakeholders are involved in making the objective of disaster risk reduction a reality in both present and future hazard occurrences. This is in line with the theory of change systems thinking. The theory is premised on the aspect of inclusiveness, depicting households and communities as significant cogs in achieving disaster resilience through ensuring an enabling environment to promote their inherent capacities. The inherent capacities are the absorptive capacity, the adaptive capacity, and the transformative capacity. Nonetheless, on their own, households and communities proffer inadequate initiatives to deal with the eminent hazards. Béné et al. [16] point out that the development of each of these resilience capacities lead to different outcomes as the capacities are interlinked with the intensity of the shock or stressor. For example, if the shock is of low intensity, a particular community is able to absorb its impact, and this reflects the absorptive capacity of that community system. On the other hand, adaptation or change may be invoked if the normal absorptive capacity of the system against a disaster is overwhelmed. In this case, the system will incorporate new strategies to counter the disaster and this is described as the adaptive capacity of the system. It is noted that incorporating new ways as counters of disaster requires a certain degree of flexibility. A transformative capacity is a descriptive of change that is realised when the new strategies of dealing with the disaster are no longer incremental but have reached a certain level whereby the disaster will have diminishing impact due to the counter measures. This framework of resilience is illustrated in Fig. 2 that also depicts that as the intensity of change increases, more expenses will also be incurred.

4 Literature Review

The United Kingdom Department for International Development (DFID) defines disaster resilience as: "the ability of countries, communities and households to manage change, by maintaining or transforming living standards in the face of shocks or stresses—such as earthquakes, drought or violent conflict—without compromising their long-term prospects" [17]. This is a working definition that links disaster risk

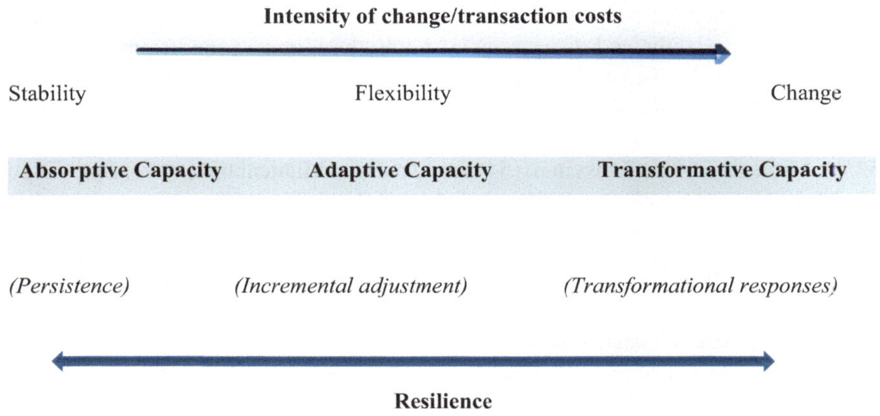

Fig. 2 Resilience framework. *Source* Adapted from Béné et al. [16]

management with development. In this regard, the definition emphasizes that disaster resilience allows for developmental progress since the intensity of a disaster is reduced through the reduction of exposure and vulnerability to hazards. Such a situation is determined by the degree of sensitivity to disturbances and the corresponding capacity to counter it. As such the DFID resilience framework depicts that resilience is when the reaction to the disturbance is described as "bounce back better". This reaction by a system implies that a system, or community in this case, has developed some incremental measures to counter the impact of disasters such that even putting into consideration the changing landscape of disaster, focussing on their increasing frequency and intensity and emerging disasters, the community development initiatives will not collapse but capitalise on the pending disaster challenges to change them into opportunities of development, thereby enhancing the already existing development capacities.

Nonetheless, it is significant to note that "A key determinant of exposure, sensitivity and adaptive capacity is the set of resources and assets that can be utilised in the face of a stress or shock" [17: 11]. This observation corresponds with the perspective that poor communities and households are less resilient because they lack the means and resources to countermand the imminent hazards. Combaz [18: 15] reveals that "Evidence consistently confirms that the capacity of an individual, family or community to prepare for, withstand and respond to a hazard or crisis is enabled or constrained by social status, income and ethnicity". As such, poor communities such as those in Matabeleland South in Zimbabwe remain trapped by regular disasters without means to leap back [13].

The Asian Development Bank (ADB) is among the major stakeholders for resilience building in South Asia. As such, ADB [19] argues that resilience building must take into consideration that climate change leads to dynamic disaster effects as the phenomenon itself is not static. As such, if less effective disaster risk management strategies are not put into operation, disaster losses will replicate. This will be due to the more intense extreme weather patterns such as cyclones and floods. Unfortunately, this comes when developing countries are experiencing such demographic excesses such as overpopulation and rapid urbanization. The rapid increase of people poses a threat to efficient resource utilization, and this pressure results in environmental degradation problems such as global warming and deforestation. Such situations further exacerbate the worsening impacts of disasters as they trigger increased intensity of climate change. By and large, if concerted actions are not taken to manage disaster risk in the context of wider development, the increasing losses will diminish previous achievements in development and put the future at risk [19].

The concerted effort in resilience building needs to include various stakeholders such as planners, financial institutions, and political offices. Streamlining disaster risk management strategies to involve all relevant stakeholders increases the chances of active participation in resilience building through understanding disaster risks. In this regard, the stakeholders will be able to determine "acceptable levels of risk and their appropriate management using a combination of measures to (i) reduce risk to acceptable levels through structural and non-structural investments; and (ii) manage the residual risk through disaster preparedness measures, including actions to avoid loss of life (e.g., early warning) and ensure adequate and timely financing arrangements for disaster relief, early recovery and reconstruction (e.g., disaster reserves, contingency budget, and insurance)" [19: 04]. In other words, disaster risk management strategies are planned beforehand, making resilience building a proactive approach to managing disasters. A proactive approach to disaster risk management promotes a transformatory capacity to resist the adverse effects of changing disaster risks.

The United Nations Food and Agricultural Organization (FAO) has a global mandate to ensure that United Nations member countries are food secure and receive both financial and technical assistance in times of emergencies. Within this ambit, the organization oversaw the drafting of the Southern Africa Resilience Strategy (2018–2021) for the Southern African Development Community (SADC) member countries [20]. The region is affected by natural hazards that include droughts, cyclone-induced floods, and livestock deaths and these are proving to be interconnected [20]. According to the SADC 2017/18 Outlook, 25.8 million people were food insecure in the 2017/18 agricultural year within the region [21]. This is an indicator that the region is being adversely affected by the disasters. Therefore, the drafted Strategy by FAO was meant to guide and inform the Southern African member states to practise resilience building in agriculture so as to achieve food and nutrition security so as to achieve the goals of the United Nations 2030 Agenda for Sustainable Development.

However, the success achieved towards the Sustainable Development Goals are being eroded by the low resilience witnessed within the region's rural areas. For example, the International Organization for Migration (IOM) submits that the most regular disaster affecting Southern Africa are droughts, with 46 of them being recorded to affect almost 74 million people and caused losses of USD 2.1 billion since the year 2000 [22]. As such, it is on record that there is a rising trend of late-onset and high variability of rains as a persistent reality in Southern Africa, resulting in cyclical episodes of food insecurity caused by droughts and mid-season dry spells as caused by the El Niño phenomenon [23, 15]. Owing to the 2015/2016 regional drought, in 2016 SADC declared a regional emergency and launched a humanitarian assistance appeal of USD 2.4 billion [24]. However, whereas such packages are necessary as immediate responses to disturbances, they tend to derail the initiative for proactive disaster risk management and community participation, a problem that is compounded by a lack of political will to adopt a proactive approach to disaster risk management [25].

The rationale of the FAO Resilience Strategy for Southern Africa is to promote resilience building on agricultural livelihoods against disturbances, particularly droughts and floods. As such, basing on the qualitative approach to measuring resilience, the Strategy has four outcomes: risk-informed policy, regulatory, institutional and investment frameworks developed and supported; disaster risks monitored and early warning information provided for potential, known, and emerging threats; vulnerability to crises reduced and disasters prevented; and preparedness for and response to crises and disasters improved through effective sub-regional coordination [20]. In other words, the Strategy calls for a collaboration of all relevant stakeholders within the region to invest and engage resilience building as informed by risk assessments and monitoring and evaluation. As such, both South-South and North–South engagements are necessary for resilience building.

The Common Market for Eastern and Southern Africa (COMESA) is the largest of the eight Regional Economic Communities (RECs) recognized by the African Union as it consists of twenty-one-member countries. According to COMESA [26: 03] "More than 90% of agriculture in the COMESA region is rain-fed and therefore very vulnerable to impacts of climate change such as droughts, erratic rainfall, flooding, rising sea levels and salt water ingress into coastal areas." These disasters result in food and nutrition insecurities and this is compounded by the erosion of livelihoods due to low capacities of the people to deal with the disaster challenges. However, despite the bloc assisting its member countries through the COMESA Climate Change Unit and supporting most of the member countries in designing their National Climate Change Response Strategies and Climate Smart Agriculture Investment Frameworks, disasters within the region are rising. As such some of the problems that affect the member countries include the following: lack of a standalone policy/strategy on resilience in most Member States and this results in an ad hoc implementation of resilience building strategies [27].

According to UNOCHA [10], Zimbabwe has a Resilience Framework draft that was proposed by United Nations Development Program (UNDP) in 2015 and the country is in the process of drafting the Zimbabwe Recovery and Resilience Framework (ZRRF) with technical support from the World Bank, UN and European Union [10]. On the other hand, the Government of Zimbabwe and UN [28: 24] reveals that "A National DRM Policy has not yet been developed in Zimbabwe, however a "civil protection policy statement" exists, stating that "every citizen of Zimbabwe should assist where possible to avert or limit the effects of a disaster"". This gap in policy frameworks has contributed to the fragmented, and thus ineffective, implementation of strategies against disasters and has contributed to the implementation of interventions often times to be in silos [26].

5 Methodology

The study is based on a desktop study involving interrogation of literature and documents mined from Google Scholar, Ebsco and websites with news and policies, statutes, and related materials. However, as Zimbabwe does not have a standalone resilience framework or a disaster risk management policy, the study made use of a case study approach to reveal the strategies that are being implemented in a bid to achieve disaster resilience in Zimbabwe at local levels. As such, convenient sampling of areas to be included in the study was undertaken and this is justified by the fact that some areas in Zimbabwe do not have the strategies they are implementing to enhance their capacities are not captured in literature. In this regard, relevant literature from the data sources was engaged systematically, being guided by key words like disaster risk management and resilience building. By and large, for data analysis, the study mainly engaged thematic analysis whereby the main ideas which were being propagated in the literature were pursued, and thus were instrumental in coming up with the results.

6 Results

6.1 Masvingo Province

The province is divided into seven districts and these are: Mwenezi, Gutu, Zaka, Bikita, Chivi, Chiredzi, Masvingo rural, and Masvingo town. Masvingo province is among the areas most vulnerable disasters in Zimbabwe. According to ZIMVAC [15], Masvingo had the highest average number of shocks in the period 2020/2021 and among the shocks were crop pests and water logging. Moreover, the province was among the areas with highest mortality rates of livestock of above 30%. According to Shamano [29], the major disasters affecting Gutu district are droughts and floods, with the area receiving erratic rainfall averaging less than 600 mm per year. On the other hand, Simba [30] reveal that the south eastern part of Masvingo province is in natural region V, with an aridity index of between 0.2 and 0.5, meaning that the region is semi-arid. Oxfarm [31] observes that Tropical Cyclone Idai affected the province through flooding. Therefore, the area is disaster-prone and some strategies to build disaster resilience are needed.

Some resilience building strategies are being undertaken in Masvingo province. Chingarande and Matondi [32] point out that resilience programming in the province is anchored by the Zimbabwe Resilience Building Fund (ZRBF). This is a long-term development initiative with an overall objective of increasing the capacity of communities to protect development gains in the face of recurrent shocks and stresses, and enabling them to contribute to the economic development of Zimbabwe. Noteworthy is that the ZRBF is supported by the Ministry of Lands, Agriculture, Water, Climate, and Rural Resettlement (MLAWCRR), the EU, the Embassy of Sweden, the UNDP, and the DFID. Moreover, USAID [33] point out that from 2013 to 2020, Masvingo province was a beneficiary of the World Vision interventions towards resilience building. The interventions included the following strategies: conservative agriculture, pest and disease management, irrigation water management, and post-harvest handling and storage. These strategies promoted sustainable food and nutrition security through technologically complementing the traditional ways of farming in the area.

However, Simba [30] points out that governmental interventions towards resilience building were not evident in the province, pointing out that the early warning systems in the province were ineffective. Moreover, "Drought management/coordination committees, composed mainly of government ministries such as agriculture, health and NGOs, are constituted largely to coordinate targeting, aid distribution and vulnerability assessment, the registration of beneficiaries under NGO and government schemes such as seed and food distribution" [29: 76–77]. This implies that the disaster risk management strategies, as championed by the government, are reactive. Noteworthy is the fact that ZIMVAC [15] captured that Masvingo province has the highest number of irrigation schemes that are non-functioning. In this regard, to one of the proactive strategies that may be adopted by the government agencies is to resuscitate the irrigation schemes such as the following: the Manjinji Irrigation Plan in Malipati, Mushandike Irrigation, Charumbira, Citrus Estate in Chiredzi, and Panganayi Irrigation plan [32].

6.2 Matabeleland South Province

According to ZIMSTATS [34], the province consists of the following provinces: Beitbridge, Bulilima, Mangwe, Gwanda Rural, Insiza, Matobo, Umzingwane, Gwanda, Beitbridge, and Plumtree. These districts may be regarded as mostly rural as the towns such as Beitbridge and Plumtree are small with only 12% of the province's population living in the urban areas. According to Phiri et al. [35], Matabeleland South is among the most vulnerable provinces to climate change effects. This may be due to the fact that the province, owing to colonial legacies, has always been marginalized in terms of development and the communities have resorted to emigration into South Africa and Botswana as an adaptation measure [36, 37]. Furthermore, agricultural production in the province is affected by long dry spells, hot temperatures and erratic rainfall patterns owing to climate change [13]. The ZIMVAC [15] report points out that Matabeleland South was among the top provinces in the country that reported livestock deaths (43% of the households), with the majority of them indicating that the cause of death was drought. The same report revealed that the province was projected to have 35% of the households facing cereal insecurity during the peak hunger period (January to March 2022).

In this regard, the Province has adopted both individual and concerted effort resilience measures to enhance their livelihoods. Ndlovu et al. [38] points out that people in Gwanda conserve soil moisture through pit planting and mulching. This is a modified indigenous way of surviving in dry areas or deserts. Moreover, through the use of indigenous knowledge systems, some people in the province are able to predict the onset of the rainy season through studying the characteristics of certain indigenous trees. However, some of the smallholder farmers adopt crisis coping strategies such as selling the productive assets like livestock at very low prices, and this is common in Umzingwane communities [13].

On the other hand, at a community level, the Government of Zimbabwe and UN [28] posit local level resilience building initiatives in Matabeleland South. The report points out that Umzingwane district benefitted from the ZRBF whereby the program supports interventions addressing the relations between hazards, existing social practices, poverty, agriculture and food security and their impact on health, nutrition and access to basic services as a way of building resilience. Additionally, a community-based disaster risk management (CBDRM) programme was launched in the province by the Catholic Relief Services, Caritas Masvingo and Caritas Bulawayo. The programme is "an example of integrated community level disaster risk reduction project aimed at increasing the resilience of communities to the impact of cyclical and repetitive extreme weather phenomena (drought and floods)" [28: 36]. As such, the programme was meant to strengthen community, district, and provincial capacities to prepare for and mitigate hydro-meteorological hazards by developing a culture of risk assessment and contingency planning, coupled with improvements to livelihoods systems and community WASH practices that increase their resilience to extreme weather events and thus improve the communities' adaptive capacities. Subsequently, two Early Warning Systems (EWS) for Matobo and Beitbridge districts for any type of disaster with 500 mobile phones were established and thus proved to be a success story for community-based resilience building. This proves that the non-governmental organizations fill the gap that is normally meant for government agencies in resilience building.

6.3 *Muzarabani District*

Muzarabani district is located within the Zambezi valley in areas below 400 m. As such, the area is susceptible to floods. Both seasonal and cyclone-induced floods are common in Muzarabani. During the peak of the rainy season, floods may be caused by backflows from Cabora Bassa and inflows from the Zambezi River [11]. On the other hand, the area is semi-arid and this implies that it is susceptible to droughts and mid-season dry spells [39]. As such, most lucrative crops that are grown in the area include cotton and other drought tolerant crops such as sorghum, pearl millet, and finger millet, although some people still grow maize despite evident climate change [40].

Muzarabani suffers from the vagaries of climate change. From a study carried out by Nyahunda, Tirivangasi, and Chibvura in 2021, it emerged that climate change is manifest in Muzarabani through unpredictable weather patterns and lower average annual rainfall and long mid-season dry spells [41]. Nonetheless, various governmental agencies have been noticed in Muzarabani and these include the Ministry of Environment, Water and Climate, the Ministry of Health and Child Welfare, agritex officers, officers from the CPU, EMA, and the Meteorological Services Department. However, some of the agencies do not take an active role in disaster risk management and resilience building [40]. According to Mavhura et al. [11: 41], "A District Civil Protection Coordinating Committee (DCPCC) exists in Muzarabani but it is inactive and has limited capacity to prepare for, respond to, or mitigate the effects of flooding." This is confirmed by the World Bank, the Government of Zimbabwe and GFDRR [7] report that reveals that the Zambezi Basin does not have automatic river stations to enhance flood risk management. Nonetheless, the report alludes to the fact that four real time stations are in the process of being implemented in the Mbire and Muzarabani districts, pointing out that these are more suitable than automated river stations for flood monitoring [7]. However, the report submits that in Zimbabwe "Currently no flood forecasting model of repute is available" (ibid.: 82). Nonetheless, Manyani and Bob [40: 78] point out that "the government through its various ministries and organs provides services to the inhabitants of Chadereka [Muzarabani Ward] ranging from agricultural support in terms of inputs and extension services to disaster relief like floods and drought relief whenever they occur." This is a revelation to the reactive approach to disaster risk management adopted by the government.

As such, it is noted that governmental initiatives in the district are limited. Nongovernmental organizations have been instrumental in implementing measures to enhance the resilience capacities of this area through the provision of drought-tolerant crops. Moreover, the people in this district make use of indigenous knowledge systems to enhance their lives and this is exemplified by growing crops in flood plains during the dry season as they are reservoirs of moisture and fertile soils [11]. On the other hand, Nyahunda et al. [41] argues that the Muzarabani communities have adopted both internal and external migration as a measure to counter flood risks in the district with over 3000 people forcefully displaced by the floods to high grounds between 2007 and 2015. On the other hand, Mudavanhu [39] posits that Mozambique, Botswana, and South Africa have become the most common migration destinations for the people from Muzarabani, the majority being men, who seek to circumvent the adverse effects of flood disasters in the district.

7 Discussion

The government of Zimbabwe, although it is a signatory to various international resilience frameworks, does not have its own standalone and consolidative resilience framework. This has impacted negatively to disaster risk management strategies that have remained fragmented and sectorial. The discourse herein reveals that it is imperative for all stakeholders in disaster risk management to be actively involved and participate towards resilience building. However, in the Zimbabwean context, although the community is involved by default, as it is the inherent victim of disasters, its role is limited by resources. The resilience framework by Béné et al. [16] depicts that as the capacity to resist the effects of disasters is enhanced, the transactional costs also rise. As rural communities are inherently incapacitated due to limited livelihoods, this implies that without the assistance of other stakeholders, their initiative alone is not enough. In this regard, as the government is the major custodian of resilience building within a nation [42], its role need not be distorted to become obscure. It is noted that the non-governmental organizations overshadow the role of the government. However, since this is not by design on the part of the nongovernmental organizations, the role they play to enhance resilience building is significant in Zimbabwe.

The ripple effect of the Civil Protection Act of 1989 is reflected in how the government deals with disaster effects in Zimbabwe. Mavhura [3] argues that the Act is no longer suitable to harness resilience building to civil protection and therefore it needs to be aligned to global resilience frameworks such as the Sendai Framework that is proactive. The proactive aspect of disaster risk management is a significant aspect to resilience building in that it serves human lives and infrastructure, and evidence has been amassed that points out that a proactive approach to disaster risk management leads to less expenses than a reactive one [19]. Unfortunately, the government of Zimbabwe, due to a dearth of a relevant regulatory framework and political will, is mainly reactive to disasters in the country. This is reflected by the preoccupation of relief programmes in the country, particularly in Matabeleland South and Masvingo provinces [12, 15]. Relief programmes are necessary in the advent of a sudden disturbance such as cyclones or floods. However, such programmes as cash transfers and food for work only cater for the recovery aspects in the aftermath of a disaster. On the other hand, resilience building strategies need to be proactive in the sense that the communities are prepared even before the disaster hits. This may only be realised if the communities are capacitated through assert provision to transform themselves to the better even when disasters strike.

Climate change is a significant determinant of disasters. This has compelled the government to initiate various drives towards reducing its effects. However, although climate change immensely contributes to the occurrence of natural disasters, it is not the common denominator for all disasters, and thus a holistic framework of resilience building cannot be substituted by such climate change initiatives. Nonetheless, if the climate change initiatives are robustically implemented, such strategies will go a long way towards resilience building. However, it seems the government only dwells on the initiatives as long as they influence international recognition, a habit observed by Olowu [6] that African governments are not sincere in policy implementation as long as it does not lead to endearing them to the electorate. As such, referencing to the case of Muzarabani, it is noted that, even though the regulatory framework for disaster risk management is archaic, the government agencies do not adhere to it in resilience building. The result for such erring in disaster risk management policies are the increasing frequency of disasters and their worsening impacts.

Migration has become a lucrative option in the face of disasters. This may be forced or voluntary migration. Displacements in Muzarabani are an example of forced migration. Such involuntary migration is an indicator of a collapsed system in trying to mitigate the impact of a disaster, unless it is temporary. On the other hand, voluntary migration of people enhances resilience building through complementing household assets that may be used for the implementation of effective disaster risk management strategies. In this vein, Mpandeli et al. [43], posits that migration is an effective way of outsourcing assets that may be used for resilience building, and this is particularly significant in Matabeleland South where emigration is a livelihood strategy for the young people and thus has evolved into transmigration [44]. As such, disaster risk management, through migration, is also exported to destination countries through the migrants, implying that their behaviour and contact in the foreign lands will also determine the reaction of their households to imminent disasters. This brings to the fore the impact of such barriers to effective migration, as xenophobic attacks, to the achievement of a transformational adaptation to disasters.

8 Conclusion and Recommendations

The government of Zimbabwe needs to adopt a political will to change its stance of a reactive approach to disaster management to a proactive one. If there is a strong political will, the redundant regulatory framework as enshrined in the Civil Protection Act of 1989 will be changed to have a comprehensive resilience framework. In this vein, resilience building would not be ad hoc and fragmented as implemented by different non-governmental organizations but will have stipulated responsible authorities who will be mandated to coordinate resilience building at all levels, from the community level to the national level. Without such a framework, other stakeholders will be lax in their roles since there is no legal accountability. Moreover, the evident dependence on nongovernmental initiatives will be done away with since the government will bear the full brunt of resilience building responsibility. In this regard, instead of local authorities hiding behind private stakeholders like non-governmental organizations, they will become more active. This encourages a bottom-up model of resilience building.

References

1. Manyena B (2016) After Sendai: is Africa bouncing back or bouncing forward from disasters? Int J Disaster Risk Sci 7(1):41–53
2. De Smet H (2017) Disaster management within the framework of a changing disaster landscape. Datawyse/Universitaire Pers Maastricht. https://doi.org/10.26481/dis.20171116hds
3. Mavhura E (2016) 'Disaster legislation: a critical review of the Civil Protection Act of Zimbabwe', natural hazards. Springer Netherlands 80(1):605–621
4. Mavhura E, Manyangadze T, Aryal KR (2021) A composite inherent resilience index for Zimbabwe: an adaptation of the disaster resilience of place model. Int J Disaster Risk Reduction. https://doi.org/10.1016/j.ijdrr.2021.102152

5. Oxfam International (2019) Tropical cyclone Idai in Zimbabwe: an analysis of policy implications for post-disaster institutional development to strengthen disaster risk management, Oxfam Briefing Paper—November 2019. Oxfam House, Oxford
6. Olowu D (2010) The Hyogo framework for action and its implications for disaster management and reduction in Africa. Ja'mba´ J Disaster Risk Stud 3(1):303–320
7. World Bank, Government of Zimbabwe and GFDRR (2019) Zimbabwe rapid impact and needs assessment (RINA) report, May 2019
8. ReliefWeb (2019) Zimbabwe Flash Appeal, January–June 2019 (Revised following Cyclone Idai, March 2019). Available online: https://reliefweb.int/report/zimbabwe/2019-zimbabwe-flash-appeal-january-june-2019-revised-following-cyclone-idai-march
9. Mazzeo J (2011) Cattle, livelihoods, and coping with food insecurity in the context of drought and HIV/AIDS in rural Zimbabwe. Hum Organ 70(4):405–415
10. UNOCHA (2019) Zimbabwe humanitarian appeal revision: February 2019–April 2020. Government Printers, Harare
11. Mavhura E, Manyena SB, Collins AE, Manatsa D (2013) Indigenous knowledge, coping strategies and resilience to floods in Muzarabani, Zimbabwe. Int J Disaster Risk Reduct 5:38–44
12. ZIMVAC (2019) Lean season monitoring report. Government Press, Harare
13. Nyahunda L, Tirivangasi HM (2021) Barriers to effective climate change management in Zimbabwe's rural communities. In: African handbook of climate change adaptation, pp 2405–2431. https://doi.org/10.1007/978-3-030-45106-6_251
14. Zimbabwe Vulnerability Assessment Committee (ZIMVAC) (2021) Rural livelihood report. Government Printers, Harare
15. ZIMVAC (2021) Rural livelihoods assessment report. Government Press, Harare
16. Béné C, Wood RG, Newsham A, Davies M (2012) Resilience: new utopia or new tyranny? Reflection about the potentials and limits of the concept of resilience in relation to vulnerability reduction programmes. IDS Working Papers 2012(405):1–61 (Wiley/Blackwell (10.1111))
17. Department for International Development (2011) Defining disaster resilience: a DFID approach paper. Department for International Development, London
18. Combaz E (2014) Disaster resilience: topic guide. GSDRC, University of Birmingham, Birmingham, UK
19. Asian Development Bank (2018) Understanding disaster risk for advancing resilient development: knowledge note, Metro Manila
20. FAO (2018) Southern Africa resilience strategy 2018–2021. Rome, p 32 (Licence: CC BY-NC-SA 3.0 IGO)
21. SADC (2018) Southern African development community regional vulnerability assessment and analysis programme. 2017/18 Outlook, Gaborone
22. International Organization for Migration (2017) Spaces of vulnerability and areas prone to natural disaster and crisis in six SADC countries. Disaster risks and disaster risk management capacity in Botswana, Malawi, Mozambique, South Africa, Zambia and Zimbabwe
23. FAO (2016) Southern Africa El Niño Response Plan (2016/17)
24. Regional Inter-Agency Standing Committee (RIASCO) (2017) Report on the RIASCO action plan for the El Niño-Induced drought in Southern Africa 2016/2017. Available online: https://drmims.sadc.int/en/documents/database/report-riasco-action-plan-el-nino-induced-drought-southern-africa-20162017
25. Van Niekerk D, Coetzee C, Nemakonde L (2020) Implementing the Sendai framework in Africa: progress against the targets (2015–2018). Int J Disaster Risk Sci 2020(11):179–189
26. COMESA (2020) COMESA regional resilience framework
27. COMESA (2021) OMESA member states meet to validate regional resilience framework. Available online: https://www.comesa.int/comesa-member-states-meet-to-validate-regional-resilience-framework/
28. Government of Zimbabwe and UN (2017) Capacity assessment of the disaster risk management system in Zimbabwe (Capacity for Disaster Reduction Initiative, CADRI). https://www.cadri.net/sites/default/files/Zimbabwe-Report-May-2017.pdf

29. Shamano N (2010) An investigation into the disaster risk reduction (DRR) efforts In Gutu District (Zimbabwe): a focus on drought early warning systems. Dissertation, University of Free State
30. Simba FM (2018) Zimbabwe's preparedness in managing meteorological disasters: a case of applying disaster risk management in managing impacts of climate change. J Geogr Nat Disast 8:231. https://doi.org/10.4172/2167-0587.1000231
31. Oxfarm (2019) Cyclone Idai in Zimbabwe—an analysis of policy implications for post-disaster institutional development to strengthen disaster risk management. https://www.oxfam.org/en/research/cyclone-idai-zimbabwe
32. Chingarande D, Matondi P (2020) Zimbabwe stakeholder mapping report: Masvingo Province. Research Technical Assistance Center, Washington, DC
33. Bureau For Humanitarian Assistance/USAID (2020) Development & disaster risk reduction: Southern Africa. Fiscal Year (FY) 2020
34. ZIMSTATS (2012) Provincial report Matabeleland South. Government Press, Harare
35. Phiri K, Ndlovu S, Mpofu M, Moyo P, Evans HC (2021) Addressing climate change vulnerability through small livestock rearing in Matobo, Zimbabwe. In: Leal Filho W (ed) African handbook of climate change adaptation. Springer, pp 1–20
36. Maphosa F (2007) Remittances and development: the impact of migration to South Africa on rural livelihoods in southern Zimbabwe. Dev South Afr 24(1):123–136
37. Sisulu E, Moyo B, Tshuma N (2007) The Zimbabwean community in South Africa. State of the nation: South Africa.In: Buhlungu S, Daniel J, Southall R, Lutchman J (eds). Human Sciences Research Council Press, Pretoria, pp 552–573
38. Ndlovu JN, Lind J, Patlán AB, Upadhaya N, Leku MR, Akellot J, Tol WA (2024) Integration of psychological interventions in multi-sectoral humanitarian programmes: a systematic review. BMC Health Serv Res 24(1):1–18
39. Mudavanhu C (2014) The impact of flood disasters on child education in Muzarabani District, Zimbabwe. Jàmbá J Disaster Risk Stud 6(1):1–8
40. Manyani A, Bob U (2018) Stakeholder participation in climate change adaptation in chadereka ward 1 of Muzarabani Rural District in Zimbabwe. J Humanit Soc Sci 23(2):72–85 (Ver. 3)
41. Nyahunda L, Tirivangasi HM, Chibvura S (2021) Migration as a determinant for climate change adaptation: implications on rural women in Muzarabani communities, Zimbabwe. In: Leal Filho W et al (eds) Handbook of climate change management. Springer Nature Switzerland AG
42. UNISDR (2015) Sendai framework for disaster risk reduction 2015–2030. Available online: https://www.undrr.org/publication/sendai-framework-disaster-risk-reduction-2015-2030
43. Mpandeli S, Nhamo L, Hlahla S, Naidoo D, Liphadzi S, Modi AT, Mabhaudhi T (2020) Migration under climate change in Southern Africa: a Nexus planning perspective. Sustainability 12(11):4722
44. Maphosa F (2010) Transnationalism and undocumented migration between rural Zimbabwe and South Africa. In: Crush J, Tevera D (eds) Zimbabwe's exodus: crisis, migration, survival. Southern African Migration Programme (SAMP), Kingston and Cape Town, pp. 346–362
45. Mavhura E, Manyena B, Collins AE (2017) An approach for measuring social vulnerability in context: the case of flood hazards in Muzarabani district, Zimbabwe. https://doi.org/10.1016/j.geoforum.2017.09.008

The Ubiquity of Politics and Its Effect on Disaster Resilience in Zimbabwe

Albert Maipisi, Oripha Chimwara, Pathias Paradzayi Bongo, and Samuel Kusangaya

Abstract The role of politics in developing countries makes it ubiquitous in all major aspects of national development. In Zimbabwe, international and national politics broadly stands at the center of household vulnerability. There is therefore a huge nexus between development and resilience. A plethora of factors that have direct impact on the nature of resilience at the household level also exist. Key amongst them includes, national political ideologies, nature of institutions, political competition and political culture. The nature and trends of the aforementioned structural and ideological factors to a greater extent reduces national resilience at the household level. This chapter examines the role of politics against resilience building in Zimbabwe. Practically, the country is a politically contested territory and allegorically presents a game of chess in which the masses constitute the cannon fodder. Their resilience has been consciously or unconsciously expended for contestants' survival. The Pressure and Release Model and in particular the Progression to Vulnerability Model is adopted in explaining the conundrum of politics and resilience in Zimbabwe. This chapter is a result of reviewed literature that was randomly selected. The literature provided data from academic journals, books, reports and studies from national and international institutions. It is concluded that politics in Zimbabwe is vastly reproducing vulnerability. Hence, possible solutions for reversing the trend are advanced.

Keywords Politics · Dasters · Resilience · Vulnerability and livelihoods

A. Maipisi (✉) · P. P. Bongo
Women's University in Africa, Harare, Zimbabwe
e-mail: albamaipisi@gmail.com

O. Chimwara · S. Kusangaya
University of Zimbabwe, Harare, Zimbabwe

R. Kasimba and I. Chirisa (eds.), *Disaster Risks and Hazards*, Advances in 21st Century Human Settlements, https://doi.org/10.1007/978-981-96-4577-0_6

1 Introduction

Resilience is the hallmark of sustainable human and national development [21]. Promotion and achievement of national and household resilience is however underpinned by national and international politics [33]. In explaining the resilience challenge in Zimbabwe, the Pressure and Release Model thinking and specifically the Progression to Vulnerability Model was used. This chapter comprises of the study background, overview and objectives, methodology, the theoretical framework; study findings; recommendations and finally the conclusion. The overall resilience of Zimbabwe and its citizens has been weakening for over two decades ago due to poor political and economic systems. These have created a new culture among citizens. The culture includes lack of national patriotism and creativity, while politics has become one of the most coveted jobs in a society where professional capacities and capabilities are generally limitedly rewarded; creating political and economic uncertainty for the majority of citizens. Other citizens have already opted to migrate abroad due to local adverse conditions with politics at the center of the challenges. Some of the migrants are hosted where they have very limited rights for facilitating meaningful human development and resilience.

2 Background and Overview

Amartya [1] supported by Oxfam [34] acknowledges the centrality of politics in development. That is confirmed in their noting that what people positively achieve is significantly influenced by their access to economic opportunities, political liberties, social powers among other related enabling factors. Those factors are anticipatively advanced to unlock conditions for good health, basic education, encouragement for further achievement and cultivation of the spirit of initiation and unlimited discovery [28] that has been weakening in Zimbabwe [42]. In many cases however, the key to such access lies in politics that significantly determines who 'gets what, when and how' [9, 22, 29] before, during and even after crunch moments. The level of access influences level of individual and community resilience. In that sense resilience is defined as competencies and capacities that cause a positive functioning of a community as the outcome whenever that community passes through adversity [3]. Unfortunately, due to its nature, politics can control and intervene even where professionals authoritatively find it difficult and unreasonable to tread [5, 15]. Confirming Honhold et al. [15] and Carresi's [5] claim above, Foley [11] and Buckle [4] adds that governments can even deny vulnerable communities, access to externally sourced disaster related aid, citing political diplomatic stances. Consciously and unconsciously consequently exposing citizens to unacceptable risks that can further weaken individual, household, community and overall national resilience.

Whereas respect for human rights is key to resilience building, in some instances the distinction between 'human rights violations' claimed by non-state actors and 'defence of human freedoms and national sovereignty' postulated by national governments have been very difficult to comprehend globally, yet contributory to weakening resilience in-state. In a utopian state though, institutional arrangements relevant for regulating access to opportunities should respect fundamental human rights and freedoms as basis for resilience of citizens. Those freedoms encompass realistic participation in critical social choices and ability to influence public decisions through means enshrined in the given society's social contract or simply put, national constitutions [37]. All things being equal, the institutions are man-made to advance human welfare in communities in which they exist. The institutions are important, noting that violation of one individual's rights advances opportunity for further violation of other people's rights with negative effects on broader national resilience [8, 14]. Such boomerang effect has been witnessed in many countries globally, including in the United Kingdom after the death of George Floyd and the rise of groups that agitate for equity and equality for example in national resource allocation. Vulnerability that is the opposite of resilience is therefore hinged on significance of resource accessibility and utilisation effectiveness with an effect on overall national capacities and capabilities development before, during and after disasters. That is why it can be acceptable that vulnerability is lack of access to political systems and processes for resource allocation, that is as good as saying vulnerability is limited development [2, 42].

Proceeding from the discussion above, it is clear that politics influences governance. Governance is the linchpin for effective vulnerability reduction that is translatable to resilience building. Understanding that governance is part of politics [18], the preceding assertion is true because national governments broadly through legal frameworks and development blueprints, secretly and explicitly constrict or support resilience building in nations. In that respect Cloete and De Coning [6] advances a quad-criteria for evaluating the degree of governance in a given society. This measurement criteria classifies governments as good or bad and/or authoritarian or democratic. The major evaluation points considered include the perceptive level of trust stakeholders have in the government under review, the government's responsive behaviour to civil society demands and/or voice, degree of accountability to voters and lastly the nature of government authority exercised over citizens. These views are likely contestable when considering the correlation between political philosophies and their relationship to sustainable resilience building in given communities or countries in general. Authoritarian governments have however, usually yielded poorer resilience in the longer term when compared to more democratic ones. Authoritarian as opposed to democratic systems, usually in the same sense fails to sustain peace in their territories [6, 30]. On the other hand, to achieve sustainable resilience building, countries transitioning to democracy may face many challenges and are to a larger extent embracing the 'dictatorship of democracy' like the case with China or gradually start with lesser democracy forms then gravitate to more acceptable extremes with time. Otherwise alternatively, an intelligently concealed push for a one-party state in order to achieve reduced national political power contestations is also opted for. This has been common where political incumbents viewed national unity, peace and coherence as the missing ingredients to their nation's development [25].

Both national and international politics have also worked against resilience. In fact, to a large extent, international politics can stimulate national political behaviours [31]. That is being witnessed globally, especially, where the national status quo is viewed as unfriendly to envisaged global powers and national government's interests, planned resource allocation, exploitation models, national strategies and plans. In that regard the disaster shock doctrine theory proponents for example, expounds that every 'stressful' situation offers an opportunity for disaster capitalists or vulture countries and institutions to reconfigure society to their advantage. In that context, national development especially in poorer countries, always rests at the mercy of some external creatures' hands especially where internationally sourced development finance's fungibility is concerned nationally [23]. The disaster capitalism idea has roots in the University of Chicago's School of Economics and Naomi Klein [19] in her book The Shock Doctrine: The rise of disaster capitalism. Staunch advocates of extreme capitalism like the late Thomas Friedman sees the need to capitalise on times of disasters or shocks, as the best way for advancing socio-economic changes that favours growth of specific groups' grip on national economies. In that theoretical view, a disaster is perceived as a community blackout that ushers in new opportunities for changing conditions that could have been prevailing before a disaster. In other words, the idea is, what is unacceptable when the community's conditions are normal and stable is thought to be easily acceptable when the same community is in a state of shock that can be economic, political or a result of any other disaster that can lead to government's desperation. Those who hold those views assume that vulnerable people have a tendency to accept anything offered to them especially immediately after a disaster as long as that is presumed capable of easing the prevailing difficult conditions [36]. COVID-19 vaccines manufacture and global supplies to many countries is a contemporary case in point. Such procurements are causing more debt burdens currently blanketed by national head immunity targets. Nevertheless, whereas the disaster capitalism assumptions have been proved to be correct in some contexts, those efforts have not always yielded expected results immediately and without contestation.

Disaster capitalists advance the point that opportunists always want to take advantage of prevailing shock and even eagerly await if not aiding creation of such moments so as to reconfigure the society according to their plans. Such persons are the ones who strive to also win contracts and/or tenders to reconstruct the disaster ravaged systems. Those involved, capitalises on new business niches and usually introduce formerly resisted ideologies during that time when local competitors, community gatekeepers and negotiators are usually in their frailest condition. That is why it remains very important for government authorities to always analyse and assess interventions and interventionists in order to make sure they prove their innocence before implementation of proposed plans and programmes. Without safeguards, those who covet disaster events as chance or moments for advancing own ideologies especially where they are resisted under normal circumstances, will always try to smuggle them back to fore even if that would lead to increased overall community vulnerability

[19]. If those disaster capitalist elements are not checked, they can compromise the long-term security of communities' resilience. Thence, national governments are sometimes justified when they closely monitor, control, regulate and coordinate perceived resilience building interventions in a state. A national government itself can nevertheless play the disaster capitalist role in some circumstances and for different reasons [24]. Whilst contrariwise, the global political systems are equally very complicatedly configured and highly invested in, for sabotaging other nations' development endeavours and for the devourer's survival. Thereby, broadly increasing vulnerability mostly in the developing world [35]. This chapter examines the role of politics on resilience in Zimbabwe in the objectives section that follows.

Objectives

- To examine the relationship between politics, development and resilience building at household level in Zimbabwe, and;
- To identify the factors that impact on household resilience in Zimbabwe.

3 Methodology

This study adopted the qualitative research methodology using the case study design. According to Yin, "... a case study research design examines a contemporary phenomenon within real life context." This method enabled researchers to examine the relationship between politics, development and resilience at household level in Zimbabwe. Data was analysed using thematic analysis that involved the categorisation of the same data under a particular theme as it emerged from the findings.

4 Theoretical Framework: The Pressure and Release Model

The Pressure and Release Model like other models, is basically a simplified representation of a complex reality, meant to describe, prescribe and explain relationships between variables and processes around them [6, 39]. In that view, a disaster is explained as the point at which vulnerability meet a hazard or a compound of them [13]. The crunch point brings the unbearable experience generally known as the disaster that Wisner et al. [39] posits theoretically as an act of two forces that would have progressed towards each other to the extent that where and when they meet, they form the crisis (crunch) situation [13] as illustrated by the Pressure and Release Model in Fig. 1. Other scholars have used the Sustainable Livelihoods Frameworks among others, to explain and tackle issues of livelihoods which are usually viewed as linked to vulnerability and resilience.

According to the illustration by Wisner et al. [39] above, vulnerability is a creation of limited access to power, structures and resources among certain groups of people

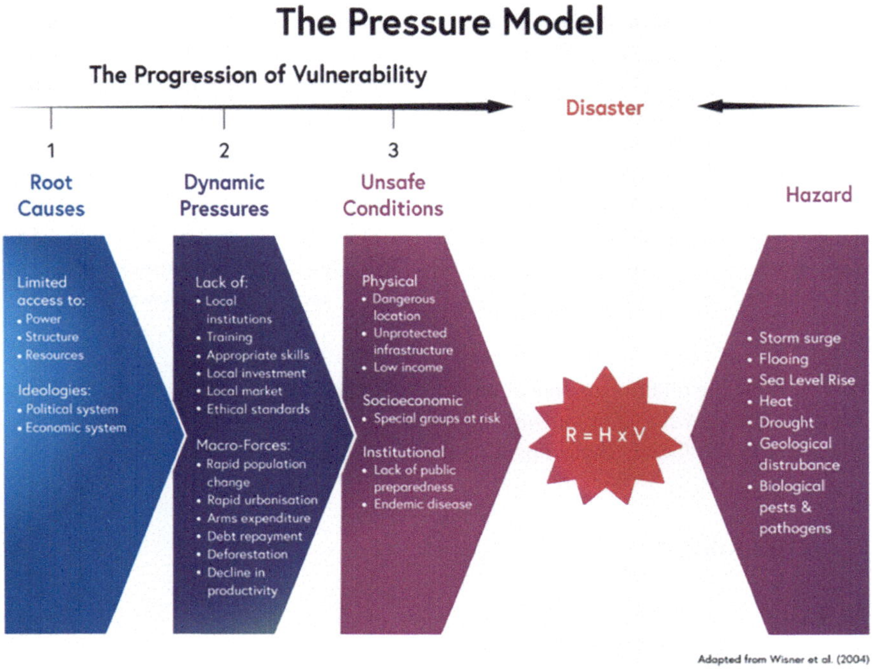

Fig. 1 The pressure and release model. *Source* [39]

in a particular society. Those challenges are further attributably rooted in ideologies that drive the referred society's political and economic systems. Hammer et al. [12] adds precarity, poverty and inequality to the list of the mentioned root causes. That is nevertheless not adding much substance than matter explicability because poverty and inequality are both a result of global, national, historical and prevailing ideologies and existing political and economic systems [10, 20, 40].

Those root causes are advanced in the model as creating dynamic pressures namely lack of local institutions, training, skills, local investment and ethical standards in public life. These factors are also coupled with population growth, urbanization and the increasing national burden of debt repayment [39]. Kruger et al. [20] however argues that no matter what, no single society can possess total capacity to completely address overall root causes of its vulnerabilities and in their view, vulnerability is found in all societies although what varies are mostly its levels. Having noted the claims by Wisner et al. [39], the researchers also note that, vulnerability is not only a result of lack of access but also overly access for example, to power, not only due to existing political and economic systems, but a combination of their porosity, abuse and an unaccountable leadership.

Due to those degenerative processes, citizens find themselves in unsafe conditions as are broadly prevailing in Zimbabwe [42]. In that respect, the Pressure and Release Model exposes four categories of unsafe conditions. The first one is fragile physical environments comprised of dangerous locations, unprotected buildings and national infrastructure. The second category of unsafe conditions are provided as a fragile local economy that leads to lower disposable income levels and increased livelihood exposures to risk (2003). Third, a vulnerable society grows, comprising of special groups at risk without local institutions for serving them. Fourth, public actions for civilian security such as disaster preparedness also ceases to be prioritized. People nevertheless find themselves living in those spaces because of individually justifiable reasons [9, 20, 26].

Reversal of progression against a disaster is postulated as achievable through addressing root causes of vulnerabilities and the foundational factors leading to hazards build-up [31]. Hazard reduction, for example through tackling climate change, has remained a contentious global issue among countries. Equally with vulnerability, certain international and national political decisions are broadly at the center. Actions for reducing vulnerability usually strengthens resilience [3]. Nevertheless, the role of politics in national vulnerability is measly acknowledged, while economics that is a lower tier cause of vulnerability is unfortunately viewed more as a standalone cause. Silence on the role of politics in key institutional reports is sometimes more revealing [42].

Disaster stricken communities are unable to easily bounce back or forward (recover) when they are caught up in a disaster unless their surplus inherent energies (resilience) can exceed their support needs or until they get proportionate external support. Without that, it will be difficult for them to transition from the pain to the pleasure economy [36]. The period of pain or falling down before bouncing back and/or forward is only shortened by the individual victim and his or her greater society's capacities. The two should join hands in adopting actions that reduces the hazard and/or vulnerability, a behaviour that Sorokin [38] views as the real duty of men. Noting that hazards, vulnerabilities and resilience are social constructs. Hence, mitigation of negative disaster effects should be possible through reformation of national and international governance and resource distribution models and structures [20]. Zimbabwe has not been spared from this call although there has been different arguments and counter arguments to the national question internally and externally.

On the other side presented in the model, there is a list of hazards threatening the society hence putting people and their assets at risk. Hazards in this model are mostly given as 'natural' but other types of hazards such as viruses, diseases and pests are generally included in this coterie particularly in the second version of the model [12]. Risk increases where vulnerability levels are high. The level of damage experienced by a community is also directly attributable to resilience levels of the community at risk and the inevitable hazard strength. Even in the same community nevertheless, households and individuals uniquely experience different vulnerability and resilience levels. Noting that resilience and resistance are different, yet, understanding their variance is important in the conceptualization of resilience.

Given the discussion above, this model also has weaknesses that the authors already acknowledged in the later version of their book "At Risk" [39]. Of significant note is that largely, the model builds on the centrality of the role of government in vulnerability creation but in some instances, access to power and economic capacities has destabilised communities, leading to increased vulnerability in the given society. The unjust distribution of and access to political and political power among others, has been used to justify advancing instability in given societies, leading even to a further rise in vulnerability. Others who have had access to political and economic power, structures and resources, have also not managed to transform their livelihoods to sustainable levels, which to some extent questions the claims by this model, which also presents a completed work than a thinking tool that should allow development of new knowledge through it. That is however acceptable with models, as they mostly reflect a static position when compared to theories. This model nevertheless provides a relevant framework for understanding the relationship between politics, development and resilience building at household level in Zimbabwe and, a framework for identifying the factors that impact on household resilience in Zimbabwe.

5 Political Ubiquity and Resilience in Zimbabwe: Study Findings

The key factors with a direct impact on the nature of resilience at national and household level are political ideologies, nature of institutions, political competition and political culture as postulated by the Pressure and Release Model [39]. The political ideology that was being driven by the post-colonial government in Zimbabwe and particularly during the first decade of independence was arguably more socialist inclined. Redistributive and restorative politics culminated to the Land Reform Programs in the country for example, leading to the subsequent chaos and heightened diplomatic fallouts that eventually saw Zimbabwe alienated from the broader international community. It is however famous that during and post the liberation struggle, nationalists have emphasized national unity and peace among citizens, as key building blocks for national development [17, 25]. Due to contestations for national resources with power being the key resource arbiter among the different interested groups, limitedly existing underlying conditions for national peace and unity could not be sustained among the different contenders and their co-contestants. This challenge further degenerated into an economic quagmire as a result of international economic sanctions that slammed the country whose economic system was heavily dependent on foreign as opposed to local capital and industrial produce market network arrangements. That legacy has seen the collapse of the economy and its social safety nets [41]. Politically, the national government become exposed to civil society pressure for reforms that had to indirectly accommodate power contenders that were out of government, as a guarantee for balancing the ruling government's possible extreme ideologies. The plan to balance the ruling party's political ideologies through contenders did not yield much success but was further weakening the national economy among others. The government has however committed to independently implement key policy reforms outlined in the first National Development Strategy (NDS) [42] with some of the contenders indirectly co-opted through the established national Political Actors Dialogue platform (POLAD). Whereas access is at the centre of the Pressure and Release model (Fig. 1), it is unfortunately revealing that, access to foreign resources for example, from the country's international aggressors by the local political elite that has mainly graduated from the middle-class, has sustained disunity among political contenders in Zimbabwe. Such international access as opposed to local access, has also reduced the cliques' vulnerability at the expense of the whole of society in the short term. This latter view is against the access to national systems thesis advanced in the Pressure and Release model.

The legacy of the political system in Zimbabwe became very visible in the nature of key institutions that then adapted to the prevailing political and economic system. The ruling government stood by its loosely defined and implemented socialist ideology with roots in the pre-independence nationalist liberation ideology [25]. Due to the rise in political competition for power, the ruling government used extra judicial tactics to retain control of political power. That saw growth in reduced meritocracy in the allocation of resources and a decline in confidence towards government structures mostly by the middle class and some elite among the population. Rural population and particularly the poorer ones who were resettled in farms and those in communal areas, were presented with rent seeking tokens in exchange for voting loyalty. The tokens were however not building resilience of the communities but in fact, weakened the same. With the focus mostly on clinging to political power, the national economy shrunk, companies closed and corruption became rife while appointments to key institutions such as those managing national elections were made to those with unquestionable loyalty. Hence, capable of entrenching perpetuation of hold to power by the incumbent. Those acts constricted state efficiency and sustainable public administration, further brooding deepening corruption at all government tentacles.

Confidence in the national political system went down nationally and internationally with time including due to reactionary government policy inconsistencies. The inconsistencies reduced investor and citizens' confidence in key national formal economy institutions. As a result, the informal economy thrived more, with negative consequences to the overall national formal economy. The national purse became dry, with minimum capacity to spent on development and maintenance of existing key infrastructures. International economic support depleted and the choice of potentially responsible investors to the country became further restricted. Local economic opportunities for the ordinary citizen and recent college graduates were also more gradually reduced [32]. That has been leading to negative coping strategies among Zimbabwe's relatively high human capital, slowing down the pace and inclusivity of economic growth [42]. Workers' bargaining powers depleted and they generally lived at the mercy of the employer, with minimum surplus for personal development and access to social services [41]. In addition, rifts among political parties widened while an increase in social entrepreneurship in the sense of foreign funded local civil society organisations (CSOs) articulating human rights and other social ills was witnessed. In that respect Dhafana [7] notes that Non-Governmental Organisations were around less than 200 in the 1990s but in a 2010 National Association of Non-Governmental Organisations (NANGO) Report, they had risen to over 1000. That nevertheless presented another difficult conundrum for the government of Zimbabwe, with negative effects on broader citizenry resilience. The rise in human rights defenders as interventionists can be attributed to such thinking advanced by the Pressure and Release model (Fig. 1) that views the centre of societal vulnerability as the national governance system.

Due to the length of time spent by the political contestants holding onto their radical views including as a source of political and economic survival of individual contestants as legal personas, the coming in of the New Dispensation has not yet managed to fully deal with the challenge. The negative political projection of government by the media and the civil society to the international community as it tries to unlock national economic and political potential, is a matter that the government still grapples with. This trend has been witnessed in the country's political participation frame post-independence. Hence, the 1987 Unity Accord and the 2008–2013 Government of National Unity (GNU) that sought to reduce competition against the ruling party from fellow political contestants. The continuous pursuit of political power consolidation by ZANU PF has also delayed full national economic growth and emancipation since any national solution should leave the party as the key broker. Many authors in Mandaza and Sachikonye [25] agrees that national peace and unity have been identified as important elements for Zimbabwe's development. They actually advanced the view that even the nationalist movements pre-independence grappled with the challenge of achieving a common voice including within the same movement. Contestants survive on their ideological positions and groupings as sources of livelihoods hence the challenge for them to demobilize or change strategic focus in their contests even where it reasonably appears necessary. The nature of politics and the political culture in the country has therefore weakened national resilience broadly and particularly among most of its poorer citizens.

Practically, the country is a politically contested territory and has allegorically presented a game of chess in which the masses have continued to constitute the cannon fodder against their resilience [41]. While the country's economy has been in abeyance, Zimbabwe failed to service its debt for over many years. Hence, interest on debt has been building up, further locking up potential access to future and more fungible development financing [23]. The battle of political ideologies saw Zimbabwe broadly under the European Union (EU) and American sanctions. Sanctions can deny ordinary citizens of their development opportunities and they can also harden the recipient government's ideological stances. Governments' blue-prints that guide such agendas are however usually unalterable until after many years [27]. In that respect, it remains for the reader to judge whether the national resilience being expended for contestants' survival is consciously or unconsciously being done by these political power contenders. Also, the nature of international developing financing to Zimbabwe has been shifting to the current extent where government fungibility is restricted. The reasons dovetailing to fear of financial diversion by the international development partners. That duplicates existing government structures through unnecessarily justified employment of people in the different CSOs, for independent implementation of projects to achieve same goals existing government structures are established to achieve, with some negative effects on financial use efficiency and effectiveness [23]. Therefore, the nature and trends of the aforementioned structural and ideological factors to a greater extent reduces national resilience and overall household welfare level [42] although it should be noted that some ideological factors have linkages to global ideologies due to different institutional survival reasons.

Sustainably dealing with the nature of power contest in Zimbabwe is the best way towards achieving lasting national resilience. The path the country has traversed since independence and even in the New Dispensation post 2018 elections, provide many lessons. One of the big questions on Zimbabwe is whether the governance philosophy should follow without due regard to its contextual characteristics. Indications are that the current Zimbabwe is following the Chinese model of democracy that is Socialism with Chinese characteristics [16]. Debate on the possibility for a one-party state has been adequately dealt with in Mandaza and Sachikonye [25] but the matter the debate sought to address has remained. Mopping of power contestants through co-option appears to be the new trajectory adopted for fostering national unity and peace as the basis for sustained economic growth. On the other end, once contestants' sponsors have their national interests taken care of, that would reduce other national political contestants' financial strength currently sustaining their existence even without drawing the political party finances from the national government. Whether this will be a success and other alternatives that can lead to national consensus, are key to national development and sustained future resilience building and vulnerability reduction.

6 Conclusion

Politics has a central role towards the achievement of national resilience and in the Zimbabwean context, it has weakened national resilience. Economics has been hugely hampered by national and international politics in response to national conditions. Fundamental local conditions for sustained national growth have not been fully addressed over the years. As a result, investor hesitancy due to policy inconsistencies has also been witnessed. Weak property ownership rights were abused for vote buying without securing the interests of the beneficiaries in the long term. That was experienced particularly where urban residential and redistributed agricultural land were concerned. Abuse of property rights was mostly used by those seeking political office as carrot and stick for the vulnerable citizens. Political ideologies to a greater extent influenced the nature of institutions, political competition and political culture that now exist in Zimbabwe and the level of citizen access to power, structures and resources. National political consensus is the key to the envisaged national development and the government of Zimbabwe have tried many ways to achieve that objective. From the ruling party's perspective, these pathways included the establishment of a one-party state, use of a legal framework to entrench national patriotism, ongoing co-option of contestants towards unity in diversity and constant monitoring, evaluation and gradual constriction of the civil society's space to ensure they realign their priorities as a guarantee for their long-term existence. From some of the opposition parties and CSOs views nevertheless, democracy is the best way for addressing the politics in Zimbabwe if the country has to achieve sustained national development. There is overall consensus that the key to national development lays in national politics and whether international sponsors to some of the national contenders will continue to fund them to a larger extent depends on whether the stakes to achieve through that means are still high. Unfortunately, they have been getting lower in the shorter term than in the longer term.

In that respect, politics in Zimbabwe has constantly reproduced vulnerability among citizens. There is therefore need for lasting solutions towards reversing that trend. Politics have shuttered the hopes and aspirations of most of the middle and lower-level groups of the society. Economic policy uncertainty and inconsistencies have further weakened the resilience levels of the people of Zimbabwe for more than two decades decade. This anomaly requires a political solution that to a greater extent is overall national political convergence towards a consensus that secures a sustainable political and economic solution towards consistent human and economic development. National unity and peace have been identified as the major anchors of national development since independence but these have been disturbed from time to time due to political expedience with the generality of Zimbabweans bearing the brunt that left them even poorer. Access to key institutions and services such as hospitals has no longer been guaranteed while the quality of service has been deteriorating creating an even wider rift between the urban and rural development divide especially around access to social services. Resilience has therefore been impacted negatively by the national politics and in particular the prevailing national political culture that continue to weaken the broader citizenry's capacities and capabilities important for resilience building.

7 Recommendations

Political contestants in Zimbabwe need to tolerate ideological diversity and foster national unity and peace as building blocks for national development and subsequent sustained resilience building in line with Kennedy [17]. That should reduce disenfranchisement of the majority of citizens from national power, resources and structures. In the same context, that would restore national pride and restore confidence in government by the citizens. Where that is achieved, collective definition of the national political and economic system would further reduce the politics of animosity, replacing it with advancement of peace, national unity and patriotism. Political solutions are further important in addressing vulnerability where communities are living with risk. There are areas where vulnerability is perennially reproduced implying that there is no immediate solution in sight for those groups particularly living in common disaster-prone areas. The government and practitioners have to therefore join hands in consulting those communities on the optimum possible options towards resolving the exposure challenge because of any reason that could be proximity to the hazards, among others. A political culture that further reinforce creation of legislative certainty, policy consistency, respect for human dignity, the right to development, equity and equality should further be promoted in the country to achieve a sense of belonging among citizens. That sense promotes improved citizen participation, initiative and scaling-up of local economic investments. Finally, political consensus has to be deliberately invested in by the government and both short-term, medium term and longer-term strategies for advancing in that direction should be developed and implemented to ensure sustainable national growth that generally translates to national and household resilience.

References

1. Amartya S (1999) Development as freedom. Oxford University Press, Oxford
2. Bankoff G, Frerks G, Hilhorst D (eds) (2004) Mapping vulnerability: disasters. Development & People. Earthscan, London
3. van Breda AD (2018) A crucial review of resilience theory and its relevance for social work. Soc Work 54(1):1–20
4. Buckle P (2012) Preparedness, warning and evacuation. In: Wisner B, Gaillard JC, Kelman I (eds). The Routledge handbook of hazards and disaster risk reduction. Routledge Taylor and Francis Group. New York
5. Carresi AL (2012) Emergency management principles. In: Wisner B, Gaillard JC, Kelman I (eds). The Routledge handbook of hazards and disaster risk reduction. Routledge Taylor and Francis Group. New York.
6. Cloete F, De Coning C (2015) Improving public police: theory, practice and results, 3rd edn. Van Schaik Publishers, Pretoria
7. Dhafana A (2017) Corporate governance and accountability of NGOs. The Catholic University of Eastern Africa, Nairobi
8. Enarson E (2012) Women confronting natural disasters: from vulnerability to resilience. Lynne Rienner Publishers Inc., London
9. Esnard A, Sapat A (2014) Displaced by disasters: recovery and resilience in a globalizing world. Routledge Taylor & Francis Group, New York
10. Flanagan BE, Gregory EW, Hallisey EJ, Heitgerd JL, Lewis B (2011) A social vulnerability index for disaster management. J Homeland Secur Emergency Manag 8(1), Article 3
11. Foley C (2007) Mozambique: a case study in the role of the affected state in humanitarian action—humanitarian working group (HWG) working paper. Overseas Development Institute (ODI). London
12. Hammer CC, Brainard J, Innes A, Hunter PR (2019) (Re-) conceptualising vulnerability as a part of risk in global health emergency response: updating the pressure and release model for global health emergencies. Emerg Themes Epidemiol 16(2):1–8
13. Hansford B (2011) Roots 9: reducing risk of disaster in our communities. Tearfund Roots resources, London
14. Hirsch S (2016) The states and multinationals: common interests, conflicting interests and public policies. Transnational Corporations Review 8(1):17–27
15. Honhold N, Douglas I, Geering D, Shimshoni A, Lubroth J (eds) (2011) Good emergency management practices: the essentials. FAO Animal Production and Health Manual Number 11. Rome
16. Jinping X (2014) The Governance of China. Foreign Languages Press, Beijing

17. Kennedy JF (1963) Why JFK was assassinated—the speech that cost him his life Part 1. (https://www.youtube.com/watch?v=Y_cPQiLBrUA) Viewed on 4 July 2021 at 2256hrs
18. Keping Y (2017) Governance and good governance: a new framework for political analysis. Fudan J. Hum. Social Science (Online)
19. Klein N (2007) The shock doctrine: the rise of disaster capitalism. Henry Holt and Company, New York
20. Kruger F, Bankoff G, Cannon T, Orlowski B, Lisa E, Schipper F (2015) Cultures and disasters: understanding cultural framings in disaster risk reduction. Routledge Taylor and Francis, London
21. Kumah MM, Shah SA (2015) A multistep model of resilience development. Int J Indian Psychol 3(1)7:99–103
22. Lasswell HD (1936) Politics: Who gets what, when and how? Whittlesey House, New York
23. Lehman HP (2010) Japan and Africa: globalisation and foreign aid in the 21st century. Routledhe Taylor & Francis Group, London
24. Maipisi A, Jordaan A, Restas A (2016) Locating the faultlines of flood disaster vulnerability and susceptability: lessons to governments, development agencies and practitioners in developing countries
25. Mandaza I, Sachikonye L (1991) The one-party state and democracy: the Zimbabwe Debate. SAPES Trust. Harare
26. Marsh GP (1864) Man and nature or physical geography. John F. Trow & Co., New York
27. Matutu V (2014) Zimbabwe agenda for sustainable socio-economic transformation (ZIMASSET 2013–2018) a pipeline dream or reality: a reflective analysis of the prospects of the economic blue print. Researchjournali's J Public Policy 1(1)
28. Mitchell JK (ed) (1999) Crucibles of hazards: mega-cities and disasters in transition. United Nations University Press. Tokyo
29. Moran M, Rein M, Goodin RE (eds) (2006) The Oxford handbook of public policy. Oxford University Press, Oxford
30. Moyo JN (1992) Politics of the national purse: public budgeting as public policy in Zimbabwe. SAPES Books, Harare
31. Moyo J (2015) The Wikileaks saga: national, regional and international implications for Zimbabwe. SAPES Books, Harare
32. Newfarmer R, Martha DP (2015) Trade in Zimbabwe: changing incentives to enhance competitiveness. Directions in development. World Bank, Washington, DC
33. Organisation for Economic Co-operation and Development (2008) Concepts and dilemmas of state building in fragile situations from fragility to resilience. J Dev 9(3):1–82
34. Oxfam (1995) The Oxfam handbook of development and relief Volume 1. Oxfam Publishing, London
35. Perkins J (1945) Confessions of an economic hitman. Berrett-Koehler Publishers, Inc., San Francisco
36. Prince SH (1920) Catastrophe and social change: based upon a sociological study of the Halifax disaster. Columbia University, New York
37. Rousseau JJ (1752) The social contract: on principles and of political right G. Putnam's Sons. New York
38. Sorokin PA (1942) Men and society in calamity: the effects of war, revolution, famine, pestilence upon human mind, behaviour, social organisation and cultural life. E. D. Dutton & Co., Inc., New York
39. Wisner B, Blaikie P, Cannon T, Davis I (2003) At Risk: natural hazards, people's vulnerability and disasters, 2nd edn. Routledge Taylor & Francis Group, London
40. Wisner B, Walker P (2005) Beyond kobe: a proactive look at the world conference on disaster reduction 18–22 January 2005, kobe, Japan. A report for the Swiss department of humanitarian aid. Feinstein International Famine Centre. Boston
41. World Bank (2016) Zimbabwe Public Expenditure Review: Volume 5, Social Protection. The World Bank. Washington D. C.

42. World Bank (2021) Zimbabwe economic update: Overcoming economic challenges, natural disasters, and the pandemic: social and economic impacts. Issue No. 3 (June 2021). The World Bank Group. Washington

Albert Maipisi holds a PhD in Disaster Management from the University of Free State in South Africa, PhD taught courses in Public Administration from the National University of Public Service in Hungary, Msc. In Disaster Management from the National University of Science and Technology (NUST) in Zimbabwe as well as a Bsc. Hons. Degree in Administration from the University of Zimbabwe (UZ). He is currently a Senior Lecturer with the Women's University in Africa domiciled in Zimbabwe and the national Institutional Development Component Lead for the New York State University Rockefeller Foundation's Parliament Support Programme in Zimbabwe. Immediately before joining the University, he held the position of Committee Clerk/ Coordinator for the Parliamentary Portfolio Committee on Mines and Mining Development under the Parliament of Zimbabwe and also have introduced and currently coordinates the Master of Science Degree in Disaster Risk and Livelihoods Management Programme at the Women's University in Africa. Research interests are disaster risk reduction, poverty reduction, parliamentarism, nationalism and public policy.

Oripha Chimwara is a lecturer in Department of Governance and Public Management, University of Zimbabwe. She holds an MSc International Relations, BSc Political Science (University of Zimbabwe). Her research interests are focused on international security, governance, women and youth studies.

Samuel Kusangaya is a physical geographer with a PhD in Hydrology (UKZN) MPhil in GIS & Remote Sensing (Cambridge), MSc in Land and GIS (UZ), and BSc in Geography & Mathematics (UZ). He is currently employed as a senior lecturer at the University of Zimbabwe lecturing in applied hydrology, water resources management, GIS, remote sensing, and climate change and climate modelling. Dr Kusangaya has a strong research background with over 50 articles in referred journals, 10 book chapters and over 60 papers published in conference proceedings. His research interests include but not limited to geo-science applications in hydrology, geo-science application in hydrology, water resources management and disaster management.

Gender and Disaster Resilience

Sandra Bhatasara, Manase Kudzai Chiweshe, and Naume Zorodzai Choguya

Abstract This chapter utilises an intersectional approach to analyse the gendered dimensions of disaster resilience. An intersectional approach identifies that people have different individualities, wants, priorities and capabilities that are not static, and will shift and change over time—disturbing their ability to arrange for, cope with and respond to natural hazards and climate variability. Intersectional approaches offer a way to understand and respond to the ways different factors, such as gender, age, disability, class and ethnicity, intersect to shape individual identities, thereby enhancing awareness of people's needs, interests, capacities and experiences. In understanding resilience, it is also important to analyse how localised responses to climate shocks are also gendered whereby women and men may have different forms of agency, capacities and knowledge related to everyday practices of resilience. Social inclusion in resilience also needs to understand the patriarchal nature of communities and how these social systems influence everyday lives including livelihood strategies and opportunities.

Keywords Disasters · Resilience · Gender · Intersectionality · Ecological disasters

1 Introduction

This chapter interrogates gendered dimensions of disaster resilience in Zimbabwe focusing on the emerging patterns and tendencies; dominant conceptualisations and distortions; silences and missing linkages in as far as gender struggles with and responses to disasters are concerned. Ecological disasters such as climate change are increasingly becoming critical challenges confronting Zimbabwe. The country's dependence on rain-fed agriculture and natural resources makes it particularly vulnerable to climate related disasters such as droughts and floods. In this context, gender and women are increasingly receiving attention. However, trawling through literature

S. Bhatasara (✉) · M. K. Chiweshe · N. Z. Choguya
Faculty of Social And Behavioural Sciences, University of Zimbabwe, Harare, Zimbabwe
e-mail: sandrabhatasara@gmail.com

R. Kasimba and I. Chirisa (eds.), *Disaster Risks and Hazards*, Advances in 21st Century Human Settlements, https://doi.org/10.1007/978-981-96-4577-0_7

on climate change in Zimbabwe, this chapter makes several observations. Firstly, it is unquestionable that the dominant literature on ecological disasters has so far not fully appreciated multi-layered gender struggles and perspectives. Secondly, although the few gender specific researches make important contributions to our understanding of the ecological disasters such as climate change, they also contribute to an unnecessarily narrow understanding of gender because of a fixation on ‘impacts’ that are material and measurable [50]. Again, research specifically interested in women’s struggles has predominantly viewed these struggles through a narrow understanding of gender thereby missing the intersectionality of experiences. As espoused by Oyewumi [68], gender has become one of the most important analytic categories in the academic enterprise of describing the world and the political business of prescribing solutions.

More so, although there are some researches that recognise women’s agency in adaptation, much has not been done to move beyond the entrenched disempowering discourse and reductionist portrayals of women (particularly rural ones) as merely vulnerable and victims. Lastly, there is also a glaring bias towards rural women in ecological disasters and climate change studies. This understandably, has been underpinned by the notion that the country is largely rural and rainfall-dependent rural livelihoods are increasingly becoming precarious under climate change. However, this is a self-limiting trajectory as women in other spaces in Zimbabwe also have to contend with various challenges posed by ecological disasters. Subsequently, all these facets have moulded specific discourses and perhaps modalities of disaster risk reduction. Our argument in this chapter is built on the increasing need to mainstream gender within resilience scholarship and programming using intersectionality perspective. We argue that there is a need to go beyond simplistic and generalised understanding of women and men as homogenous groups. We refer to Jerneck’s [46] approach that goes beyond the dichotomies of victim or heroine that characterise some studies on gender and climate change. We argue, that intersectionality should be the starting point to understand the varied experiences, struggles and agency of women and men, young and old, persons with disabilities, rural and urban (among others) emerging under ecological disasters in Zimbabwe.

2 Methodology

The research utilised a desk research approach focused on collecting data from a variety of existing sources. Desk research involves the analysis of data or information that was either gathered by someone else (e.g., researchers, institutions, NGOs, etc.) or for some other purpose than the one currently being considered, or often a combination of the two. It involves the summary, collation or synthesis of existing research rather than primary research where data is collected from subjects. Sources of secondary information include research journals, newspapers, study books, and annual reports from companies, internet, policies, and publications of governmental and non-governmental organizations. The desk research utilised a systematic review method. Systematic review is appropriate in identifying, appraising and synthesizing research-based evidence and presenting it in an accessible format [63]. According to the Cochrane Handbook, a systematic review attempts to collate all empirical evidence that fits pre-specified eligibility criteria in order to answer a specific research question. It uses explicit, systematic methods that are selected with a view to minimizing bias, thus providing more reliable findings from which conclusions can be drawn and decisions made. In general desk research suggests published reports and statistics, however in a wider context it includes all sources of information that do not involve a field survey [74]. For this chapter we conducted a systematic review of literature from the following: libraries sources, online journals, research papers, newspapers, commentaries, non-governmental organisations' reports, government and policy documents.

3 Literature Review and Theoretical Framework

3.1 Nature and Patterns of Ecological Disasters in Zimbabwe

Studies have already demonstrated that Zimbabwe is experiencing the effects of climate change, notably rainfall variability, extreme events and warming trends that are expected to render land increasingly marginal for rain-fed agriculture [6]. Zimbabwe has over the years experienced multiple ecological hazards such as droughts, floods, veld fires, storms [75]. Chikoto and Sadiq [16] provide a detailed catalogue of disasters in Zimbabwe's history. They argue:

> Between 1980 and 2010... 35 natural disaster events, that resulted in 6,448 deaths, averaging 208 deaths from disasters annually were documented. Of the 35 natural disasters, 6 were drought occurrences, 7 were floods, 2 were storms, and 20 were epidemic occurrences.

These occurrences highlight Zimbabwe's vulnerability to shocks and hazards such as floods, droughts, epidemics and populations displacements due to climate change or natural or human induced emergencies [99]. Below we provide more detail on the nature and impacts of these shocks and disasters.

3.1.1 Droughts

Colonial and post-colonial Zimbabwe has a long-documented history of droughts. Chigodora [15] highlights that drought have become common occurrences in Zimbabwe. The Gender Equality and Youth Inclusion Strategy and Action Plan (2020), argues that Zimbabwe is experiencing the effects of climate change characterised by rainfall variability and extreme events such as droughts, heatwaves, long dry periods and heavy rains, accompanied by flash floods. In the last 30 years, the trend indicates changes in the beginning of rains and upsurges in the occurrence

Table 1 Ecological disasters in Zimbabwe

Incident	Dates	Extent of damage	Affected areas
Eline	Mid Feb 2000	Widespread damage: 2.7 million people affected, 91 deaths, 357 injuries, 59,187 huts were destroyed	Manicaland, Matabeleland South, Midlands, Masvingo
Japhet	3 March 2003	6 deaths	Masvingo, Midlands South, Manicaland South and parts of Mashonaland East
Dineo	March 2017	136 people dead and 59,184 houses damaged 20,000 livestock killed and damaged 230 dams	Manicaland
Idai	March 2019	Widespread damage; 270,000 people affected; over 340 dead	Manicaland, Masvingo, Matabeleland South

Sources Brown et al. [12], Chatiza [14]; https://reliefweb.int/report/zimbabwe/cyclone-dineo-pounds-zimbabwes-southern-regions

and intensity of localised heavy rainfall occasions, upsurges in the intensity of mid-season dry spells and/or droughts occurring back-to-back in the same season. While the incidence and extent of dry spells during the rainy season have amplified, that of rainy days has declined. In explaining the causes of droughts in Zimbabwe, Mazzeo [55] points towards the effects of the El Nino-Southern Oscillation.

3.1.2 Floods

Flooding in Zimbabwe has been mainly related to the effects of tropical cyclones. Table 1 highlights the various tropical cyclones that affected Zimbabwe post 2000. Disasters such as Tropical Cyclone Idai in 2019 have increased the need to focus on social vulnerability and community resilience especially in rural spaces. There are emerging trends in disaster-induced internal displacement. Invariably, there is growing interest in how inequitable governance interventions in the aftermath of climate-related disasters can "add insult to injury" [51], exacerbating social, cultural, economic, environmental and political dimensions of injustice in a myriad of often under-recognized ways.

3.1.3 Other Ecological Disasters

Other ecological disasters occur at a smaller scale such as earthquakes in Zimbabwe. Clark for instance note that there have been 3400 earthquakes in Zimbabwe before 2000, with the largest occurring beneath Lake Kariba in 1963, measuring 5.8 on the Richter scale. Another ecological challenge facing Zimbabwe is veld fires. The 2016 fire season recorded 1652 incidences resulting in 1,197,335.52 ha of land being destroyed by fire while 1881 incidences and 1,336,746.77 ha were burnt in the 2015 fie season [22]. The Fire Assessment Report by EMA in the same year highlighted that Mashonaland West Province had the largest area burnt with a total 631,622 ha. Mashonaland Central (188,219 ha), Matabeleland North (157,550 ha), and Mashonaland East (112,729 ha). A total of 1,653,822 ha were burnt by veld fires in 2014 compared to 1,179,273 in 201 [23]. EMA [24] observes that farming land destroyed by veld fires in Zimbabwe amounted to 950,905 ha in 2009, 1,152,413 ha in 2010, 713,770 ha in 2011 and 1,320,325 ha in 2012.

3.2 *Resilience to Disasters in the Zimbabwean Context*

Growing evidence shows that adaptation is the vanguard of building resilience to ecological disasters, particularly those induced by climate change disasters. Table 2 summarises some of the adaptation measures in some of the critical livelihood sectors in Zimbabwe.

Adaptation strategies in various local communities have also been documented. For instance in her work in Mutoko [4] alluded to such strategies as conservation agriculture, water conservation, micro-irrigation and re-organisation of farming systems through increasing on-farm crop and livestock diversification. Brazier [11] also noted that communities have to diversify livelihoods into climate resilient-areas, including animal husbandry, horticulture and off-farm activities as new ways of adapting. She also noted that the water sector was high susceptible to climate change hence the need for adaptation in that sector. Besides, the strategies shown in the water in the table above, Ndlovu et al. [66] reported that smallholder farmers in drought-prone regions were using pit planting and mulching in-field rainwater harvesting techniques to reduce crop vulnerability. In Mbire district, smallholder farmers smallholder farmers have adapted to climate change through multiple strategies including planting improved and drought resistant crops, cultivating fewer plots, mixed cropping, keeping more livestock, applying agrochemicals and local ecological knowledge [1]. Communities in Middle- Zambezi river basin have also been adapting to droughts and flooding through vegetable farming and crop production in the floodplain, taking on local jobs that brought in wages, planting late and livestock disposal and out-migration [8]. Small grains production is also one of the key adaptation options that are utilized by smallholder farmers in response to climate change in Tsholotsho [72].

3.3 *Resilience to Disasters: An Intersectionality Theory*

Resilience to disaster is a growing scholarly field and policy arena. Morecroft et al. [57] noted that it has become a common objective of climate change adaptation across the whole range of human activities. The commonly used terminology, such

Table 2 Adaptation measures in Zimbabwe

Water	• Improved monitoring and analysis of available national surface and groundwater reserves • Improved water supply to communities from groundwater sources • Protecting catchments especially wetlands • Preventing of deforestation • Managing grazing areas • Reducing fires • Water harvesting • Conservation agriculture • Agroforestry • • Planting cover crops • Mulching • Planting windbreaks • Drip irrigation • Using underground water stores rather than dams
Land	• Gulley reclamation and slope protection • Reducing burning and land clearance • Improved grazing management • Protection of forests and planting on windbreaks • Conservation agriculture • Crop rotation, inter cropping, compost, and mulch
Vegetation	• Crop diversification • Encouraging sustainable use of forests • Controlling fires • Promoting alternatives to firewood • Promoting agroforestry • Encouraging seed banks • Improving post-harvest storage • Encouraging integrated pest and disease management

Source Brazier [11]

as climate change impacts and vulnerability, is now being increasingly replaced by a preference for language with more positive connotations as represented by resilience and a focus on the 'strengthening' of local communities [56]. Resilience has multiple meanings. For instance, it refers to the amount of disturbance an ecosystem can withstand without changing self-organised processes and structures; and also the return time to a stable state following a perturbation (see Gunderson [41]). Broadly, resilience describes the abilities of systems (such as agroeconomic systems), organisations and people to recover from an adverse impact [57]. Whilst disaster resilience has become popular with policy makers and scientists, the notion of resilience has also been found wanting. Based on the understanding that climate change disaster risk reduction requires new forms of theorising and communicating people's practices, resilience alone is not sufficient. There is need for a more nuanced understanding of resilience, offered by such theories as intersectionality. McEvoy et al. [56] highlighted that the major weakness for the resilience frame is its inability to adequately account for social phenomena. This means the notion fails to bring out the complexities of adaptation because its 'scientific basis' does not capture how different actors behave and actions take place in power structures, interactions and processes. The traditional view of resilience does not question the power relations, the unequal distribution of natural or economic resources, how risks are distributed or the possibilities of participation and political decision-making [91].

Therefore, we adopt a critical stance to the notion of resilience because as currently applied it marginalises multilayered-struggles and fails to account for how and why peoples' experiences have been primarily reduced to gender. In short, the chapter attempts to resist marginalisation, simplification, homogenisation, aggregation, distortion and disempowerment of peoples' experiences. A strong argument is made that intersectionality should be the starting point to understand the varied experiences and struggles of women and other groups emerging under ecological disasters in Zimbabwe. As espoused by Kaijser and Kronsell [48] an intersectional analysis of climate change illuminates how different individuals and groups relate differently to climate change, due to their situatedness in power structures based on context-specific and dynamic social categorisations.

Crenshaw [20] provides a theoretical orientation that explains the interplay of various factors in explaining an individual's position. This theory also analyses how various social and culturally constructed categories interact on multiple levels to manifest themselves social inequalities [5]. Intersectional thinking challenges 'one size-fits-all' approaches—often perceived to be the best way to quickly reach most people in emergencies—to offer a framework for better integrating social heterogeneity, by exposing explicit and implicit assumptions about predefined social categories. It provides an analytical tool for understanding and responding to the ways in which individual factors or identities intersect with others, to enable more nuanced understanding of people's needs, interests, capacities and experiences. Gender inequalities deeply shape people's vulnerability and resilience to risks, but sometimes other social differences, such as ethnicity, access to education and geographic location, are equally or even more important to understand who is at risk and who is able to gain protection [13]. Identifying gender as a single analytical category does not adequately capture the vulnerability and resilience of all women and men [13, 43]. Gender intersects with other identities [42, 47], yet Iniesta-Arandia et al. [44] found that the intersectionality of gender with other identities was nearly absent in analysis of vulnerability, adaptation and resilience, although there are notable exceptions to this—for example work by van Aelst and Holvoet [89] on the intersections of gender and marital status in accessing climate change adaptation and that of Gaillard et al. [37] on the role of gender minorities in disaster risk reduction. What also makes intersectionality ideal in understanding resilience is also that it invokes issues of power, social justice and equity (see Terry [81], Tuana [82]).

4 Discussion of Findings

4.1 Ecological Disasters and Women

There has been a growing recognition of the need to identify gender-differentiated impacts of climate change [83]. The main arguments are not only that climate change will be experienced by men and women differently, but also that women will be more severely hurt by the impacts than men [50]. There has been a growing recognition of the need to identify gender-differentiated impacts of climate change [84]. In Zimbabwe, studies that recognise how men and women perceive climate change [62], how gender roles result in different impacts (see Dube et al. [21], Phiri et al. [73]) and how women are adapting (see Chimanikire [17]) have emerged.

Though historically construed as gender-neutral phenomenon, since the 1990s, research has shown that gender inequalities, combined with other factors, can increase women's vulnerability to natural disasters [46–33]. These contributions highlight situations of high vulnerability of women and their need to be taken into consideration in policies and practices aimed at reducing disaster risk. The studies also show that the different vulnerabilities are socially constructed and linked to the social roles historically assumed by men and women. This further deepens the social inequality they already suffer from. Additionally, women's pre-disaster vulnerabilities like sociocultural political and economic factors also determine the impacts of disasters and vulnerability among them [3]. Women's vulnerability is connected to their generally socio-economic status. Often, they do not hold land or property rights, have fewer educational opportunities, less political voice and mobility due to cultural restrictions. Cultural norms, biological conditions and the sociopolitical environment intersect and add to women's vulnerability by restricting their access to support services for recovery [3]. Thus disasters expose their vulnerability greatly.

Yonder et al. highlights four impacts endured by women in disasters: (1) loss of productive employment outside the home (2) loss of household production and income (3) decrease in homebased work that reduce the family's daily expenditures and also sometimes brings in some income (4) other economic damage due to outstanding debts existing in the pre-disaster period that become a bigger crisis after disaster induced losses. Disasters also lower the life expectancy of women more than that of men. More women than men are shown to die and those that survive may also experience a decline in sexual and reproductive health, increased gender-based violence, disruption to education or are forced into harmful coping mechanisms such as child marriage or transactional sex [90]. On average natural hazards and their associated impacts kill more women than men or kill women at an earlier age than men [67]. Women tend to suffer more injuries and die more frequently in most disasters. Women and children (boys and girls alike) are fourteen times more likely than men to die in disasters [10]. This is attributed to a number of factors such as restrictive clothing that deters them from running fast, the inside the house workspaces cause higher vulnerabilities in earthquakes, maternal and reproductive aspects render them vulnerable to infections. They are often not the first to be attended to when injured, and also tend to take more risks to save children and household belongings during disasters among other factors. We ought to appreciate that women's vulnerability differs according to the nature and intensity of the disaster. Disasters resulting in intense damages cause serious concerns in the post-disaster situation as greater number of resources for rebuilding are required and usually the resultant rehabilitation process is slow. For most middle- or lower-income countries rebuilding after a disaster becomes a challenge. Thus, women continue to face multiple difficulties in their personal, family and community lives. While 'she survives a disaster such as a flood or earthquake, a woman will likely have fewer options to recover' (United Nations Women [86]: 22).

Disaster-induced displacement results in women living in amps and temporary shelters. These are not designed to contain an influx of people for a longer period and end up as unhealthy cramped living conditions wherein women face multiple problems. Women's biological need to maintain privacy during menstruation causes a high level of stress among them. Defined by an absolute loss of personal home space, camps afford them neither privacy nor comfort. Therein women are subjected to higher rates of culturally inappropriate exposure [45]. Their reproductive health issues are intricately linked with psychosocial stress resulting in higher psychosomatic stress levels being reported. In these spaces adolescent youths regardless of gender fall into the trap of high-risk adolescent behavior causing conflict and concern in families and communities. Though important in disaster studies, gender must be considered in its particular context, because the specificity of women's vulnerability and/or capacity/resilience may differ. It is important to pay attention to the fact that causes are not necessarily universal; women's tactical and strategic choices are made in different contexts, times, and spaces. Perhaps through more empirical case studies in the Zimbabwean context in particular, we will be able to find out how and in what contexts women of different ethnic, religious, marital and disability status encounter multi-layered and dynamic ecological hazards. Currently climate change studies not only fail to unpack those aspects but are generally not inadequate.

4.2 Men, Masculinities and Ecological Disasters

A disaster is "a highly gendered social event" [26]. Disasters are a disruption of the taken-for-granted order of everyday life [80]. Thus, they are considered as providing a window of opportunity for change of the established social order [71, 7], affecting the gender order as well [9, 34]. Nonetheless, empirical research suggests a resistance of the social order against disasters. At the same time feminist disaster scholars have also observed a perpetuation, or even exaggeration, of the pre-existing gender patterns as well [9, 2, 69]. Disasters also contain moments of deviation and ruptures of established power relations (and, more recently, as crises of masculinity. "Real" men are often elided with physical strength, competitiveness, assertion, courage, power, dominance, and control, hence, the opposite of feminine traits [93]. These inform, how gendered actors are portrayed, perceived and expected to behave in the disaster setting. "Real" men are assumed to actively engage in the disaster fight and take on the role of a heroic rescuer. As this requires the existence of passive and helpless victims, this is largely assigned to women [28, 29]. However, what is really going on in disasters does not necessarily adhere to the images and narratives of gender [79].

Often, men are found to refuse asking for help, since this portrays them as needy, incapable, and dependent [28], or are more prone to risk-taking behaviour or the refusal of protective actions, such as evacuation [85]. As outlined by Bradshaw [68], such a risk-taking behaviour is far away from a simple expression of machismo, but rather a result of men's socially imposed responsibility to protect their family and livelihoods. Austin [2] observed the emergence of a "disaster masculinity" in the wake of disasters. It depicts the extreme demonstration of masculinity through toughness, courageous behaviour, physical empowerment, or the use of violence. Austin [2] argues that disasters constitute "crises of masculinity" by rupturing institutional arrangements that underpin the masculine domination. Against this background, the exaggeration of "manliness" purposes to restore masculine status. This sheds new light on the often observed increase in gender violence in the aftermath of disasters as well [2, 28, 30, 76].

4.3 Young People in Disaster Contexts

Peek [70] highlights that children and youth are among the most susceptible to the negative impacts of disasters including psychological morbidity, physical injury, sanitation-related illnesses and other health concerns, death, and the adverse impacts of losing child-friendly spaces (e.g., home, schools and playgrounds) including educational decline and school dropout. Despite this recognition of their vulnerabilities in disaster, relatively little research has explored their specific needs or their perspectives on what supports recovery and resilience. However, a growing body of research is beginning to explore youth adaptation to disasters and other traumatic events [52, 88]. Other studies on children and youth identified a range of interacting vulnerabilities (risks, stressors, and exposure), and protective or promotive factors for recovery and resilience at the personal, relational, environmental and cultural levels [74–78].

Progress has been made over the past decade to identify the disaster-related vulnerabilities specific to children and youth in disasters [70] and paediatric-specific needs in emergency response and longer term recovery. Studies have identified a range of post-disaster-related psychological and behavioural issues, including the potential for higher rates of stress and/or traumatic stress [92, 96], increased criminal activity, lower grades and higher rates of high school dropout, increased use of alcohol and other substances, and other behavioural and psychological outcomes [53, 35]. Advances in research include an intersectional analysis of children and youth that explores questions of which *children* and/or *youth* in which *contexts* are most vulnerable and have the highest capacity for coping and recovery [53, 39].

This consist of identifying how and that effects intersect with other ascribed and achieved categories, such as age and stage of development of the child or youth, race and ethnic group membership; household and community level income and wealth; physical and mental ability; and other characteristics, to influence risk and resilience for disaster-affected youth.

However, few studies exist in Zimbabwe that pay attention to the multifaceted risks young people faced based on their sex, class, age, location and disability. It has been noted that young girls in specific cultural contexts maybe vulnerable to early and forced marriages in exchange for food in times of droughts due to increased poverty, and patriarchal practices that devalue women. For instance, in Muzarabani an area of low levels of socio-economic development, girls normally leave school as early as 14 years of age because of cultural expectations such as marriage, and are often given in marriage during disaster situations for family security [61]. Also speaking on early marriages in the same areas others have also noted that during droughts (leading to poor harvests) girls are pulled out of school more promptly than boys, either to provide additional productive work or for subjection to early marriage [98]. Floods and droughts destroy crops resulting in loss of income and food insecurity, forcing some families to withdraw their children from school [54]. A study in Muzarabani, it was observed that about 80% of the children had tried to cross flooded rivers, missed school, and experienced separation from parents during flooding [61]. Other challenges include children being exposed to Malaria and Cholera in flood zones, low retention and performance, social and psychological stress and malnutrition from poor harvests and food scarcity. Similarly, the 2016/17 agricultural season recorded heavy downpours throughout the country, destroying approximately 18% of the country's schools, and affecting the education of approximately 500,000 children [98]. The incidence of school dropouts increases in cases where dwellings are destroyed, children cannot cross bridges to attend school, and people are moved to safer areas, leading to withdrawal of children from schools. Where there is insufficient water and sanitation, young girls miss school during menstruation (ibid.).

4.4 Disaster Resilience: An Intersectional Intervention

Nonetheless, from a feminist standpoint, the language used seems to be disempowering to women. In most instances, women are portrayed as the 'weaker sex', 'more vulnerable' and the group at the worst point and risk of being affected by climate change. Most feminist scholars highlight that climate change affects women as much as it affects men especially in developing nations. In light of this, it can be noted that most articles have focused much on how women have been more prone to poverty a result of climate change [58, 59]. Women's fight and resilience to climate change has been underreported whilst these have done much more and continue to fight against the odds to keep families afloat in times of poverty from the scourge of climate change in Africa. Supporting this fact, Mutami and Chazovachii [64] postulates that focus should be put on the efforts that women are putting in the fight against the ravaging effects of climate change and ways to improve these rather than focusing mostly on their vulnerabilities. In as much as there are contested notions of the feminization of poverty as a result of climate change, the masculinisation of poverty and how this connects to resilience is not given so much attention.

The intersectionality framework also attacks the problem of homogenizing women and their agency [94]. In as much as the discourse has done fairly enough in documenting issues around women's vulnerability to climate change, the extent of attention to agency has been far-fetched in many cases. From an intersectional perspective women's response to disasters are diverse based on their locality, race, ethnicity, class, status and power.

Entrenchment of power dynamics and cultural hegemony over women in issues of climate change has been noted as one of the strangulating impacts of the climate change discourse with regards to women. Patriarchal power dynamics have been entrenched as these are seen through how international conferences such as the IPCC, UNFCCC and also most protocols signed such as the Kyoto Protocol and Paris Climate Change Convention all being male dominated. Besides that, power dynamics and cultural hegemony has been noted in the women's community on their own where adaptation and resilience beneficiation is structured and layered based on gerontocracy amongst the women. In food distributions, farming process and other livelihood strategies being employed by women, the basis of demarcation is the age factor thereby making young women tend to be left out.

Moreover, another critique is on gender adaptation in climate change interventions. Gendered adaptation approaches by various intervention groups may target women mostly leaving men. As highlighted above, the concept of the feminisation of poverty in Zimbabwe is fast losing value yet interventions carried by most NGOs are gendered in nature as these are targeting women. These have proffered feminised systems of adaptation strategies such as basketry, weaving pottery and small livestock. Mubaya and Mafangoya [60] noted that women have even been restricted by culture in more lucrative livelihood options as the livelihood options they choose are tied to the home so as to attend to roles of taking care of the family. These simply resonate with cultural ascribed roles of women, and do not recognises women's capacities and build resilience. Women's agency beyond gender roles is hardly taken into consideration.

The climate change discourse in Africa can been accused of further domesticating women in the name of climate change adaptation and resilience. Scholars such as Oyewumi [68] argue that the goal of feminism is to make sure that women have been removed from the hegemonic dominance of men and shackles of patriarchy that have ostracized them form the public sphere and also the control of their environment. This has been done through the control of the environment of production such as agriculture and also natural resources by men. This has been done through the domestication of women via the adaptation agenda that keeps women in the private sphere in climate change adaptation mechanisms. These mechanisms include growth of small crops, small livestock production, basketry and weaving that are done on the farm with little contact with the public sphere. With this in mind it can be noted that women's agency is heavily subdued in adaptation as most initiatives are done for them by the men who control the narrative and also implementation of what needs to be done.

Additional research is needed to identify the factors that promote resilience and to explore the roles that different groups such as the youth play in identifying risks, needs and vulnerabilities, and their potential contributions to the design and implementation of effective risk mitigation and community resilience interventions [70]. Understanding the social ecology of youth through theirown eyes is critical to enhancing resilience [87]. Involving youth directly in risk and resilience education and activities increasing their sense of agency and self-efficacy when disasters occur, two psychological factors shown to be associated with resilience [49]. Although the role of youth in community resilience is only beginning to be explored [65, 77], there is a growing recognition that the youth have the capacity to contribute not only to their own, their families and communities' wellbeing in disasters [35].

5 Conclusion

Our chapter argues that an intersectional approach to building resilience is urgently required if policies and programmes are to fully respond to the wide array of gendered needs in ecological disasters. Resilience is gendered in multifaceted ways that require a nuanced approach that places emphasises social positionality to understand how women and men cope with and respond to disasters. Resilience programming tends to view people within a binary (male or female) of homogenous groupings. The chapter has argued that such thinking that promotes a 'one-size-fits-all' approach cannot provide a holistic response to the unique needs of people. It argues for an intersectional approach that provides an analytical tool for understanding and responding to the ways in which individual factors or identities intersect with others, to enable more nuanced understanding of people's needs, interests, capacities and experiences. Intersectional theory contends that social groups are neither homogenous nor static, and recognise this complexity by taking historical, social, cultural and political contexts into account. Intersectional approaches help therefore to understand the differentiated nature of vulnerability and resilience.

References

1. Asare-Nuamah P, Mandaza MS, Amungwa AF (2021) Adaptation strategies and farmer-led agricultural innovations to climate change in mbire districtof Zimbabwe. Int J Rural Manage. 0973005221999913
2. Austin DW (2015) Hyper-Masculinity and disaster. The reconstruction of hegemonic masculinity in the wake of calamity. In: Enarson E, Pease B (eds) Men, masculinities and disaster. Routledge, New York, pp 45–55
3. Bhadra S (2017) Women in disasters and conflicts in india: interventions in view of the millennium development goals. Int J Disaster Risk Sci 8:196–207

4. Bhatasara S (2018b) Understanding adaptation to climate variability in smallholder farming systems in Eastern Zimbabwe: a sociological perspective. Rev Agric Food Environ Stud 99(2):149–166
5. Bhatasara S, Chiweshe MK (2017) Beyond Gender: interrogating women's experiences in FTLRP in Zimbabwe. Africa Rev 9(2):154–172
6. Bhatasara S (2018a) Climate variability in local scales: narratives and ambivalences from Mutoko District. In: Helliker K, Chiweshe MK, Bhatasara S (eds) The political economy of livelihoods in contemporary Zimbabwe. Routledge, London and New York, pp 139–153
7. Boin A, Hart P, McConnell A (2009) Crisis exploitation. Political and policy impacts of framing contests. J Eur Public Policy 16(1):81–106
8. Bola G, Mabiza C, Goldin J, Kujinga K, Nhapi I, Makurira H, Mashauri D (2014) Coping with droughts and floods: a case study of Kanyemba, Mbire District, Zimbabwe. Phys Chem Earth 67–69:180–186. https://doi.org/10.1016/j.pce.2013.09.019
9. Bradshaw S (2015) Engendering development and disasters. Disasters 39:s54–s75
10. Bradshaw S, Fordham M (2013) Women, girls and disasters—a review for DFID. GSDRC Applied Knowledge Services. DFID, London
11. Brazier A (2015) Climate change in Zimbabwe facts for planners and decision makers. Konrad-Adenauer-Stiftung, Harare
12. Brown D, Rance Chanakira R, Chatiza K, Dhliwayo M, Dodman D, Masiiwa M, Muchadenyika D, Prisca Mugabe P, Zvigadza S (2012) Climate change impacts, vulnerability, and adaptation in Zimbabwe. IIED Climate Change Working Paper No. 3, October 2012
13. Carson M, Remling E, Johanssen A (2013) Institutionalizing gender equality in disaster risk reduction. MSB and SEI, Stockholm
14. Chatiza K (2019) Cyclone Idai in Zimbabwe: an analysis of policy implications for post-disaster institutional development to strengthen disaster risk management. Oxfam. https://doi.org/10.21201/2019.5273
15. Chigodora J (1997) Famine and drought: the question of food security in Zimbabwe. Drought Network News 1994–2001
16. Chikoto GL, Sadiq A (2012) Zimbabwe's emergency management system: a promising development. In: McEntire DA (ed) Comparative emergency management: understanding disaster policies, organizations, and initiatives from around the world (Chapter 20). FEMA, U.S. Department of Homeland Security
17. Chimanikire DP (2013) Climate change and gender in Zimbabwe. World Social Science Report. UNESCO, Paris
18. Clausen L (1992) Social differentiation and the long-term origin of disasters. Nat Hazards 6:181–190. https://doi.org/10.1007/BF00124622
19. Clausen L (1994) Krasser sozialer Wandel. Leske und Budrich, Opladen
20. Crenshaw K (1991) Mapping the margins: intersectionality, identity politics, and violence against women of color. Stanford Law Review 43:1241–1299
21. Dube T, Itauno S, Moyo P, Phiri K (2019) The gender-differentiated impacts of climate change on rural livelihoods labour requirements in Southern Zimbabwe. J Hum Ecol. https://doi.org/10.1080/09709274.2017.1316958
22. EMA (2016) Fire assessment report. EMA, Harare
23. EMA (2014) Assessment of the state of the environment report. EMA, Harare
24. EMA (2011) Are veld fires a threat to the environment, viewed 06 June 2012, from http://www.ema.co.zw/index.php/ema-newspaper-articles/116-veld-fires
25. Enarson E (1998) Through women's eyes: a gendered research agenda for disaster social science. Disasters 22(2):157–173
26. Enarson E (2006) Women and Girls Last? Averting the Second Post-Katrina Disaster in SSCR. https://items.ssrc.org/understanding-katrina/women-and-girls-last-averting-the-second post-katrina-disaster. 29 May 2020
27. Enarson E, Scanlon J (1999) Gender patterns in a flood evacuation: a case study of couples in Canada's Red River Valley. Appl. Behav. Sci Rev 7/2

28. Enarson E (2001) Promoting social justice in disaster reconstruction: guidelines for gender-sensitive and community-based planning. Report from Gujarat, India. Disaster Mitigation Institute, Ahmedabad
29. Enarson E, Meyreles L (2004) International perspectives on gender and disaster: differences and possibilities. Int J Sociol Soc Policy 24:49–93
30. Enarson E, Fothergill A, Peek L (2017) Gender and disaster: foundations and new directions for research and practice. https://doi.org/10.1007/978-3-319-63254-4_11
31. Enarson E, Fothergill A, Peek L (2007) Gender and disaster: foundations and directions. In: Rodríguez H, Donner W, Trainor J (eds) Handbook of disaster research. Springer, New York, NY, pp 130–146
32. Fordham M (1998) Making women visible in disasters: problematizing the private domain. Disasters 22(2):126–143
33. Fothergill A (1996) Gender, risk, and disaster. Int J Mass Emerg Disasters 14(1):33–56
34. Fothergill A (1996) The neglect of gender in disaster work: an overview of the literature. Int J Mass Emerg Disasters 14
35. Fothergill A, Peek L (2006) Surviving catastrophe: a study of children in Hurricane Katrina. In: Learning from catastrophe: quick response research in the wake of Hurricane Katrina. Natural Hazards Research and Applications Information Center, University of Colorado Boulder, CO, pp 97–129
36. Furr JM, Comer JS, Edmunds JM, Kendall PC (2010) Disasters and youth: a metanalytic examination of posttraumatic stress. J Consult Clin Psychol 78:765–780
37. Gaillard JC, Sanz K, Balgos BC, Dalisay SNM, Gorman-Murray A, Smith F, Toelupe VA (2017) Beyond men and women: a critical perspective on gender and disaster. Disasters 41(3):429–447
38. Gibbs L, Block K, Harms L, MacDougall C, Baker E, Ireton G, Forbes D, Richardson J, Waters E (2015) Children and young people's wellbeing post-disaster: safety and stability are critical. Int J Dis Risk Red 14:195–201
39. Gibbs L, Mutch C, O'Connor P, MacDougall C (2013) Research with, by, for and about children. Lessons from disaster contexts. Glob Stud Childhood 3(2):129–141
40. Goundar S (2012) Research methodology and research method. Victoria University of Wellington. https://prallagon.com/wp-content/uploads/2021/05/Research-Methodology-A2.pdf
41. Gunderson LH (2000) Ecological resilience—in theory and application. Annu Rev Ecol Syst 31:425–439
42. Hackfort S, Burchardt HJ (2018) Analysing socio-ecological transformations—a relational approach to gender and climate adaptation. Crit Policy Stud 12(2):169–186
43. Huynh PTA, Resurrección BP (2014) Women's differentiated vulnerability and adaptations to climate-related agricultural water scarcity in rural central Vietnam'. Clim Dev 6:226–237
44. Iniesta-Arandia I, Ravera F, Buechler S et al (2016) A synthesis of convergent reflections, tensions and silences in linking gender and global environmental change research. Ambio 45:383–393
45. International Federation of Red Cross and Red Crescent Societies (2015) Unseen, unheard: gender-based violence in disasters. IFRC and RCS, Geneva. Filho WL, Luetz J, Ayal D (eds) Handbook of climate change Mmanagement. Springer, Cham
46. Jerneck A (2018) What about gender in climate change? Twelve feminist lessons from development. Sustainability 10(3):627. https://doi.org/10.3390/su10030627
47. Jordan JC (2018) Deconstructing resilience: why gender and power matter in responding to climate stress in Bangladesh. Clim Dev 1–13
48. Kaijser A, Kronsell A (2013) Climate change through the lens of intersectionality. Environmental Politics 23(3):417–433. https://doi.org/10.1080/09644016.2013.835203
49. Luthar SS (2006) Resilience in development: a synthesis of research across five decades. In: Cicchetti D, Cohen DJ (eds) Developmental psychopathology: risk, disorder, and adaptation. Wiley, pp 739–795
50. MacGregor S (2010) Gender and climate change: from impacts to discourses. J Indian Ocean Region 6(2):223–238

51. Marino E, Ribot J (2012) Special issue introduction: adding insult to injury: climate change and the inequities of climate intervention, vol 22. Elsevier, pp 323–328
52. Masten AS (2014) Global perspectives on resilience in children and youth. Child Dev 8:6–20
53. Masten A, Narayan A (2012) Child development in the context of disaster, war, and terrorism: pathways of risk and resilience. Ann Rev Psychol 63:227–257
54. Mavhura E (2017) Building resilience to food insecurity in rural communities: evidence from traditional institutions in Zimbabwe. Jamba: Disaster Risk Stud 9(1):1–9
55. Mazzeo J (2011) Cattle, livelihoods, and coping with food insecurity in the context of drought and HIV/AIDS in rural Zimbabwe. Hum Organ 70(4)
56. McEvoy D, Fünfgeld H, Bosomworth K (2013) Resilience and climate change adaptation: the importance of framing. Plann Pract Res 28(3):280–282
57. Morecroft D, Crick HQP, Duffield SJ, Macgregor NA (2012) Resilience to climate change: translating principles into practice. J Appl Ecol 49:547–551
58. Mtekwa V (2009) Climate change impacts and adaptation in the agricultural sector: the case of smallholder farmers in Zimbabwe. J Sustain Dev Africa 11:237–256
59. Mubaya CP, Mafongoya P (2017) Local-level climate change adaptation decision-making and livelihoods in semi-arid areas in Zimbabwe. Environ Dev Sustain 19:2377–2403. https://doi.org/10.1007/s10668-016-9861-0
60. Mubaya CP, Mafongoya PL (2017) Local-level climate change adaptation decision-making and livelihoods in semi-arid areas in Zimbabwe. Environ Dev Sustain 19:2377–2403
61. Mudavanhu C et al (2015) Taking children's voices in disaster risk reduction a step forward. Int J Disaster Risk Sci. 6:267–281
62. Mudzengi KB, Simba FM, Murwendo T, Mdlongwa T (2013) Perspectives in climate change and gender issues: a case study of Masvingo Province in Zimbabwe. Russ J Agric Soc-Econ Sci 3(15):1–10
63. Mulrow CD (1994) Rationale for systematic reviews. Br Med J 309(6954):597–599
64. Mutami C, Chazovachii B (2012) Effective livelihood strategies in distressed environments: the case of Mudzi District of Zimbabwe. Curr Res J Social Sci 4(5):362–371
65. Mutch C, Garwith L (2014) The role of schools in engaging children in emotional processing of disaster experience. Pastoral Care Educ 32(1):54–67
66. Ndlovu S, Mathe B, Phiri K, Nyathi D (2020) Factoring water harvesting into climate change adaptation: endogenous responses by smallholder farmers in Gwanda district. Zimbabwe. https://doi.org/10.1080/23311886.2020.1784652
90 Neumayer E, Plümpe T (2007) The gendered nature of natural disasters: the impact of catastrophic events on the gender gap in life expectancy, 1981–2002. Ann Assoc Am Geogr 97(3):551–566
68. Oyewumi O (2004) Conceptualizing gender: eurocentric foundations of feminist concepts and challenge of African epistemologies. CODESRIA, Dakar, pp 1–8
69. Pacholok S (2013) Into the fire. disaster and the remaking of gender. University of Toronto Press, Toronto
70. Peek L (2008) Children and disasters: understanding vulnerability, developing capacities, and promoting resilience-an introduction. Child Youth Environ 18:1–29
71. Pelling M, Dill K (2010) Disaster politics: tipping points for change in the adaptation of sociopolitical regimes. Prog Hum Geogr 34:21–37
72. Phiri K, Ndlovu S, Mpofu M, Moyo P (2020) Climate change adaptation and resilience building through small grains production in Tsholotsho, Zimbabwe. In: Leal Filho W, Luetz
73. Phiri K, Ndlovu S, Chiname TB (2014) Climate change impacts on rural based women: emerging evidence on coping and adaptation strategies in Tsholotsho, Zimbabwe. Mediterr J Soc Sci 5(23):2545–2552
74. Pickard AJ (2007) Research methods in information. Facet publishing, London
75. PreventionWeb (2012) Zimbabwe—disaster statistics. Accessed 9 March 2012, <http://www.preventionweb.net/english/countries/statistics/?cid=193>
76. Rezwana N, Pain R (2021) Gender-based violence before, during, and after cyclones: slow violence and layered disasters. Disasters 45(4):741–761

77. Ronan K, Johnston D (2005) Promoting community resilience in disasters: the role for schools, youth, and families. Springer Science & Business Media
78. Sapienza K, Masten AS (2011) Understanding and promoting resilience in children and youth. Curr Opin Psychiatry 24:267–273
79. Scanlon J (1998) The perspective of gender: a missing element in disaster response. In: Enarson E, Morrow BH (eds) The gendered terrain of disaster, pp 45–52
80. Stallings RA (1998) Disaster and the theory of social order. In: Quarantelli EL (ed) What is a disaster? perspectives on the question. Routledge London, pp 127–145
81. Terry G (ed) (2009) Climate change and gender justice. Practical Action Publishing in Association with Oxfam GB, Rugby, UK (Toronto Public Health (2009). Climate Change)
82. Tuana N (2019) Climate apartheid: the forgetting of race in the anthropocene. Crit Philos Race (Special issue: Race and the Anthropocene and race, immigration, and refugees) 7(1):1–31
83. Tuana N (2014) Gender and climate change. Hypatia: J Feminist Philos Rec 29(3):533–540
84. Tuana N (2013) Gendering climate knowledge for justice: catalyzing a new research agenda. In: Margaret A, Whittenbury K (eds) Research, action and policy: addressing the gendered impacts of climate change. Springer, Dordrecht, pp 3–16
85. Tyler M, Fairbrother P (2013) Bushfires are "men's business": the importance of gender and rural hegemonic masculinity. J Rural Stud 30:110–119
86. UN Women (2016) Women and sustainable development goals. UN Women Eastern and Southern Africa Regional Office, Nairobi
87. Ungar M (2011) Community resilience for youth and families: facilitative physical and social capital in contexts of adversity. Child Youth Serv Rev 33(9):1742–1748
88. Ungar M (2011) The social ecology of resilience: addressing contextual and cultural ambiguity of a nascent construct. Am J Orthopsychiatry 81:1–17
89. Van Aelst K, Holvoet N (2016) Intersections of gender and marital status in accessing climate change adaptation: evidence from rural Tanzania. World Dev 79:40–50
90. Van der Gaag N (2013) Because I am a girl: the state of the world's girls 2013' report. In: Double Jeopardy: adolescent girls and disasters. Plan on 11 October 2013. Plan International
91. Vargas CA, Romero SH, Sicard TL (2020) Key points of resilience to climate change: a necessary debate from agroecological systems. Climate Dev 12(6):564–574
92. Weems CF, Watts SE, Marsee MA, Taylor LK, Costa NM, Cannon MF, Carrion VG, Pina AA (2007) The psychosocial impact of Hurricane Katrina: contextual differences in psychological symptoms, social support, and discrimination. Behav Res Ther 45(10):2295–2306
93. Willer R, Rogalin CL, Conlon B, Wojnowicz MT (2013) Overdoing gender: a test of the masculine overcompensation thesis. Am J Sociol 118(4):980–1022
94. Winker G, Degele N (2011) Intersectionality as multi-level analysis: dealing with social inequality. Eur J Womens Stud 18:51–66. https://doi.org/10.1177/1350506810386084
95. Wolmer L, Hamiel D, Laor N (2011) Preventing children's posttraumatic stress after disaster with teacher-based intervention: a controlled study. J Am Acad Child Adolesc Psychiatry 50(4):340–348, 348.e1-2
96. Yelland C, Robinson P, Lock C, La Greca A, Kokegei B, Ridgway V, Lai B (2010) Bushfire impact on youth. J Trauma Stress 23:274–277. https://doi.org/10.1002/jts.20521
97. Yoder JD, Kahn AS (2003) Making gender comparisons more meaningful: a call for more attention to social context. Psychol Women Q 27(4):281–290 (Original work published 2003)
98. Zimbabwe Human Development Report (2017) Climate change and human development: towards building a climate resilient nation. UNDP, Harare
99. Zimbabwe Vulnerability Assessment Committee (2010) Rural livelihoods assessment. FNC, Harare

Sandra Bhatasara has a background in sociology and social anthropology. Currently she is a Senior Research Fellow at the Environment, Climate and Sustainable Development Institute at the University of Zimbabwe. Her research utilizes social theory to understand gender, agrarian and livelihoods issues, climate change adaptation and resilience in African spaces. She has edited two books, co-authored a book on land and published more than 25 articles in refereed journals. Her latest publication is on climate justice in Africa published by Brill.

Manase Kudzai Chiweshe is a Senior Lecturer in the Department of Social and Community Development, University of Zimbabwe. He is the winner of the 2015 Gerti Hesseling Prize for Best Paper Published in African Studies. His work revolves around the sociology of everyday life in African spaces with specific interest in gender, livelihoods, football and sports studies.

Cyclone Damage Assessment Using High Resolution Satellite Imagery, Drone Technology and GIS: Lessons Learnt

Samuel Kusangaya, Mhosisi Masocha, Isaiah Gwitira, Pianos Gweme, Victor Mukungunurwa, Pathias P. Bongo, Albert Maipisi, and Dominic Mazvimavi

Abstract Observed increases in the frequency of extreme climatic events such as tropical storms amplify the vulnerability of communities to disasters in developing countries. For example, in developing Zimbabwe, the low adaptive capacity of the population associated with recurrent economic stressors compromises the ability of communities to cope with natural disasters. As such rapid assessment of the disruption of livelihood systems as well as damage to structures caused by disasters is critical for implementing measures to restore affected communities to the pre-disaster state. Considering that all disasters whether they are natural or anthropogenic in origin, have the following key characteristics: location, magnitude, intensity, onset, duration and frequency, the increasing availability and accessibility of earth observation data coupled with advances in geographic information systems (GIS) provide vast opportunities for developing rapid assessment of damage. Specifically, the availability of cloud computing applications and the access to high resolution satellite imagery with increased revisit time have boosted the robustness of earth observation in monitoring the intensity, frequency, duration and impact of extreme weather events which result in disasters such as cyclonic winds and torrential storms that trigger flooding. In recent years, unmanned aerial vehicles (UAVs) or drones have provided an additional platform for collecting near real time data for post-disaster damage assessment. The main advantages of using UAVs are that the critical datasets for post-disaster damage assessment can be acquired rapidly, at relatively low cost at the desired frequency. Using Cyclone Idai which hit the eastern parts of Zimbabwe

S. Kusangaya (✉) · M. Masocha · I. Gwitira · P. Gweme · P. P. Bongo
Department of Geography, Geospatial Sciences and Earth Observation, University of Zimbabwe, Harare, Zimbabwe
e-mail: kusangayas@yahoo.com

V. Mukungunurwa
Zimbabwe National Geospatial and Space Agency, Harare, Zimbabwe

A. Maipisi
Women's University in Africa, Harare, Zimbabwe

D. Mazvimavi
University of the Western Cape, Cape Town, South Africa

R. Kasimba and I. Chirisa (eds.), *Disaster Risks and Hazards*, Advances in 21st Century Human Settlements, https://doi.org/10.1007/978-981-96-4577-0_8

particularly Chimanimani district in March of 2019 and was declared a national disaster, as a case study. This chapter describes a simple and repeatable change detection approach for integrating satellite and drone imagery in a GIS framework to undertake comprehensive cyclone damage assessment. The methodology described in this chapter is ideal for giving a quantitative measure of the type and extent of physical damage for disaster management practitioners and emergency rescue personnel operating in resource-constrained environmental settings.

Keywords Cyclone · Idai · Post-disaster · Unmanned aerial vehicles

1 Introduction and Context

All places on earth are prone to natural disasters such as earthquakes, cyclones and droughts. Therefore, post-damage assessment is an integral part of disaster relief and recovery. In fact, damage assessment is one of the most powerful tools in the disaster response tool kit, since no significant or systematic response can be mounted until the impact of a disaster is properly assessed [26]. Formally, damage assessment entails the systematic collection and analysis of data concerning the disaster's impact on target elements at risk such as buildings, bridges and crops [48]. To date, a suite of methods ranging from expert-based systems, statistical analysis, mathematical modelling, and remote sensing have been developed for post-disaster damage assessment [23, 41, 42, 62]. For example, Murlidharan et al. [45] presented a knowledge-based system for damage assessment and vulnerability analysis of structures subjected to cyclones targeting the coastal areas of India. The knowledge base comprised of damage survey reports and information acquired through feedback to questionnaires administered to experts including engineers. Fuzzy logic was then implemented to handle uncertainties inherent in the knowledge base. The system they developed based on probabilistic rules was validated using expert opinions and aerial photographs of damaged structures. The assessment generated by applying the knowledge-based expert system was cross-checked with assessments made by field experts and found to be useful for facilitating decisions essential for resource allocation during repair and reconstruction work.

Goyal [23] also used a statistical approach for assessing damage to semi-engineered rural houses in India due to cyclonic wind. The procedure outlined involved calculation of conditional probability of occurrence and integrating this with the conditional probability of failure of rural houses to obtain the overall damage estimate for a specified region. Five damage states comprising openings failure, roof cover failure, secondary roof member failure, roof member to wall connection failure and mud wall failure were considered. The procedure yielded damage estimates that were deemed useful for decision making. In a related work done in Taiwan, Huang and Wang [28] built damage assessment models based on used simple mathematical functions. For example, the model for assessing crop losses for Type A typhoons was:

$$\text{Type A} = \text{EXP}(17.16572 + 0.0046109X_1)$$

where X_1 maximum daily rainfall at each weather station during the typhoon-affected period and EXP is the exponent. Similar damage assessment models were developed for deaths or missing persons, and totally destroyed houses based on other meteorological variables such as maximum central wind speeds and the radius of class seven winds. It has to be noted the models were developed for different categories of typhoons (Type A to Type E).

In recent years, remote sensing has proved a cost-effective technology for post-damage assessment. The physical basis of using remote sensing is that different materials on the earth's surface interact with electromagnetic radiation in a unique way. From this one can deduce that if a structure or an object suffers damage due to a disaster, its spectral properties change. As such change detection which is based on comparing how the same object interacted with electromagnetic radiation in the pre- and post-disaster period. Whether the source of energy for remote assessment is natural as in passive remote sensing or active as is the case with radar, change detection methods are widely used for damage assessment [49, 61]. In the context of synthetic aperture radar (SAR), change detection involves identifying differences in the state of features on the ground by observing them at different times [3]. The same approach based on discriminating areas of change on imagery holds when using remote sensed optical data collected in the visible bands of light. Later on, change detection based on integrating satellite imagery and drone imagery will be described for Cyclone Idai which made landfall on the 15th of March 2019 in Eastern Zimbabwe. In addition to change detection, polarimetric decomposition and texture-based methods are also gaining popularity for building damage assessment using SAR imagery. More details about these relatively new methods as well as an evaluation of their advantages and disadvantages can be found in Yamazaki [62],Yamazaki and Matsuoka [61] as well as Akhmadiya and Nabiyev [3]. In regard to passive remote sensing, sensors mounted on air borne or space borne platforms are also used for damage estimation. The advantage of space-borne sensors is that they allow monitoring of damage over a wide field of view spanning tens of kilometer to more than 100 kilomteres [49].

To demonstrate the applicability and value of integrating satellite data and drone imagery the chapter now focuses on a tropical cyclone Idai that made landfall in Zimbabwe on the 15th of March 2019. The damage to infrastructure such as bridges and houses was so extensive that the event was declared a national disaster. At the time of its onset, the country did not have a systematic procedure for damage assessment yet that information was required to buttress emergency rescue operations as well as recovery efforts. The starting point for the work was the existing satellite imagery taken at the date as close to the time the cyclone made landfill as possible. Multi-spectral cameras mounted on drones piloted by first responders were then used to take imagery of the state of the affected areas three days after torrential rains subsided.

2 Frequency of Cyclonic Activity in Southern Africa

Tropical cyclones are one considered some of most destructive natural disasters in the coastal and near coastal regions of Southern Africa. The first Category 5[1] tropical cyclone to form over the South-West Indian Ocean (SWIO) was observed in 1994. Since then, there have been 12 category 5 cyclones [18]. As such previous studies have indicated that extreme hydrological events are increasing in frequency and magnitude in most parts of southern Africa due to cyclonic activity, (e.g., Kabat et al. [31], Reason and Keibel [52], Kadomura [32], Patt and Schröter [46], Yanda [63], Goodess [22]). Available literature indicates that southern Africa is vulnerable to tropical storms that form in the between latitudes 5° and 30° S. Beyond the 30° S latitude, the ocean tends to be colder than 26°C and this inhibits cyclogenesis [38]. Every year about 9–10 tropical storms and cyclones form over the SWIO. Of these 4–5 develop into tropical cyclones. The cyclone season in the SWIO is from November to April, and 74% of the tropical systems occur during the December–March period. In the case of Zimbabwe, most of the cyclones that form in the Indian Ocean reach country via the Mozambican channel. The country has recorded cyclones from as far back as the early 1900s. Tropical cyclones that had an impact on southern Africa occurred in 1904, 1934, 1946, 1948, 1957–58 (Cyclone Astrid), 1959 (Cyclone Collene), 1967 (Cyclone Daphne), 1969 (Cyclone Berthe), 1970 (Cyclone Emily), 1996 (Cyclone Bonita), 2000 (Cyclone Eline), 2014 (Cyclone Hellen) and 2019 (Cyclone Idai). The advancement in satellite technology has made it possible to track and predict cyclone paths making it possible to alert communities in the path of impending danger.

Tropical Cyclone Idai originated from a tropical depression that formed over the Mozambican Channel on 4 March 2019.[2] This storm moved initially over land in Mozambique and Malawi during the 4–9 March 2019 period. It then returned to the Mozambican Channel where it made a rapid intensification to become a tropical cyclone. On the 11th March 2019 at 06.00 UTC the system had become an intense tropical cyclone with wind speed up to 165 km/hr and heading towards mainland Mozambique. The cyclone which was now a category 4 storm made landfall north of Beira during the early hours of 15 March 2019. This system weakened on the same day into at tropical depression as it moved overland. By the evening of the same day the tropical depression had moved some 230 km from the coast into Zimbabwe affecting the administrative districts of Chimanimani and Chipinge. This system moved as far as the central part of Zimbabwe and started dissipating on 16 March 2019 after moving overland for 470 km from Beira, Mozambique. Cyclone Idai lasted 12 days (4–16 March 2019) from being a tropical depression to dissipation.

---

[1] Tropical cyclones around the world are measured by the Saffir-Simpson Hurricane Wind Scale which originated from 1971 with Herbert Saffir, a civil engineer and Bob Simpson of the US National Hurricane Center. The severity of a tropical cyclone is described in terms of categories ranging from 1 (weakest) to 5 (strongest) related to the maximum mean wind speed. http://www.bom.gov.au/cyclone/tropical-cyclone-knowledge-centre/understanding/tc-info/.

[2] http://www.meteofrance.re/cyclone/trajectoire.

Table 1 Effects of Cyclone Idai[3]

Effects	Malawi	Mozambique	Zimbabwe
Number of human deaths	59	603	344
Number of persons displaced	60,880	400,000	4500
People in need	868,900	1.2 M	279,000
Cultivated areas damaged	Not available	715,378 hectares	Not available
Schools damaged	104	540	139

Tropical cyclone Idai was the third worst event in history and changed the lives and livelihoods of an estimated 270,000 people in Chimanimani District. The official deaths were between 299 and 344 while more than 300 people were reported missing. The health delivery systems and educational institutions; fresh water sources, transport infrastructure and shops; cultural and religious heritage sites, artefacts were destroyed, displaced or disrupted. The cyclone also impacted on the intangible heritage of the people. In the aftermath of Cyclone Idai, search and rescue operations were carried out, and remote sensing (satellite and drone) were used for damage assessment. This chapter therefore analyses the use of GIS, satellite and drone technology in disaster damage assessment.

3 Analysis of Impacts of Cyclone Idai

3.1 Damage Assessment Using Satellite Imagery

Cyclone Idai caused a disaster with the highest fatalities in comparison to previous cyclones. The life of over 1000 people in Malawi, Mozambique and Zimbabwe was lost including damage to houses, roads, bridges and washing of crop land (Table 1).

In some parts of Beira in Mozambique, the high wind speeds destroyed residential areas and slums. Cyclone Idai induced storm surges reached heights of up to 4.5 m along the Mozambican coastline. The rainfall associated with the cyclone was in the region of 500–1000 mm which contributed to flooding. High resolution satellite image analysis by the European Space Agency and NASA showed that most of the low-lying areas with an elevation of less than 20 m stretching from the coast to about 100 km along both the Pungwe River and the Buzi River, were inundated creating a lake with water depths of up to 6 m in some parts (Fig. 1). Flooding occurred on about 3000 sq km in the Manica and Sofala Provinces of Mozambique.[4] An

[3] https://reliefweb.int/sites/reliefweb.int/files/resources/joint_national_sitrep_1_mozambique_10_may_2019_final_whoinsdps_for_trans.pdf

https://reliefweb.int/report/mozambique/unicef-cyclone-idai-and-kenneth-post-impact-situation-may-2019.

[4] https://reliefweb.int/sites/reliefweb.int/files/resources/joint_research_centre_analysis_of_wind_rainfall_and_storm_surge_impact_09_april_2019.pdf.

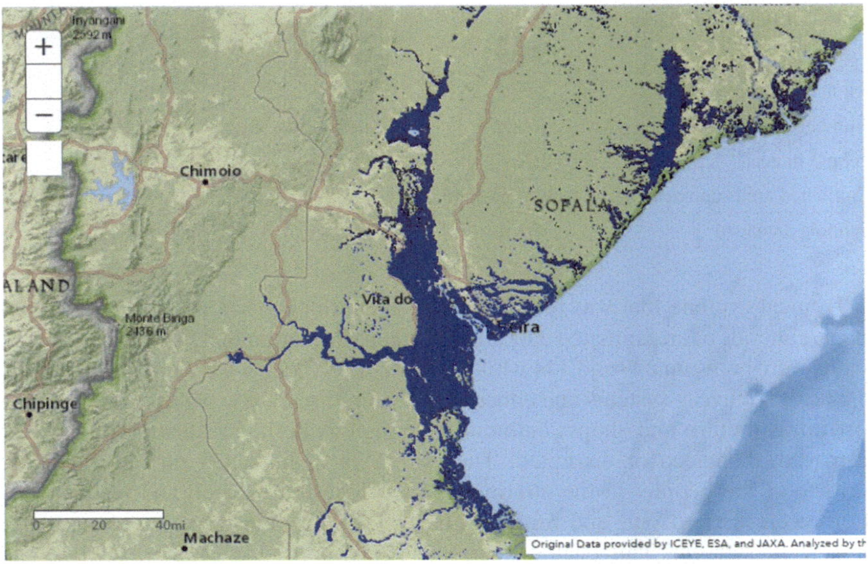

Fig. 1 Flood extent in Mozambique mapped using from 18–23 March 2019 images by the Advanced Rapid Imaging and Analysis (ARIA) team at NASA's Jet Propulsion Laboratory using modified JAXA ALOS-2 data (2019), Copernicus Sentinel data (2019), ICEYE data (2019) (https://maps.disasters.nasa.gov/)

estimated 715,378 hectares of cultivated lands with crops ready for harvesting were inundated in these two provinces. In addition, fisheries infrastructure, which was a main source of livelihood along the coast and Pungwe River, were damaged. The damage to agricultural lands just before the harvesting period caused severe food shortages. Estimates by humanitarian organizations are that about 1.1 million people were made food insecure.

The channel followed by Cyclone Idai from Mozambique is characterised by mountain ranges aligned in an east–west direction with an altitude of 1000–2000 m above sea level. This resulted in the cyclone being forced to ascent these mountains which enhanced rainfall formation. This region has very steep slopes and high rainfall in a short period caused flash floods, landslides and rockfalls (Figs. 2 and 3). The high rainfall induced landslides and mudslides which resulted in massive destruction of infrastructure. At the Dzingire Centre 26 km south of Chimanimani in Zimbabwe, over 80 houses located within the then floodplain of the Rusitu River, were washed away (Figs. 4, 5 and 6). The remote sensing image data for this study were taken from Sentinel-2A. The satellite images in this study were obtained from

---

http://unosat-maps.web.cern.ch/unosat-maps/MZ/TC20190312MOZ/UNOSAT_A3_TC20190312MOZ_FloodExtent_Dombe.pdf.

http://unosat-maps.web.cern.ch/unosat-maps/MZ/TC20190312MOZ/UNOSAT_A3_Natural_Portrait_TC20190312MOZ_cumulative.pdf.

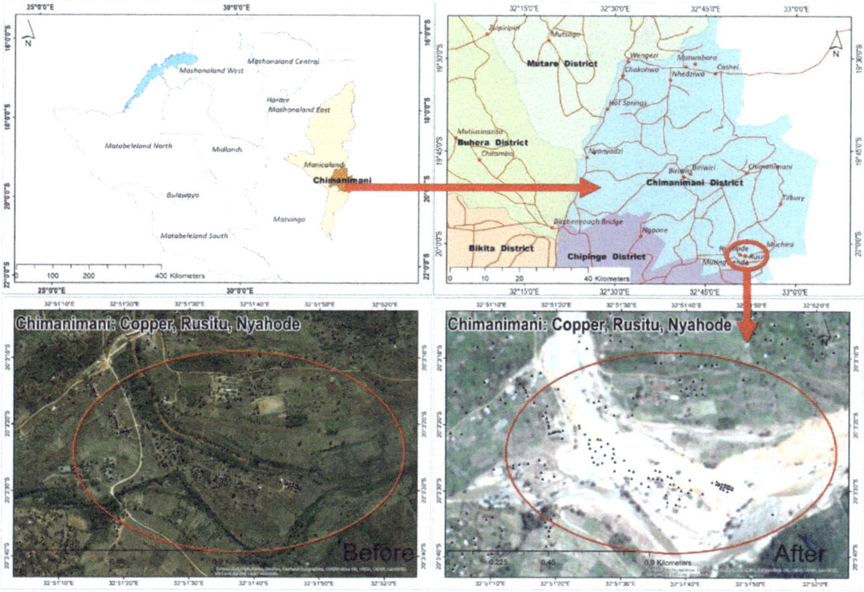

Fig. 2 Satellite imagery showing cyclone damage at the confluence of Rusitu and Nyahode rivers in Chimanimani. *Source* Sentinel 2 images from the European Space Agency

the website of the European Space Agency (ESA) (https://scihub.copernicus.eu/). The Sentinel-2 data used in this study were Level-2A products and all the bands were resampled to 10 m via the S2 Resampling Processor in the software SNAP (the bicubic method was used for resampling). Sentinel-2 images were atmospherically corrected using the Sen2Cor algorithm (http://step.esa.int/main/third-party-plugins-2/sen2cor), which processes ESA's Level-1C top-of-atmosphere reflectance to atmospherically-corrected bottom-of-atmosphere (BoA) reflectance (Level-2A).

Most of the infrastructure such as roads and bridges were damaged by uprooted trees and landslides which made most parts inaccessible. Power lines and mobile phone base stations were also destroyed contributing to the inaccessibility of affected areas.

3.2 Damage Assessment Using Drone Imagery

A few days after Cyclone Idai hit Chimanimani District and specifically on the 22nd and 25th of March, three small drones namely the Easy UAV and DJI Mavick Pro and DJI Phantom 4 were flown over five selected sites to acquire images of the damage from the cyclone. The drones were calibrated to take geo-tagged images with a spatial resolution of 4 cm. The images were processed separately for each site

Fig. 3 A comparison of satellite images before (24/6/2018) and after (20/3/2019) Cyclone Idai at Dzingire Centre in Zimbabwe located at the confluence of three rivers, Nyahode River from the north joining Rusitu River which flows from west to east, which is joined by the Chipita River from the south. About 65 buildings disappeared while 39 were damaged. *Source* Images processed and obtained from SERTIT, http://sertit.u-strasbg.fr/index_en.htm

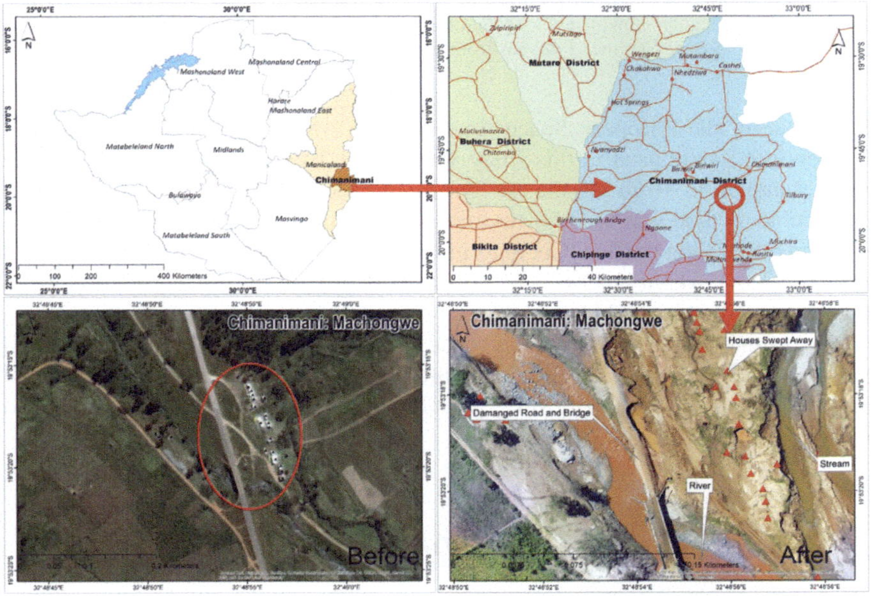

Fig. 4 Sentinel satellite imagery showing damage at Machongwe Business Centre. *Source* Sentinel 2 images from the European Space Agency

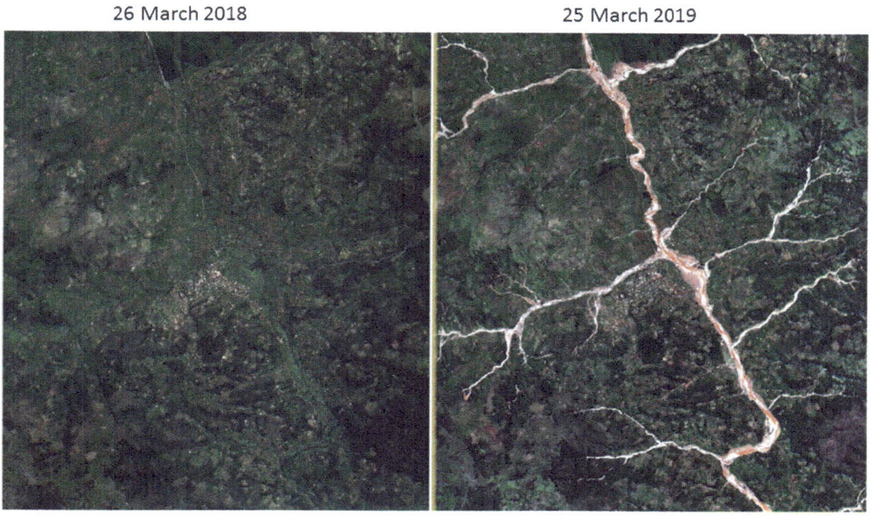

Fig. 5 Landslides and widening of the Nyahode River 13 km south of Chimanimani due to rainfall and floods during the Cyclone Idai period. *Source* Sentinel 2 images from the European Space Agency

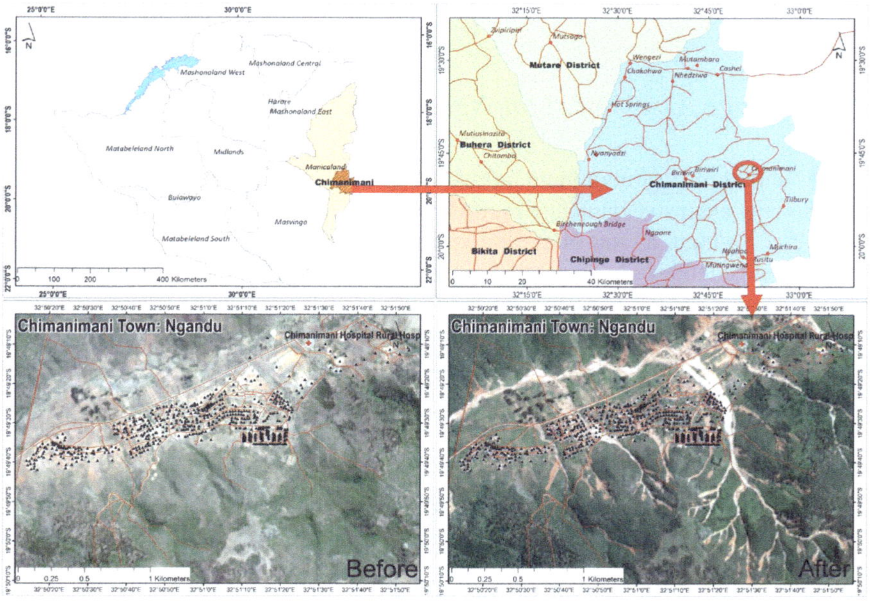

Fig. 6 Landslides and widening of streams passing through Ngangu Township in Chimanimani due to rainfall and floods during the Cyclone Idai period. *Source* Sentinel 2 images from the European Space Agency

Fig. 7 Satellite and drone imagery of Ngangu showing infrastructure damage before and after Cyclone Idai

to determine the state of the infrastructure and landscape within ten days of Cyclone Idai. Thematic layers showing the state of the infrastructure before Cyclone Idai hit the area were overlaid on top of the drone imagery. This allowed the comparison of the pre-disaster and post-disaster images and allowed assessment of damage from the cyclone. In addition, the areal extent of flooding was mapped using drone images as the base layer and overlaid on top of maps showing location and type of damage to infrastructure at Ngangu and Rusitu to help interpret the magnitude of the damage.

3.3 Damage Assessment for Ngangu

Figure 7 shows the layout of main infrastructure at Ngangu before and after Cyclone Idai, showing extensive damage to infrastructure (Fig. 8). The results in Fig. 7 indicate that at Ngangu, a total of 103 buildings were completely destroyed while 461 buildings were damaged. The destroyed and damaged buildings were mostly sited close to streams and rivers. Four wide channels were opened up at Ngangu and these were responsible for transporting large rock boulders from the mountains (Fig. 8). In combination with landslides, flooding was mainly responsible for destroying and damaging infrastructure at Ngangu.

3.4 Damage Assessment for Rusitu Valley

Prior to Cyclone Idai, the settlement at Rusitu Valley was dominated by properties belonging to Agricultural Technical and Extension Services (Agritex) and the Zimbabwe Republic Police (ZRP). Over 90% of the settlement was flooded resulting in 87 houses being completely destroyed and 70 building structures including two

Fig. 8 Stones and boulders observed at Ngangu

bridges severely damaged. The main mechanism behind the damage was transportation and deposition of massive boulders (in excess of 5 tonnes—Fig. 9) at the settlement.

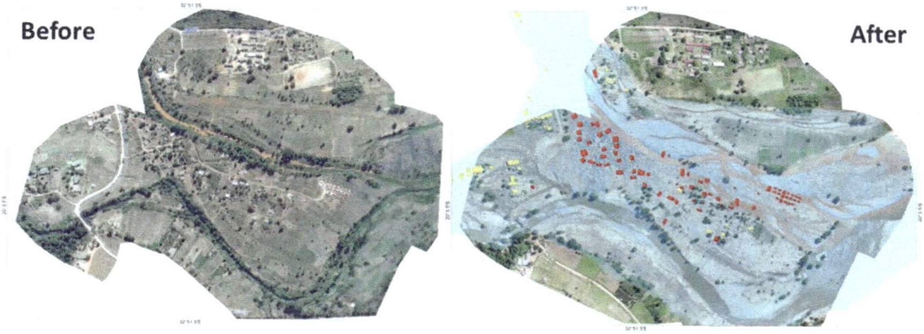

Fig. 9 Satellite (before) and drone (after) images of Rusitu Valley showing infrastructure damage due to Cyclone Idai

Fig. 10 Satellite (before) and drone (after) images of Machongwe showing infrastructure damage due to Cyclone Idai

3.5 Damage Assessment for Machongwe

Figure 10 shows the damage assessment at Machongwe Business Centre, indicating the area that was flooded and the damaged infrastructure. Eight building structures were completely destroyed and 40 were damaged at Machongwe.

3.6 Damage Assessment for Peacock Business Centre

Figure 11 shows the location of shops and other infrastructure in the area before and after cyclone Idai. All 12 shops at Peacock Business Centre were completely washed away. The road and the bridge at Peacock were also extensively damaged (see Fig. 11).

Fig. 11 Satellite (before) and drone (after) images of Peacock Business Centre showing infrastructure damage due to Cyclone Idai

Fig. 12 Satellite (before) and drone (after) images of Skyline Junction showing infrastructure damage due to Cyclone Idai

3.7 Damage Assessment for Skyline Junction

Cyclone Idai resulted in widespread landslides around the Skyline Junction. Figure 12 shows the Skyline Junction landscape before the cyclone. After the cyclone the landscape was completely transformed by landslides. The landslide scars are visible on the drone imagery (Fig. 12). The landslides destroyed settlements and sections of the Chimanimani-Wengezi main road resulting in the two communities being disconnected.

4 Discussion and Conclusion

Despite the utility of earth observation techniques such as high-resolution satellite and drone imagery, the impact of cyclone Idai in Chimanimani was a culmination of several factors. Chief among them are the ones outlined below:

4.1 Response to the Disaster

The Department of Civil Protection in Zimbabwe coordinated the relief work. However, due to the severe damage to most of the roads, the most severely affected Chimanimani District was not accessible during the first 3 days of the cyclone. Continuous rainfall and misty weather made access by helicopters problematic during the first 3–5 days after the cyclone. The Department of Civil Protection had never planned for a disaster at the scale of Cyclone Idai. In affected communities, there was no suitable equipment for retrieving persons buried in damaged houses or heavy boulders. The problems of accessibility soon after a cyclone as indicated above, have been widely reported elsewhere (see e.g.: Alam and Collins [4], Chatiza [11],

Chanza et al. [10], Misra et al.[44]). Alam and Collins [4] reported the same challenges in Bangadesh, for coastal communities repeatedly affected by cyclones. They thus recommended the involvement of local communities as active participants in response preparedness and strategies. For Zimbabwe, the same challenges as outlined above were also reported by Chatiza [11] and Chanza et al. [10]. Chatiza [11] recommended that policies and agencies need urgent improvement to frame more proactive disaster risk management, characterised by stronger social protection systems, enhanced DRM funding, and better settlement systems regulation. Chanza et al. [10] also advocates for better involvement of local players in disaster response. They highlighted the importance of grassroots level support as they react to disasters first before they call for external assistance. However, Troy et al. [56] and Ramsbottom et al. [51], noted that an important component of community engagement is the availability of local resources, timely information dissemination which can enable local communities to respond to a disaster whilst addressing the gaps due to the inevitable delay in formal disaster management assistance, since the first response always comes from local actors [33].

4.2 Cyclone Idai Amplified Pre-existing Development Challenges

The major lesson is that Cyclone Idai amplified pre-existing human development challenges and high vulnerability of large sections of the society arising from weak national and local governance, communities largely depending on near subsistence livelihoods, lack of access to basic services and inequalities at national and international levels. The affected communities have very low resilience to any form of disaster, and Cyclone Idai unfortunately demonstrated this. Our findings were similar to what was reported by Kmoch et al. [36] who showed that, in western Myanmar, Cyclone Komen's aftermath was exacerbated by endemic poverty and pre-existing inequalities. In particular, the complex interplay of households' pre-existing vulnerabilities, persistent farming challenges, extensive disasters' cascading effects were noted to disparately affect lowland and upland communities.

4.3 Need for More Localized Early Warning Systems

RMSC La Union and the national meteorological organizations were able to track Cyclone Idai from formation, to intensifying intense tropical cyclone, heading towards mainland, and finally dissipation. Thus, remote sensing information was available about an impending disaster in both cases. In Mozambique evacuation plans and measures for protecting some buildings were put in place. In Zimbabwe cyclone alerts were provided as part of the usual weather bulletins. The early warnings provided were not effectively acted upon by district and local authorities including those in the path of the cyclone and those affected in various ways. With readily available satellite data some effort should have been made to provide location specific warnings to community leaders, schools, local police stations, and clinics. The need for localised early warning systems is widely reported in literature. For example: Gwimbi [24], Mallick and Rahman [39], Alam and Collins [4], Ahmed et al. [2]. The basic conclusion from all these studies is that the absence of appropriate forecasting and early warning systems exacerbates the cyclone effects on the already vulnerable communities.

4.4 Need for Capacity Building and Clearly Defined Roles of Local Authorities

During the period Cyclone Idai was affecting parts of Mozambique and Zimbabwe, local authorities who are close to affected communities were not prominent in leading search and rescue operations. In relation to this, Bongo and Manyena [7] refer to the important role of leadership for effective disaster governance. Bongo and Manyena (ibid) present an interesting case study about a veldt fire incident in one of the districts in the semi-arid part of the country. They relate that upon hearing about the veldt fire, the DA, in the company of two officials, rushed to the scene of the outbreak. The group arrived at the scene of the veldt fire, disembarked from their vehicle and stood at akimbo as the fire spread before their eyes. At the district level, civil-protection leadership, had at least gone to 'assess' the fire outbreak. It was clear that, as the leader, the DA had to be seen to be doing something about the fire, but he actually did not do what was needed. This incident glaringly portrayed the lack of empowerment, resources and initiative to deal with the fire. There also seems to be very little investment in disaster management and dissemination of early warnings at the local authority level. Even within communities, the roles of representatives in coordinating disaster management does not seem to be well defined, except to act as lead persons when external humanitarian organizations arrive. Thus, investment in disaster management at the local authority level should be made a priority. National governments should allocate adequate resources for disaster management to local authorities. There is a desperate need to effectively equip and train disaster management teams at district levels.

4.5 Unavailability of Disaster Response Options Due to Poverty

In Mozambique due to the frequent occurrences of tropical storms, there is some level of awareness about the possible effects of these weather systems. However, due to poverty that affects the majority of the population, they often do not have many options for responding. The same is true for rural communities that were affected in Zimbabwe. Even if they had ample information about the impending disaster, they didn't have resources and capacity to move out of harm's way on time. Stocker [54] reported that developing countries often face extreme impacts of climate change due to poor adaptation abilities and restricted natural disaster coping capacities due to poverty. As reported in Kusangaya et al. [37] the African continent is still particularly vulnerable to climate change due to its low adaptive capacity [29] and vulnerability [9] with the southern African region regarded as the most vulnerable owing to lack of financial, institutional and technological capacity [6], low adaptive capacity [53], endemic poverty [15] low technology uptake, and dependence on rain fed agriculture [63]. This in total limits the countries capacity to respond to any disaster as was noted during cyclone Idai. Similarly, previous research by Hülssiep et al. [30] showed that socio-economically poor people are marginalised in most aspects of life.

4.6 Public Unawareness About the Nature of Impending Disasters and Appropriate Responses

In Zimbabwe, Cyclone Idai demonstrated the low level of public awareness about disasters such as landslides, rockfalls, uprooting of large trees, and flooding arising from cyclones. Almost all the affected were not expecting major problems. In addition, there had never been any information disseminated about the appropriate response as part of disaster management plan. This has also been reported elsewhere, for example [47] reported that in Kinnaur District of Himachal Pradesh in the Indian Himalaya, for most of the people, disaster protection was a low priority with immediate, more tangible needs dictating perceptions of risk hence response to any disaster.

4.7 Integration of Disaster Reduction in Land Use Planning

Analysis of high resolution and drone imagery showed that considerable proportion of the communities affected by Cyclone Idai resided in unplanned settlements, or settlements planned in marginal areas vulnerable to flooding and landslides in both rural and urban areas. Thus, there is a need to integrate management of flooding and landslides in land use planning in order to reduce the impacts of these disasters. The use of remote sensing technologies in this demarcation is fundamental to this effect extensive drone mapping is very useful, pre and post disaster occurrence. Several studies have looked at the importance of linking disaster management and land-use planning (e.g., Wang [60], Godschalk et al. [21], King et al. [35], Glavovic

[20]). These studies indicated that land-use planning is a critical measure and tool to promote community safety.

4.8 Need for Improved Coordination

During the aftermath of Cyclone Idai, Zimbabwe witnessed a huge effort by citizens in and outside the country who voluntarily offered assistance in various ways. Private sector organizations in the major centres also participated in providing assistance, e.g., transporting donated goods. Medical personnel and other experts also moved in rapidly on a voluntary basis to provide assistance. Local humanitarian NGOs based in Zimbabwe were also actively involved. This demonstrates that there is a potential for citizens and the local NGOs in contributing to significant disaster management. However, these efforts were not coordinated at national and local levels. Eller et al. [13] highlighted the need for improved coordination of voluntary and disaster management efforts by the public. McLennan et al. [43] proposed implementation of the Emergency Volunteering—Community Response to Extreme Weather (EV CREW) as a best-practice model for centrally coordinating spontaneous volunteers who respond during emergencies. Such an approach can be implemented for Zimbabwe.

4.9 Need for Systematic Use of Satellite Data

Satellite data for monitoring weather events and monitoring their effects such as flooding and landslides are now readily available. The European Space Agency and NASA made some of the information available. However, in Zimbabwe, as evidenced during cyclone Idai, there were no effective institutional frameworks for coordinated and systematic utilization of satellite data and products for timely managing disasters. However, several studies have shown that today's highly developed geospatial technology have the potential to be highly effective for disaster management including disaster response [57, 1, 59]. For example, Van Aardt et al. [1] highlighted the capabilities offered by geospatial portals (UN inter-agency OneResponse website, the Sahana Free and Open-Source Disaster management System, and the crowdsourcing platform Ushaidi for deployment, data collection and information dissemination during Haiti Earthquake disaster management. These platforms allowed the public to provide information about missing persons or to track the location of missing people. As such use of satellite data can help to minimize the impact of a catastrophe, vulnerability assessment, readiness, and in building resilience in disaster management. Additionally, unmanned aerial vehicles (UAVs are increasingly being used in damage assessment; natural hazard management and infrastructural damage assessment [14, 17, 25, 34]. For example, studies by Feng et al. [16], Popescu et al. [50]

showed that UAVs can be used to identify disaster damage and serve humanitarian response services during disaster events.

5 Use of Unmanned Aerial Vehicles (UAVs)

Damage assessment using UAVs is possible if the area to be assessed is relatively small. In addition, the manoeuvrability of UAVs permits remote data acquisition in environments dangerous to human life and/or inaccessible to direct examination (e.g., flooded area, thick forests and mountainous areas). UAVs provide potential for acquiring remote data more rapidly and at lower cost and at the desired frequency, which is critical for areas during and after a disaster. UAVs are capable of covering large areas much faster than people on the ground especially during search and rescue operations. Overall, the use of UAVs may be a much better method since it provides accurate, flexible and cost-effective techniques [5, 55]. Colomina and Molina [12] and thus are flexible, allow near real time data acquisition and are non-destructive. Images with resolutions much higher than those offered by satellites can be obtained even with the UAV flying at high altitudes. In turn, this makes it possible to detect problems before they become widespread. In addition, UAVs have become much easier to operate because flight missions can be entirely automated using offline flight planning. Finally, image processing and machine learning tools have evolved to a point in which visual cues contained in an image can be successfully extracted and transformed into useful information for management of disaster.

However, the greatest limitations of multi-rotor UAVs are the relatively short flight time, lower payload and the sensitivity to weather [19, 64]. In addition, the efficiency of computer data processing needs a robust computer (high-performance computing) and dedicated software to enable rapid and accurate data processing and analysis. For example, the geometric correction and radiometric calibration constitute the most demanding work for UAV remote sensing data processing, which is time-consuming, hence the need for a high-performance computer. At minimum a high processing computer (HPC) is recommended to ensure efficient image data processing.

Furthermore, there is need to use UAVs with the right sensors for the job at hand. The type of sensor plays an important role in the ability that UAV-based monitoring has to detect the stresses of interest [8]. Conventional RGB cameras as used in Chimanimani, can deliver very high spatial resolutions, and they tend to be much more affordable, but their spectral information is very limited. On the other hand, multispectral cameras can deliver high spatial resolutions and detailed reflectance information for a few bands with relatively wide bandwidths but tend to be considerably more expensive than RGB cameras. In addition, hyperspectral sensors offer a very high-resolution spectral profile containing reflectance information for hundreds of narrow spectral bands allowing for the detection of finely localized spectral alterations [40, 58], but this type of sensor (hyperspectral) also tends to be the least affordable. Thus, choosing the right sensor for a given application is a complex task that should always take into account the trade-off between the potential gain in accuracy and the cost associated with more sophisticated sensors. For damage assessment, multi-spectral sensors (which cover the visible as well as near infra-red range) are recommended provided a bigger drone is available to carry the payload.

As was observed during the cyclone Idai, there are other several challenges and opportunities associated with use of UAVs. These challenges can be classified as follows:

(i) *Technical challenges*: The deployment of UAVs is hampered by challenges which range from lack of technical experts who can fly the drone and process the data, to the limited size of the payload. One of the key challenges in Zimbabwe is the availability of licensed UAVs operators. Secondly, lightweight UAVs systems have limited capabilities to carry several sensors. Most UAVs systems have a short battery lifetime resulting in a short flight time ranging from 30 min to one hour, which hinders progress when doing damage assessment mapping for large areas since batteries must be changed regularly.

(ii) *Policy and regulations issues*: This is also one of the key challenges in the use of UAVs in most developing countries. Most countries do not have regulations and policies that can be used as a guide by UAVs users, while others have unsuitable regulatory frameworks. Based on the reviewed literature, it was be observed that there is low usage of UAVs systems in most developing countries because they are mainly used for military operations. Therefore, there is a need to craft some related legislation and national/regional rules that will be used to regulate the users and application of UAVs in different countries.

(iii) *Cost*: UAV technology is an important tool for damage assessment. However, purchasing a super UAVs system with a good balance of endurance, stability, durability and optional ability to fly autonomously is costly. In addition, the use us UAVs systems require experts and skilled personnel who are able to operate the systems and process the data acquired.

(iv) *Ethics and Privacy*: The use of UAVs systems in other places for mapping and monitoring might lead people to feel as if there is a privacy invasion. This can be regulated by improving public awareness.

6 Conclusion

After Cyclone Idai, many essential services such as hospitals and schools were disrupted while roads were damaged. It is necessary to focus on the affected areas and rebuild the lives of affected communities. This chapter has demonstrated the utility of satellite imagery, drone technology and GIS for post-disaster assessment. As opined by Griffin (2015) the community's attitude and perceived risks will also influence the potential use of UAVs for disaster management. It is essential that officials address these concerns and build in safeguards to protect against any breaches of privacy and individual rights. Even though this chapter has demonstrated the use of satellite imagery and UAVs in damage assessment, they can be embedded into the entire life-cycle of disaster management. Hoque et al. [27] argue that with the substantial advancement of remote sensing and spatial analysis techniques, tropical cyclone disaster management will be more effective than in previous periods. Previously, limited spectral, spatial and temporal resolution of remote sensing data was a drawback for its wider application. At the same time, spatial data is being updating frequently with the advances in technology and rising demand level. Several international organisations are supporting and financing the access to spatial data for researchers and agencies engaged in natural disaster management. Due to continuous progress in these areas, there is a strong prospect of significant improvements in cyclone disaster management.

References

1. Van Aardt JA, McKeown D, Faulring J, Raqueño N, Casterline M, Renschler C, Gill S (2011) Geospatial disaster response during the Haiti earthquake: a case study spanning airborne deployment, data collection, transfer, processing, and dissemination. Photogramm Eng Remote Sens 77(9):943–952
2. Ahmed B, Rahman MS, Sammonds P, Islam R, Uddin K (2020) Application of geospatial technologies in developing a dynamic landslide early warning system in a humanitarian context: the Rohingya refugee crisis in Cox's Bazar, Bangladesh. Geomat Nat Haz Risk 11(1):446–468
3. Akhmadiya A, Nabiyev N (2019) Review of building damage assessment methods using SAR data. Kazakh Agro Technical University, Nur-Sultan, Kazakhstan
4. Alam E, Collins AE (2010) Cyclone disaster vulnerability and response experiences in coastal Bangladesh. Disasters 34(4):931–954
5. Banu TP, Borlea GF, Banu C (2016) The use of drones in forestry. J Environ Sci Eng B 5(11):557–562
6. Boko M, Niang I, Nyong A, Vogel C, Githeko A, Medany M, Osman-Elasha B, Tabo R, Yanda P (2007) Africa. Climate change 2007: impacts, adaptation and vulnerability. In: Parry ML, Canziani OF, Palutikof JP, van der Linden PJ, Hanson CE (eds) Contribution of working Group II to the fourth assessment report of the intergovernmental panel on climate change. Cambridge University Press, Cambridge UK, pp 433–467
7. Bongo PP, Manyena SB (2015) From "Government" To "Governance": tensions in disaster-resilience leadership in Zimbabwe. Jàmbá: J Disaster Risk Studies 7(1), Art. #188, 10. https://doi.org/10.4102/jamba.v7i1.188
8. Von Bueren SK, Burkart A, Hueni A, Rascher U, Tuohy MP, Yule IJ (2015) Deploying four optical UAV-based sensors over grassland: challenges and limitations. Biogeosciences 12(1):163–175
9. Callaway JM (2004) The benefits and costs of adapting to climate variability and climate change. Benefits Climate Change Policies 111–157
10. Chanza N, Siyongwana PQ, Williams-Bruinders L, Gundu-Jakarasi V, Mudavanhu C, Sithole VB, Manyani A (2020) Closing the gaps in disaster management and response: drawing on local experiences with Cyclone Idai in Chimanimani, Zimbabwe. Int J Disaster Risk Sci 11(5):655–666
11. Chatiza K (2019) Cyclone Idai in Zimbabwe: an analysis of policy implications for post-disaster institutional development to strengthen disaster risk management
12. Colomina I, Molina P (2014) Unmanned aerial systems for photogrammetry and remote sensing: a review. ISPRS J Photogramm Remote Sens 92:79–97
13. Eller W, Gerber BJ, Branch LE (2015) Voluntary nonprofit organizations and disaster management: Identifying the nature of inter-sector coordination and collaboration in disaster service assistance provision. Risk, Hazards and Crisis in Public Policy 6(2):223–238

14. Erdelj M, Natalizio E (2016) UAV-assisted disaster management: applications and open issues. In: 2016 International conference on computing, networking and communications (ICNC). IEEE, pp 1–5
15. Eriksen S, O'Brien K, Rosentrater L (2008) Climate change in Eastern and Southern Africa: impacts, vulnerability and adaptation. GECHS Report 2008:2. University of Oslo
16. Feng Q, Liu J, Gong J (2015) Urban flood mapping based on unmanned aerial vehicle remote sensing and random forest classifier—a case of Yuyao, China. Water 7(4):1437–1455
17. Fernandez Galarreta J, Kerle N, Gerke M (2015) UAV-based urban structural damage assessment using object-based image analysis and semantic reasoning. Nat Hazard 15(6):1087–1101
18. Fitchett JM (2018) Recent emergence of CAT5 tropical cyclones in the South Indian Ocean. S Afr J Sci 114(11–12):1–6
19. Garnica-Peña RJ, Alcántara-Ayala I (2021) The use of UAVs for landslide disaster risk research and disaster risk management: a literature review. J Mt Sci 18(2):482–498
20. Glavovic BC (2010) The role of land-use planning in disaster risk reduction: an introduction to perspectives from Australasia
21. Godschalk DR, Kaiser EJ, Berke PR (1998) Integrating hazard mitigation and local land use planning. In: Cooperating with nature: confronting natural hazards with land-use planning for sustainable communities, pp 85–118
22. Goodess CM (2013) How is the frequency, location and severity of extreme events likely to change up to 2060? Environ Sci Policy, 27:S4-S14
23. Goyal PK (2022) Cyclonic damage assessment of rural houses for the east coastal region of India. Prog Disaster Sci 15. https://doi.org/10.1016/j.pdisas.2022.100238
24. Gwimbi P (2007) The effectiveness of early warning systems for the reduction of flood disasters: some experiences from cyclone induced floods in Zimbabwe. J Sustain Dev Africa 9(4):152–169
25. Ham Y, Han KK, Lin JJ, Golparvar-Fard M (2016) Visual monitoring of civil infrastructure systems via camera-equipped Unmanned Aerial Vehicles (UAVs): a review of related works. Vis Eng 4(1):1–8
26. Hodde H (2012) The damage assessment process: evaluating coastal storm damage assessments in Texas after Hurricane Ike. Thesis for: M.S. Environmental Management.https://doi.org/10.13140/RG.2.1.4366.1845
27. Hoque AM, Phinn S, Roelfsema C, Childs I (2017) Tropical cyclone disaster management using remote sensing and spatial analysis: a review. Int J Disaster Risk Reduction. https://doi.org/10.1016/j.ijdrr.2017.02.008
28. Huang W-K, Wang J-J (2015) Typhoon damage assessment model and analysis in Taiwan. Natural Hazards 79(1). https://doi.org/10.1007/s11069-015-1858-8
29. Hulme M, Doherty R, Ngara T, New M, Lister D (2001) African climate change: 1900–2100. Climate Res 17(2):145–168
30. Hülssiep M, Thaler T, Fuchs S (2021) The impact of humanitarian assistance on post-disaster social vulnerabilities: some early reflections on the Nepal earthquake in 2015. Disasters 45(3):577–603
31. Kabat P, Schulze R, Hellmuth ME, Veraart JA (2002) Coping with impacts of climate variability and climate change in water management: a scoping paper. International Secretariat of the Dialogue on Water and Climate Wageningen
32. Kadomura H (2005) Climate anomalies and extreme events in Africa in 2003, including heavy rains and floods that occurred during northern hemisphere summer, African study of monographs. Supplementary Issue 30:165–181
33. Kapucu N (2008) Collaborative emergency management: better community organising, better public preparedness and response. Disasters 32(2):239–262
34. Kerle N, Nex F, Gerke M, Duarte D, Vetrivel A (2019) UAV-based structural damage mapping: a review. ISPRS Int J Geo Inf 9(1):14
35. King D, Gurtner Y, Firdaus A, Harwood S, Cottrell A (2016) Land use planning for disaster risk reduction and climate change adaptation: operationalizing policy and legislation at local levels. Int J Disaster Resilience Built Environ

36. Kmoch L, Palm M, Persson UM, Jepsen MR (2021) Cyclone Komen's aftermath: local knowledge shows how poverty and inequalities fuel climate risk in western Myanmar. Reg Environ Change 21(4):1–15
37. Kusangaya S, Warburton ML, Van Garderen EA, Jewitt GP (2014) Impacts of climate change on water resources in southern Africa: a review. Phys Chem Earth, Parts a/b/c 67:47–54
38. Leroux MD, Meister J, Mekies D, Dorla AL, Caroff P (2018) A climatology of southwest Indian Ocean tropical systems: their number, tracks, impacts, sizes, empirical maximum potential intensity, and intensity changes. J Appl Meteorol Climatol 57(4):1021–1041
39. Mallick F, Rahman A (2013) Cyclone and tornado risk and reduction approaches in Bangladesh. In: Disaster risk reduction approaches in Bangladesh. Springer, Tokyo, pp 91–102
40. Maslekar NV, Kulkarni KP, Chakravarthy AK (2020) Application of unmanned aerial vehicles (UAVs) for pest surveillance, monitoring and management. In: Innovative pest management approaches for the 21st century. Springer, Singapore, pp 27–45
41. Massarra, Carol C (2012) Hurricane damage assessment process for residential buildings. Louisiana State University Master's Theses, https://digitalcommons.lsu.edu/gradschool_theses/520
42. Masutomi Y, Iizumi T, Takahashi K, Yokozawa M (2012) Estimation of the damage area due to tropical cyclones using fragility curves for paddy rice in Japan. Environ Res Lett 7:014020
43. McLennan B, Molloy J, Handmer J (2016) Centralised coordination of spontaneous emergency volunteers: the EV CREW model. Australian J Emerg Manag 31(1):24–30
44. Misra S, Goswami R, Mondal T, Jana R (2017) Social networks in the context of community response to disaster: study of a cyclone-affected community in Coastal West Bengal, India. Int J Disaster Risk Reduction 22:281–296
45. Murlidharan TL, Durgaprasad J, Appa Rao TVSR (1997) Knowledge-based expert system for damage assessment and vulnerability analysis of structures subjected to cyclones. J Wind Eng Ind Aerody 72:479–491
46. Patt A, Schröter D (2008) Perceptions of climate risk in Mozambique: implications for the success of adaptation strategies. Global Environ Change 18(3):458–467
47. Pilgrim NK (1999) Landslides, risk and decision-making in kinnaur district: bridging the gap between science and public opinion. Disasters 23(1):45–65
48. Planitz A (1999) A guide to successful damage and needs assessment: south pacific disaster reduction programme (SPDRP)
49. Plank S (2014) Rapid damage assessment by means of multi-temporal SAR—a comprehensive review and outlook to sentinel-1. Remote Sensing 6(6):4870–4906. https://doi.org/10.3390/rs6064870
50. Popescu D, Ichim L, Stoican F (2017) Unmanned aerial vehicle systems for remote estimation of flooded areas based on complex image processing. Sensors 17(3):446
51. Ramsbottom A, O'Brien E, Ciotti L, Takacs J (2018) Enablers and barriers to community engagement in public health emergency preparedness: a literature review. J Community Health 43(2):412–420
52. Reason C, Keibel A (2004) Tropical cyclone Eline and its unusual penetration and impacts over the southern African mainland. Weather Forecasting 19(5):789–805
53. Solomon S, Qin D, Manning M, Chen Z, Marquis M, Averyt K, Tignor M, Miller H (2007) Intergovernmental panel on climate change, climate change 2007: the physical science basis. contribution of working group I to the fourth assessment report of the intergovernmental panel on climate change. Cambridge University Press, Cambridge
54. Stocker T (ed) (2014) Climate change 2013: the physical science basis: working group i contribution to the fifth assessment report of the intergovernmental panel on climate change. Cambridge university press
55. Tang L, Shao G (2015) Drone remote sensing for forestry research and practices. J Forest Res 26(4):791–797
56. Troy DA, Carson A, Vanderbeek J, Hutton A (2008) Enhancing community-based disaster preparedness with information technology. Disasters 32(1):149–165
57. Underwood S (2010) Improving disaster management. Commun ACM 53(2):18–20

58. Vanegas F, Bratanov D, Powell K, Weiss J, Gonzalez F (2018) A novel methodology for improving plant pest surveillance in vineyards and crops using UAV-based hyperspectral and spatial data. Sensors 18(1):260
59. Walter LS (1990) The uses of satellite technology in disaster management. Disasters 14(1):20–35
60. Wang JJ (2012) Integrated model combined land-use planning and disaster management: the structure, context and contents. Int J Disaster Prev Manag
61. Yamazaki F, Matsuoka M (2007) Remote sensing technologies in post-disaster damage assessment. J Earthquake Tsunami 1(3):193–210
62. Yamazaki F (2001) Applications of remote sensing and GIS for damage assessment. In: Corotis et al. (eds) Structural safety and reliability, swets and zeitlinger, ISBN 90-5809-197X
63. Yanda PZ (2010) Climate change impacts, vulnerability and adaptation in Southern Africa. In: Kotecha P (ed) Climate change, adaptation and higher education: securing our future. SARUA Leadership Dialogue Series, Cape Town, South Africa, pp 11–30
64. Zhang C, Kovacs JM (2012) The application of small unmanned aerial systems for precision agriculture: a review. Precision Agric 13(6):693–712

Samuel Kusangaya is a physical geographer with a PhD in Hydrology (UKZN) MPhil in GIS & Remote Sensing (Cambridge), MSc in Land and GIS (UZ), and BSc in Geography & Mathematics (UZ). He is currently employed as a senior lecturer at the University of Zimbabwe lecturing in applied hydrology, water resources management, GIS, remote sensing, and climate change and climate modelling. His research interests include but not limited to geo-science applications in hydrology, geo-science application in hydrology, water resources management and disaster management.

Albert Maipisi holds a PhD in Disaster Management from the University of Free State in South Africa, PhD taught courses in Public Administration from the National University of Public Service in Hungary, Msc. In Disaster Management from the National University of Science and Technology (NUST) in Zimbabwe as well as a Bsc. Hons. Degree in Administration from the University of Zimbabwe (UZ). He is currently a Senior Lecturer with the Women's University in Africa domiciled in Zimbabwe and the national Institutional Development Component Lead for the New York State University Rockefeller Foundation's Parliament Support Programme in Zimbabwe. Immediately before joining the University, he held the position of Committee Clerk/ Coordinator for the Parliamentary Portfolio Committee on Mines and Mining Development under the Parliament of Zimbabwe and also have introduced and currently coordinates the Master of Science Degree in Disaster Risk and Livelihoods Management Programme at the Women's University in Africa. Research interests are disaster risk reductionpoverty reduction, parliamentarism, nationalism and public policy.

The Resilience of Child Protection Systems During the COVID-19 Pandemic: Perspectives of Social Work Practitioners in Zimbabwe

Tanatswa Silvanus Chineka, Edward Muzondo, and Mulwayini Mundau

Abstract Moments of crisis are moments of truths. The COVID-19 pandemic has brought about vicissitudes that entrap children's rights and curtail the capacity for delivering protection. Concerns arise regarding the resilience of child protection mechanisms in such contexts, particularly those rendered by the Department of Social Development (DSD). Through the eyes of practising social workers, this study examines the endurance of child protection services provided by the DSD and other development partners, in the context of perturbations instigated by the COVID-19 pandemic. The study adopted a qualitative approach, were data was collected using semi-structured interviews. A sample of 12 purposively selected social work practitioners in Zimbabwe took part in the study. The study's findings reveal that the COVID-19 pandemic has reignited concerns of child protection as indirect risks for children's safety emerge. Children, particularly those in poor and marginalized households lack access to the various child protection responses instituted by both state and non-state actors. Further, the attendant economic challenges are stifling the efforts of both state and non-state actors. It is therefore concluded that such circumstances endanger and reverse the gains in child protection of the past years. Recommendations have been proffered that seek to anchor child protection services within the needs and realities of the communities in Zimbabwe.

Keywords Child rights · Child protection · COVID-19 · Social work · Resilience

---

T. S. Chineka (✉)
School of Health and Social Care, University of Essex, Colchester CO4 3SQ, UK
e-mail: tasilvac@gmail.com; tc23277@essex.ac.uk

African Center for Migration and Society, University of Witwatersrand, Johannesburg, South Africa

E. Muzondo
Department of Social Work, Eswatini Medical Christian University, Mbabane, Eswatini

M. Mundau
Department of Social Work, University of Lesotho, Roma, Lesotho

R. Kasimba and I. Chirisa (eds.), *Disaster Risks and Hazards*, Advances in 21st Century Human Settlements, https://doi.org/10.1007/978-981-96-4577-0_9

1 Introduction

The Coronavirus Disease (COVID-19) pandemic has ushered in an unprecedented global health crisis that has violently shaken the world order across the social, economic, cultural and political dimensions. Zimbabwe, in line with global trends, responded with various anti-contagion measures including national lock-downs as provided for by Statutory Instruments 76 and 99 of 2020. This brought most social and economic activities to a virtual halt, stifling an already harrowing socio-economic order [9]. Given the socio-economic ills that have long bedeviled Zimbabwe prior to the onset of the pandemic, the pressure to cope is immensely exacerbated for individuals, families and communities. Such convulsive and strenuous circumstances are known to promote unhealthy ways of coping (violence, abuse, depression) of which the most vulnerable, particularly children, bear the brunt. The concerns for child rights and child protection, under such a precarious state of affairs, are reignited in an amplified manner. For instance, rising levels of depression, anxiety, loneliness and even self-harm have been identified as concerning negative associations of the pandemic with children's mental wellbeing.

Zimbabwe has an extensive child protection framework built around the National Action Plan for Orphans and Vulnerable Children Phase III (2016–2020), as informed by the National Social Protection Policy Framework of Zimbabwe (NSPPF: 2015). As such, various state and non-state initiatives have been put in place as a means for promoting child care and protection in the pursuit of preserving children's rights. It is worth noting that the majority of these initiatives are premised around the notion of free movement of peoples, and are therefore seriously compromised in the wake of COVID-19 related restrictions on movement. This paper, basing on the views of practicing social workers, examines the resilience of these mechanisms and their capacities to safeguard the rights of children under such debilitating conditions as presented by COVID-19.

2 Background and Context

The COVID-19 pandemic declared on the 11th of March 2020, whose scale and severity is unprecedented, is impacting humanity in enormous ways. The COVID-19 pandemic has ushered in social, cultural, economic and political shocks that have fundamentally disrupted the lives and livelihoods of individuals, families and communities in catastrophic magnitudes. The loss of life and the toll on health care systems globally is ginormous. The Committee for the Coordination of Statistical Activities [3] highlights devastating statistics of reported cases as at 30 April 2020, Europe at 1,406,899, the Americas at 1,246,190, Eastern Mediterranean at 181,119, Western Pacific at 147,743, South-East Asia at 52,266, and Africa at 24,713. The World Health Organisation reported a total of 6,132,154 reported cases and 371,154 deaths due to the coronavirus disease as at 31st May 2020. France is noted to have the highest fatality rate at 18.8 deaths per one hundred infections [3].

The economic ramifications associated with the COVID-19 pandemic have been equally colossal. The global lockdown has negatively impacted on trade, both internal and external, with sharp contractions across productive and service sectors. CCSA [3] argued that COVID-19 induced immediate and sweeping shocks on the labour markets culminating in the biggest employment decline since the Second World War. The International Labour Organisation (ILO) predicted a 10.5% drop in global hours worked, an equivalent of 305 million full-time workers with a 48-h workweek, for the first quarter of 2020 [7]. The informal sector, which has a larger representation in most developing countries and Zimbabwe in particular, bears the brunt with the likelihood of permanent loss of employment. Vast resources have been channeled towards alleviating the pandemic, at the expense of productive activities.

The global child care circumstances under which COVID-19 hit are quite dire. The United Nations [19] underscores that prior to the onset of COVID-19, the global child care scenario was dire: where a child under age 15 dies every five seconds; where one in every five children is malnourished; over half (53%) of 10-year old children in low and middle-income countries can't read and understand simple stories; and one child in four under the age of 5 does not have their birth registered. Given such a background, as economies struggle and government spending for child care is restricted under COVID-19, such grim circumstances are prone to worsen.

The COVID-19 pandemic has found Zimbabwe under dire humanitarian circumstances. The United Nations Children's Fund [17] observes that the humanitarian situation in Zimbabwe has continued to deteriorate as a result of multiple natural hazards; drought and related food insecurity, flooding and the risk of epidemic/ disease outbreaks. Debilitating economic imperatives, for instance hyperinflation at 481% in the first quarter of 2020, curtail the provision of and access to basic services. UNICEF [17] further depict worrying humanitarian statistics; 6.7 million people (3.2 million children) in urgent need of humanitarian assistance, severe acute malnutrition (SAM) rate rising from 0.2% in 2018 to 1.4% in 2019, continuing cholera threat due to protracted dire water and sanitation situation and electricity shortages. The Zimbabwe Humanitarian Response Plan (2020) acknowledges that children remain highly exposed to violence, abuse, mental stress, school drop-out and exploitation.

3 A Brief Overview of the Child Protection Framework in Zimbabwe

It is not the intention nor within the scope of this paper to provide an extensive discussion of the entire gamut of the child protection framework in Zimbabwe. Rather this section briefly highlights the legislative and programmatic dimensions of the child protection framework in Zimbabwe. It is however imperative to clarify that the term 'child protection' often connotes the "prevention and response to violence, exploitation and abuse of children in all contexts" [18]. This includes providing relative services to children who are especially susceptible to these threats, such as those living without or with insufficient family care, in situations of conflict or on the streets in natural disasters.

Zimbabwe is signatory to the United Nations Convention on the Rights of the Child (UNCRC) and the African Charter on the Rights and Welfare of the Child (ACRWC) having ratified these conventions in 1990 and 1999 respectively. As such, the child protection framework in Zimbabwe is committed, in form or effect, to fundamental child rights principles: best interest of the child, survival and development, participation and non-discrimination. The domestication of the UNCRC and the ACRWC in Zimbabwe is primarily through the Children's Act (Chapter 5.06), and is however partial and unsatisfactory [2]. Muchacha [12] highlights that the Children's Act (Chapter 5:26) is largely remedial and curative and thus not commensurate with the child protection imperatives obtaining in Zimbabwe. It is however noteworthy that the Children's Act

SOS Children Villages International (2014) finds that the legislative and policy frameworks making up the child protection system in Zimbabwe are conceptually adequate to cater for needy children. Pertinent legislative and policy provisions relating to child care and protection include the National Social Protection Policy Framework of Zimbabwe (2015), National Orphan Care Policy (1999), the National Action Plan for Orphans and Vulnerable Children III (NAP for OVC, 2016–2020) and the National Case Management System for the Welfare and Protection of Children in Zimbabwe formally adopted and nationally scaled up in 2014. Notwithstanding the extensive legislative and policy framework relating to child care and protection in Zimbabwe, their coordination and effective implementation is hampered due to their scattered administrative location across different government ministries. This creates occasion for gaps in the care and protection of children in Zimbabwe.

It is worth noting that efforts to provide a coordinated, systematic and efficient response to the needs of children in Zimbabwe have been envisaged in the National Case Management System for the Welfare and Protection of Children in Zimbabwe (NCMS) [15]. Muchacha [12] identifies the NCMS as one of the central interventions of the National Action Plan for Orphans and Vulnerable Children. The NCMS is considered particularly suited to Zimbabwe as it is a hybrid of statutory, customary and community approaches [15]. The National Case Management System for the Welfare and Protection of Children in Zimbabwe facilitates inter-agency collaboration between stakeholders and provision of integrated child safeguarding services [15].

Mushunje and Mafico [13], highlight that formal educational support through the Basic Education Assistance Module (BEAM) programme is the most common form of support for orphans and vulnerable children in Zimbabwe. The NAP for OVC is a programmatic framework developed by the government of Zimbabwe to address the orphan crisis [11]. Interventions such as Child Supplementary Feeding Programme (CSFP), BEAM, NAP for OVC, the Victim Friendly Initiative, and Assisted Medical Treatment Orders (AMTO) are suffering from inadequate funding [14, 16]. This results in gaps in quality care of needy children. Extended family care (kinship care) and community care are the major grassroot resource points that addresses gaps in formal child care and protection in Zimbabwe. Persistent droughts, economic meltdown and HIV and AIDS have however compromised the viability of extended family care in Zimbabwe [13].

Non-state actors, particularly humanitarian agencies, are central actors in the arena of child protection in Zimbabwe [8]. Chikwaiwa et al. [4] identify Non-Governmental Organisations (NGOs) as particularly critical in rendering support to vulnerable children in Zimbabwe. The support rendered by NGOs include provision of basic needs and the creation of a conducive social context that promote the development of positive attitudes and emotions (psychosocial support). Further, NGOs through direct contact with communities, facilitate the generation of statistical and qualitative data that informs policy and programmatic interventions. Garutsa and Masuku [6] underscores that NGOs have provided an alternative channel for economic and social development in the wake of state limitations due to internal and external exigencies.

4 Critical Child Rights and Child Protection Issues Under COVID-19

The United Nations [19] observes that children are at risk of being the most affected by the COVID-19 pandemic, particularly as the scourge threatens their wellbeing. In this regard, the socio-economic impacts of the COVID-19 pandemic and the mitigation measures directed at it inadvertently place children at risk of lifelong harm. Ostensibly, children at greatest risk are those in the poorest countries, poorest neighbourhoods, and those in already disadvantaged and vulnerable situations, circumstances of which are prevalent in Zimbabwe. The UN [19] notes that the COVID-19 pandemic affects children in three main channels: infection with the virus itself; the immediate socioeconomic impacts of measures to stop transmission of the virus and end the pandemic; and the potential persistent effects of delayed implementation of the Sustainable Development Goals. Further, as the pandemic increased poverty levels and hunger, and decreased access to education, concerns arise regarding the risk of girls becoming child brides. Statistics from the Zimbabwe Humanitarian Response Dashboard (2021) raised concerns relating to increased food insecurity, challenges in accessing nutritious and diversified foods, correlating with the reverberations of the pandemic.

It is noted that the actual threat of life posed by COVID-19 to children is quite negligible in comparison to adults [19]. A study by Cruz and Zeichner [5] shows that children are less likely to become severely ill than older adults, with the exception of sub-populations of children with an increased risk for more significant illness. This is supported by UN [19] who note that severe symptomatic reactions are less likely in children, with exceptions relating to prior conditions. However, the loss of a parent or caregiver to COVID-19 threatens the wellbeing of children drastically. This ranges from the psychosocial, economic and cultural implications that indirectly impact on the livelihood and general wellbeing of children. Given the magnitude of loss of life suffered globally as a result of COVID-19, these indirect repercussions on children can be phenomenal.

As most countries have embarked on national lockdowns, this has brought some inadvertent risks to child safety. The United Nations [19] notes several risks on child safety and multiple impacts associated with these measures: heightened risk of children witnessing or suffering violence and abuse; significant health risks for children living in unsanitary and crowded conditions; the reliance on online platforms for distance learning increasing children's risk of exposure to inappropriate content and online predators. Further to this: 42–46 million children are expected to fall into poverty in 2020; economic hardships pose a threat to child survival and health, with potential to reverse up to 3 years of progress in reducing infant mortality. Further, malnutrition is expected to rise as some children relied on schools, which are closed, for a reliable daily nutritious meal [19]. The widespread school closures have affected 1.5 billion children and youths, whereas distance learning is inadequate as one third of the world's youngest people are digitally excluded. The risks on child mental health are therefore quite considerable as children strive to come to terms with such debilitating circumstances.

Globally, the exhaustion of health services due to the burden of treating infected patients has left children and women less able to access standard care [19]. In addition, pertinent direct child services have been partially or in some cases totally suspended: nutrition programmes; maternal and newborn care; immunization services; sexual and reproductive health services; HIV treatment; alternative care facilities; community-based child protection programmes and case management for children requiring supplementary personalised care (e.g., Children with e.g., Children disabilities, abuse victims). The heightened risk to exposure of frontline workers necessitates that their children be provided with alternative care arrangements. The pressures to adapt to the changing environment can be considerable [19].

The physical distancing and lockdown measures instituted by more than two-thirds of the countries globally have led to a significant reduction of economic activity across all major economies, resulting in a global recession [19]. Under the assumption that the pandemic recedes in the second half of the year, the International Monetary Fund (IMF) anticipates global income to contract by 3% in 2020 [10]. Such income shocks at the household level can have devastating effects on children, particularly those living in poor households with limited assets [1, 20]. Further, UN [19] argues that such shocks have localised effects for instance the household breadwinner being forced to shelter in place, or migrate back to their rural homes, abandoning their normal livelihoods. It is worth noting that a gender dynamic accompanies such shocks as historically the girl child disproportionately shoulders the attendant burdens.

5 Methods

This chapter is the result of a qualitative study undertaken during the height of the COVID-19 pandemic. Data was collected through semi-structured questionnaires conducted virtually with 12 practicing social workers based in Zimbabwe. This sample of social workers worked within the DSD or in organisations that worked directly with the DSD. Given the regulatory restrictions under which the study was carried out, electronic strategies were employed to engage participants. Specifically, telephonic interviews were employed in which semi-structured interview guides were utilised. Audio recording of the interviews was carried out after verbal consent was sought from the participants. Thematic content analysis was employed to develop a codebook from which themes were drawn, developed and refined. A rapid review of relevant literature was undertaken to better appreciate the evolving context of child protection in the pandemic environment within Zimbabwe, regionally and globally.

6 Findings and Discussion

The purpose of the research was to examine the resilience of child protection mechanisms under covid-19 from the perspective of practicing social workers in Zimbabwe. One of the key objectives of this paper was to establish awareness of and familiarity of practicing social workers with the child protection framework of Zimbabwe. This is critical as social workers are central to the delivery of child protection services in Zimbabwe as, inter alia, probation officers, child protection officers and Pre-Trial Diversion officers. Semi structured interviews were employed to engage [participants and elicit the knowledge of practicing social workers on the delivery of child protection policies and services in Zimbabwe during this crisis period. The extent of their knowledge and familiarity with such legislative and service provisions provides an insight into the ability of key child protection practitioners, such as social workers, to influence the trajectory of child protection in Zimbabwe.

6.1 Child Protection Issues and COVID-19 in Zimbabwe: Perspectives of Social Workers

The study's findings reflects on various child protection issues that are central to COVID- 19 as opined by social workers. Among these factors are the knowledge of child protection frameworks, Child Care Systems under COVID-19 and the child rights issues under COVID-19.

6.1.1 Knowledge of Child Protection Frameworks

Participant social workers demonstrated in-depth awareness of the policy, legislative and programmatic dimensions of child protection in Zimbabwe. Identified policies and legislature include the Constitution of Zimbabwe Amendment 20 (2013), Children's Act (chapter 5:06), National Orphan Care Policy (1999) and the National Action Plan for Orphans and Vulnerable Children Phase III (2016–2020). Pertinent child protection programmes identified include the Basic Education Assistance Module, National Case Management System for the Welfare and Protection of Children in Zimbabwe, the Pre-Trial diversion Programme, the Victim Friendly Initiative and Harmonised Cash Transfers. Knowledge of traditional child protection systems such as the Zunde Ramambo concept was also demonstrated. These indigenous mechanisms become critical particularly in resource constrained environments such as characterise Zimbabwe.

6.1.2 Child Care Systems Under COVID-19

Child care under the national lockdown in Zimbabwe was reported to be provided through the three primary child care systems as espoused in the National Orphan Care Policy (1999). One of the participants during the interview observed that nuclear family care, kinship care and community care were present and providing care to children under the COVID-19 national lockdown. The resilience of such modes of care, as well as other supportive child protection mechanisms was observed to be affected by the economic contraptions, inadequate funding, shortage of material resources and shortage of human personnel. Particularly, transportation challenges and limited staffing were observed to militate against case reporting and compromise investigations. Alternative care through institutionalisation and foster care was reported to be on suspension as new children could not be taken in due to the pandemic. Further, material support for institutional care was limited as key resource points such as NGOs had suspended services or realigned funds.

The operation of the National Case Management System was noted to be compromised as the members were not able to meet and discuss issues affecting children in their communities. Accessing a birth certificate was identified as one of the critical aspects of Case Management, which is hampered due to closure of services and limited movement. In the case of the Harmonised Cash Transfer (HSCT) programme, one social worker submitted through questionnaire that several logistical issues could not be addressed during the lockdown period. Cash disbursements could not be carried out, monitoring and attending to beneficiary grievances was also compromised. For example those who have challenges with their mobile money wallets could not immediately find redress.

Social work participants observed through questionnaire submissions that Non-Governmental Organisations that are critical stakeholders involved in child welfare and child protection temporarily suspended their operations in light of the COVID-19 pandemic. Inadvertently, this left a gap in addressing pertinent issues such as gender-based violence and child sexual abuse. Further children who were working and living in the streets could no longer access needed services due to dislocation. Efforts to provide remote services through such strategies as toll free lines were hailed as commendable. However, coverage was observed to be limited, particularly for children without access to mobile phones.

6.1.3 Child Rights and Protection Under COVID-19

The pandemic was observed by social work participants to inadvertently and indirectly lead to violations of children's fundamental rights. Children's right to education was observed through both interviews and questionnaire submissions to be heavily curtailed. The widespread closure of schools put learning beyond the reach of children in poor and marginalised communities. E-learning was highlighted as inadequate in meeting the needs of children in remote and poor communities. Such pertinent technological devices necessary for accessing classes or reading material on electronic platforms were argued to be lacking in such communities. Participants bemoaned the lack of resources to capacitate child protection interventions and safeguard the rights of children in such contexts.

Further, school closures were associated by the participants with militating factors against the protection of children. Participants observed that schools are critical avenues for detecting any incidences of child abuse and neglect. Their closure thus leads to a higher number of cases going unnoticed or unreported. Unintended consequences of school closure were also observed as one social worker submitted during an interview that children are now being sent for vending errands to support their family livelihoods. This exposed children to health risks as well as limited the time they had for learning. No specific child protection intervention was identified by the participant social workers as having the capacity to address the presenting gaps in learning. Participants also questioned the capacity of vulnerable households to meet the costs associated with compliance to COVID-19 regulations in the event of schools being reopened.

The COVID-19 pandemic was noted by the participants as stifling children's access to healthcare services. Three of the eight participants who took part in the semi-structured interviews argued that children's right to health was severely affected by the national lockdown. All the 42 electronic semi-structured questionnaire submissions highlighted the plight of children in regards to access to health care services as affected in one way or the other. For instance, children in need of Anti-Retroviral Treatment were noted to have met with difficulties in accessing their medication. One of the participants in the interviews observed that such circumstances gave rise to cases of relapsing. In the face of such calamities, critical services such as Assisted Medical Treatment Orders (AMTO) were observed to be ineffectual due to limited funding.

Concerns with issues of child justice emerged as central concerns during the interviews and through questionnaire submissions. The reduced operation of courts was observed to be combative to children's access to justice. Further, participant social workers noted that cases of child abuse and neglect escalate during such restricted conditions but could not be effectively addressed. In addition, as it is commonly known that the highest perpetrators of child abuse are caregivers and trusted persons, it becomes highly likely that children are locked down with their perpetrators and are therefore subjected to violence. The closure of schools further meant that the time children spend with such perpetrators becomes prolonged. The ability of child protection services to address such issues was considered as limited, given that direct services could not easily be accessed under lockdown conditions. Hotlines were argued to have some range of alleviation but issues of limited coverage combat such initiatives.

7 Conclusions and Recommendations

Social workers who participated in the study suggested various strategies that could strengthen the capacity of child protection services in Zimbabwe within the environment imposed by COVID-19. Interview participants called for the anchoring of child protection services within communities and the avoidance of elitist approaches, for instance in the use of hotlines and media platforms. One questionnaire responded submitted that given the changing trajectory towards embracing e-technologies, there was a danger of leaving out the most vulnerable children in the communities. The inaccessibility of child friendly spaces external to the home due to restricted movements was identified as necessitating alternative designs that focused on the immediate family environment. Child protection interventions were thus called upon to balance the needs for embracing e-technologies and ensuring inclusivity, as well as ensuring adherence to principles of equity and social justice. Submissions through the questionnaires alluded to the need for addressing the gaps of such e-services, especially for children who are too young or lacking resources.

Adequate budgetary allocation was identified as key in capacitating the operations of various child protection interventions. Increasing human resources and improving access to pertinent technologies were identified as key measures. Particularly, strengthening and enabling the resumption of all critical and essential staff, as well as ensuring their protection was identified by all the 42 questionnaire respondents as vital. The engagement of volunteers was highlighted as a critical strategy of addressing human resource limitations. In addition, participants in the interviews argued that funding of such disasters should be considered under critical budgetary issues and lessen reliance on donations as these are not reliable nor sustainable.

Training and capacity building of key personnel was identified as key in addressing the challenges imposed by the COVID-19 pandemic. In addition, the need for continuous research and professional development was identified as critical to enhancing the capacity of child protection services to address the needs of children under such environments. One social worker submitted during interview that, '...that's what has to be researched on since no one ever dreamt of responding to a disaster in the magnitude of Covid-19.' In particular, participants expressed interest in having awareness of how other child protection practitioners such as social workers across the globe have responded to the pandemic. Such interest lies in learning particularly in regards to the evolving child protection practice.

References

1. African Child Policy Forum (ACPF) (2019) For lack of will: child hunger in Africa. Addis Ababa, African Child Policy Forum (ACPF)
2. Bhaiseni B (2016) Zimbabwe Children's act alignment with international and domestic legal instruments: unravelling the gaps. Afr J Soc Work 6(1):3–6
3. CCSA (2020) How COVID19 is changing the world; a statistical perspective. Committee for the coordination of statistical activities. Available at https://unstats.un.org/unsd/ccsa/documents/covid19-report-ccsa.pdf
4. Chikwaiwa BK, Nyikahadzoi K, Matsika AB, Dziro C (2013) Factors that enhance intrapersonal wellness of orphans and other vulnerable children (OVC) in institutions and community-based settings in Zimbabwe. J Soc Dev Afr 28(2):53
5. Cruz AT, Zeichner SL (2020) COVID-19 in children: initial characterization of the pediatric disease. Pediatrics 145(6)
6. Garutsa T, Masuku MM (2020) Intervention measures in addressing the needs of orphans and vulnerable children in Marondera, Zimbabwe. J Soc Sci Humanit 17(2):198–207
7. ILO (2020) COVID-19 and the world of work. Rapid assessments of the employment impacts and policy responses. International Labour Organisation
8. Kurevakwesu W, Mukushi AT (2018) Charitable work in a developing region: why humanitarian work is mostly in the hands of foreign organizations in Africa. Afr J Soc Work 8(1):56–62
9. Lee Y (2009) Child rights and child well-being. In: The 3rd OECD world forum on "statistics, knowledge and policy," Busan, Korea. Retrieved from http://www.oecd.org/site/progresskorea/44137252.pdf
10. McKibbin W, Vines D (2020) Global macroeconomic cooperation in response to the COVID-19 pandemic: a roadmap for the G20 and the IMF. Oxf Rev Econ Policy 36(Supplement_1), S297–S337
11. Mtetwa E, Muchacha M (2013) Towards a national social protection policy: knowledge and lessons from a comparative analysis of the design and implementation of public assistance and harmonised social cash transfer programmes in Zimbabwe. IOSR J Humanities Soc Sci 11(3):18–24
12. Muchacha M (2015) An evaluation of the implementation of the national case management model for children orphaned as result of HIV and AIDS in Zimbabwe (Maters dissertation). University of KwaZulu Natal
13. Mushunje MT, Mafico M (2010) Social protection for orphans and vulnerable children in Zimbabwe: the case for cash transfers. Int Soc Work 53(2):261–275
14. Mwapaura K, Chikoko W, Nyabeze K, Kabonga I, Zvokuomba K (2022) Provision of child protection services in Zimbabwe: review of the human rights perspective. Cogent Soc Sci 8(1):2136606
15. Nhapi T (2020) Zimbabwe's national case management system for child protection and enhanced rights realisation for children with disabilities (CWDS). Afr J Soc Work 10(2):124–131
16. SOS Children's Villages International (2014) Assessment report of the alternative care system for children in Zimbabwe. Innsbruck, SOS Children's Villages International
17. UNICEF (2020) Zimbabwe humanitarian situation report: multihazard annual situation report January–December 2020. Final Report, UNICEF
18. UNICEF (2021) Child protection strategy 2021–2030. UNICEF

19. United Nations (UN) (2020) Policy brief: the impact of COVID-19 on children
20. United Nations Children's Fund (UNICEF) (2019) UNICEF humanitarian action for children 2020: overview. Available at https://www.unicef.org/media/62606/file/HAC-2020-overview.pdf

Mulwayini Mundau is a Senior Lecturer in the Department of Social Work at the National University of Lesotho. He has previously served as the head of the Social Work Department at the University of Zimbabwe, as well as a Council member in the Council of Social workers in Zimbabwe's Fifth Chapter. Dr Mundau holds a PhD in Social Work from the University of Hare, South Africa, Master of Social Work from the University of Fort Hare, Postgraduate Diploma in Monitoring and Evaluation from Stellenbosch University, South Africa, Bachelor of Social Work Degree from the University of Fort Hare and a Bachelor of Arts general degree from the University of Zimbabwe. His professional career hails from High school teaching experience in Zimbabwe, Social work practice in Zimbabwe and South Africa. He has also taught at various institutions of social work education and training in Zimbabwe and South Africa. These include the University of Zimbabwe and the Women's University in Africa in Zimbabwe and the University of Fort Hare in South Africa where he also did his Post-Doctoral Research Fellowship. Dr. Mundau's research interest are in community, family and child welfare, social development and climate change and livelihoods.

Zimbabwe's Disaster Preparedness, Readiness and Response: Juxtaposing the COVID-19 Pandemic to the 2019 Cyclone Idai

Tapiwa Musasa and Innocent Chirisa

Abstract This study examines how Zimbabwe handled Cyclone Idai, a disaster that occurred in March 2019 and the COVID-19 pandemic that started in December 2019 worldwide but had its first case recorded in Zimbabwe in March 2020, separating the two hazardous events by one year. The study argues that if a community is affected by consecutive events in the same manner, with more or less the same impacts, its either the community is incapacitated to improve or no learning or efforts are being made to improve on Disaster Risk Reduction (DRR), thus, many critical areas need to be revisited in terms of policy formulation, implementation, preparedness and response to disasters. Disaster management is a continuous on-going process that should improve temporally as an individual, household, community or nation utilises past events to make an improvement on forthcoming similar quagmires. Through the use of desktop research, events analysis during the disasters, and interviews, the study found out that the state of the country's disaster preparedness still need to be improved to be proactive rather than reactive, the nation responded better to COVID-19 than to Cyclone Idai leaves a lot to be desired since the institutions are always caught unaware and not ready for these catastrophic events. Recommendations are therefore made that great improvement should be done in disaster policy formulation, and implementation to involve different stakeholders from all walks of life for effectiveness so that people and infrastructure is saved from destruction.

Keywords Disaster management · Quagmire · Anthropogenic events · Indelible effects · Praxis · Accountability

---

T. Musasa (✉)
Catholic University of Zimbabwe, Harare, Zimbabwe
e-mail: tapiwa.musasa@cuz.ac.zw

I. Chirisa
Office of the Vice Chancellor, Zimbabwe Ezekiel Guti University, Bindura, Zimbabwe

R. Kasimba and I. Chirisa (eds.), *Disaster Risks and Hazards*, Advances in 21st Century Human Settlements, https://doi.org/10.1007/978-981-96-4577-0_10

1 Introduction and Background

This study intends to explore the effectiveness of the existing structures and pieces of legislation in promoting disaster preparedness, readiness and response, as well as examine the effectiveness in terms of resource allocation, resource utilisation, accountability and strategies employed to respond to the two quagmires. The impacts of Cyclone Idai, a disaster that occurred on 15–16 March 2019, killing more than 340 people and affecting more than 270,000 others, and COVID-19, a pandemic that struck the country in March 2020 when the first case was reported [5, 20], cannot be effectively compared because the two hazardous events are not of the same nature. However, the two events are similar in that they are all disasters declared by a State President, thus this study focuses on the way the Zimbabwean Government responded to the two events. In addition, both Cyclone Idai which affected four provinces in Zimbabwe namely Masvingo, Manicaland Midlands and Mashonaland East, was declared a national disaster on 15 March 2019 by the President of Zimbabwe E.D Munangagwa [20], while COVID-19 was declared a national disaster through SI/76/202 [17] on March 23 2020. As of July 2020, the pandemic had killed four people in Zimbabwe, with several others infected with the corona virus. This study argues that lessons should have been learnt so that the level of preparedness and response is raised as compared to previous hazardous events. A disaster is a natural or anthropogenic event occurring with or without warning, causing deaths and/or destruction of infrastructure, injury or disease but exceeding the capability of a community to cope on its own [1, 10, 13]. Disaster risk, in this context is considered as a function of hazard, vulnerability and capacity and each of these terms are defined herein. A risk can be viewed as the probability of something with bad consequences to happen in future. Hazards become disasters when lives and livelihoods are destroyed, and all these concepts are related but not the same. The concept of disaster management has increased over the years because there has been an increase in the rate and intensity of disasters worldwide and the focus for most governments and communities is on Disaster Risk Reduction (DRR), aiming to reduce erosion of development gains and create sustainable livelihoods for current and future generations. Currently, the world poses an unpredictable, hazardous unstable situation where disaster is imminent at any time, ranging from terrorism, climate change to possible economic collapse [1].

Disaster management therefore seeks to mitigate these inexorable events and protects society from collapse in the aftermath of these horrible events. Disaster management is a process not an event, where efforts are made to control and reduce the impacts of these chaotic situations. The major objective of disaster management is not to stop disasters but to reduce their impacts on humanity in terms of saving lives and livelihoods. Zimbabwe is a country in Sub-Saharan Africa, a region classified under Less Developed Countries (LDCs), where disasters have had more damages with an increase of 142% in the period 1994–1998 [10].

From a Marxist perspective of disaster management, the high impact of disasters on the lives and livelihoods in Sub-Saharan Africa is due to endemic poverty in the region that promotes vulnerability and lack of resilience and resistance. It is also in LDCs where there are no long term and coherent institutional arrangements to address disaster issues with a long-term vision, due to the fact that institutions work in isolation, lacking involvement of indigenous knowledge systems and local people [10]. A disaster management cycle is given herewith and the article focuses on the concept of disaster preparedness and response as shown in this cycle adapted from Khan et al. [10]: 49.

Over the years, international frameworks, like the Sendai Framework of Action 2015–2030 have continued to guide the world in approaching disaster management issues. According to Government of Zimbabwe [8], the Sendai Framework of Action has four pillars namely: Understanding risk (Pillar 1), Strengthening governance and effective legislation to manage disaster risks (Pillar 2), Investing in economic, social, cultural and environmental resilience (Pillar 3) and enhancing preparedness for effective response and building back better in recovery and reconstruction (Pillar 4).

With knowledge of international frameworks and guidelines on disaster management, the Government of Zimbabwe established the Department of Civil Protection of 1989 that was later revised in 1992 and in 2001 to coordinate and manage disasters, and implementing guidelines of the Sendai framework of action. There is much work going on within the Civil Protection Unit (CPU) in terms of awareness raising, partnering with institutions of higher learning and responding to disasters. The current pieces of legislation in Zimbabwe empower the Minister of Local Government, Rural and Urban Development to coordinate all disasters as stipulated by the Civil Protection Act no 5 of 1989 [2, 13], and Simba 2018. The point of reference in this article is pillar 4 of the Sendai Framework of action, that is a continuation from priority Action 5 in the Hyogo Framework of Action (2000–2015) emphasising preparedness, response recovery and reconstruction. of critical reference, also, is the disaster management cycle (see Fig. 1), that shows that disaster management is a continuous process that should not be broken, thus, responsible authorities should never rest or relax but continue working on a daily basis through research, partnerships, resource mobilisation and reconstruction of lives and livelihoods so that the future is better than the present in terms of disaster preparedness and response.

2 Theoretical Framework

This study engages the Weberian perspective of emergency management and the Marxist interpretation of disaster vulnerability. The Weberian perspective has it that cultural values, attitudes and practices determine the level at that a society perceives risk, thereby increasing or decreasing vulnerability due to their social constructionist views. This idea is closely related to organisational behaviour, where the same cultural values may give barriers across institutions in terms of planning, or poor emergent responses during disasters due to poor planning within and across organisations. The Marxist interpretation of disaster management holds that economic conditions and poor political organisation and powerlessness are related to disaster vulnerability. On

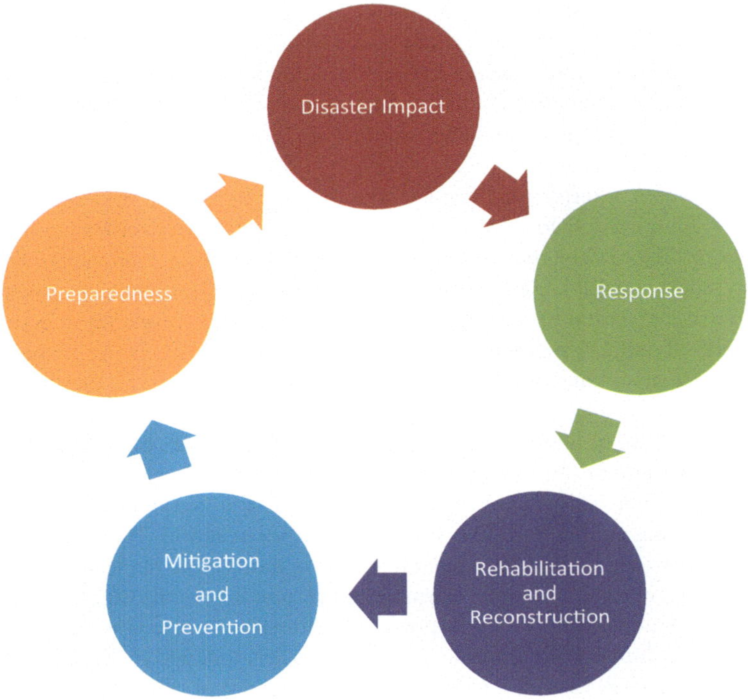

Fig. 1 Disaster management cycle. *Source* Adapted from Khan et al. [10]: 49

one hand there is a scenario where poor minorities and other marginalised groups are most likely to live in hazardous areas that are easily affected by disasters, while on the other hand progressive development leads to well-planned disaster plans with the economic theory suggesting that economic prosperity and recovery enables a nation to respond well to disasters. This is promoted by good technological development and environmental management practices, that avert natural resources depletion, creating an enabling environment for all to respond and recover quickly from disasters [23].

3 Literature Review

The Sub-Saharan Africa (SSA) region has experienced so many indelible occasions resulting from a diversity of disasters, with drought as a perennial disaster for several years in retrospect. Epidemics and famine have been the next close causes of loss of life in the region and they are linked to meteorological and hydrological conditions, that are linked to the on-going climate change process that is increasing the frequency, intensity and variability of these natural disasters [3]. Responding to international calls, guidelines and frameworks for disaster management, the greater SSA region has made efforts to establish platforms to work towards the reduction of the growing risks. At the regional level, the African Union (AU), together with the New Partnership for Africa's Development (NEPAD) Secretariat, has developed the African Regional Strategy for Disaster Risk Reduction and a Programme of Action for the Implementation of the Africa Strategy (2005–2010). The Economic Community of West African States (ECOWAS) in early 2007 approved a sub-regional Common Policy and mechanisms for DRR, an indication that DM issues have taken the top concern in most regions. Similarly, the Southern Africa Development Community (SADC) has revised its sub-regional strategy, factoring in DRR and the Economic Community of Central Africa States (ECCAS) has established a sub-regional centre for DRR in the Republic of Congo and is developing a sub-regional strategy. The Indian Ocean Commission (IOC) is already preparing to develop its sub-regional strategy, still focusing on guidelines and influence from the international arena and the greater Sub-Saharan African region.

At national level, several countries in the region are making efforts to factor in Disaster Risk Reduction as priorities in their development plans, as they work with the greater region and the sub-regions and these include Mozambique, Liberia, Angola, Madagascar, Tanzania, South Africa, Uganda and Ghana, just to name a few [3]. In Liberia, for example, the disaster management policies are not reviewed in time to be able to capture in new trends and dimensions of DRR in contemporary society. Despite the fact that there has been a paradigm shift from response to proactiveness, the country continued to use a 1976 National Disaster Relief Plan until 2012.This very old disaster plan lacked inclusion of local communities, particularly women who were only included later, in section 4.2 of the new DRR plan in 2012 [7, 9].

In Madagascar, the situation is similar in that more than 70% of the population survives on agriculture and fisheries, and the livelihoods are easily affected by the erratic weather patterns [19]. Although the legislation on DRR is well-planned from higher offices to communities, the effective implementation is absent due to lack of decentralisation, absence of specific legal framework for responsibility, accountability and liability relating to DRR [19, 22]. However, the policies and pieces of legislation in most of these countries in SSA lack financial support, enough skills, and inclusiveness in terms of gender, age, physical ability, geographical location and involvement of local people. Most of the policies are very good on paper but the practicalities leave a lot to be desired at the regional and sub-regional levels.

The situation in Zimbabwe is not isolated from other African states in terms of disaster preparedness and response. Different scholars analyse the level of preparedness in the country and concur that the policies are very good on paper but the praxis has many loopholes and weaknesses. According to Mavhura [14], Zimbabwe is prone to every type of known disasters due to its complex climatic conditions and varied geological dimensions. Although the Government has enacted the Department of Civil Protection to coordinate and manage disasters, the good policies are not matched by the practicalities on the ground leaving a lot to be desired for the institutions that should be given enough authority to execute their mandates. From a social work dimension, Mhlanga et al. [12] posit that there is need for the praxis of professional disaster-oriented social work so that common survival forms, like food security, health and education do not continue to be destroyed. Betera [2], in line with the Marxist perception of disasters, laments the way that most hydrological disasters, such as cyclones in Zimbabwe have a dire impact on the poor, incapacitated and marginalised communities. Being the majority (67% according to Zimstats 2012), development programmes should be giving priority to the rural poor but unfortunately, these communities have no savings or fall-back plans in times of difficulty. They lack sustainable livelihoods, live off-farm produce produced through rain-fed agriculture that suffers any minute changes in rainfall patterns, that has been known to be erratic in most parts of the country. This is the reason why they are easily affected by meteorological disasters and also the reason why it matters to address issue of social vulnerability and economic recovery with the attention and emergency they deserve so that the majority are protected from the impacts of natural disasters.

Mavhura [13] avers that the occurrence of disasters in not a new phenomenon in Zimbabwe since so many disasters can be cited in retrospect. The indelible effects of cyclone Eline in 2000 when 120 people died and 7.5 million US dollars in economic losses was experienced, Cyclone Japhet in 2003, Cyclone Idai in 2019 and a series of droughts over all those years are a clear indication that the frequency and intensity of disasters has increased over the years. Countries should therefore possess improved professional skills in disaster management planning, preparedness and response for civil protection. The level of preparedness for Zimbabwe remains on more or less the same level because of the way legislation and policies are implemented. Gonye [6] posits that communities are not even aware of the institutions responsible for Disaster Risk Reduction (DRR). Even students at institutions of higher learning also wonder who is responsible for disaster management in the country, an indication that disaster communication, sensitisation and awareness still lack in the country.

Mavhura [13] takes a swipe at the Civil Protection Act (Chapter 10: 06 of 1989), lamenting the lack of community involvement and participation in the structures. The centralisation of power and resources, unavailability of dedicated finances, focus on natural hazards than vulnerability and resilience are some of the key weaknesses of the act that make effective disaster preparedness and response a mirage, an elusive goal in Zimbabwe. On the same vein is Manyena et al. [11] who posit that Zimbabwe adopted a command-and-control model in terms of disaster legislation, that was derived from a militaristic system and largely a top-down approach where decisions and actions are taken at the top and those below the system have no mandate to act without instructions from the top. This scenario is hazardous because it is at the grassroots level where quick decisions are needed in terms of disaster response. Similarly, Chatiza [5] believes the legislation has weaknesses in terms of inclusiveness where women and girls are highly excluded and the systems are male dominated, thus, decisions and problems affecting women and girls are the last to be given attention to at any given time. In addition, institutions are more reactive than proactive and this eliminates the concept of disaster mitigation and preparedness.

Chanza et al. [4] emphasises that Zimbabwe is prone to much disasters and the experiences from recent disasters are still fresh on people's minds. While it is essential to get external help in disaster response and evacuation systems, the scholars argue that local disaster management regimes should be strengthened through training, financial capacitation and availability of emergency equipment and plans because of their pragmatic essence at the onset of the disaster when no one else is there to help. From the experiences of Cyclone Idai, it is the local heroes who actually saved and the external help came in later to assist in the search for bodies. These arguments reveal that time is critical in the disaster management cycle, thus, any local community should be trained to be responsive and knowledgeable to options and priorities available in case of disasters, the scholars lament the way that outside helpers may cause division and discrimination that was never there before as communities inclusively helped each other in a spirit of family hood. This was identified in the way resources were reportedly distributed on partisan lines and other discriminatory methods while the communities themselves had already managed to help each other indiscriminately in times of real need. This is real help offered on a brotherhood, sisterhood and family hood concept as opposed to outside help that, in some situations, can only be offered when the media is present to broadcast for popularity and fame.

In addition, Zimbabwe has no designated evacuation centres, and the construction industry in rural areas is hardly regulated, exposing the rural populace to residing in poor, weak disaster-prone accommodation. According to Munsaka et al. [16], Cyclone Idai exposed policy and capacity gaps in the country as well as deficiencies in risk management systems at local and national levels. This lack of preparedness was noted to have exacerbated community vulnerabilities across the whole country. In this view, World Bank [24] recommends a Zimbabwe Recovery and Resilience Framework so that a continuous mitigation and preparedness plan is always in place and funding to be provided by the bank. Mlevu [15], posits that the Civil Protection Act (Chapter 10:06) does not offer women protection from abuse, thus, the critical needs of women during such difficult times are never considered during disaster response. Ndlovu [17] laments the way public funds are being used and recommends that Government should increase transparency and accountability that helps to instil confidence in the people.

4 Methodology

The study employed desktop research, interviews, events observation, analysis and documentation to examine the way the Government of Zimbabwe responded to the two disasters. Thirty (30) participants were purposively selected and interviewed during the COVID-19 era while an additional 15 participants participated in focus group discussions meant to validate the responses obtained from the interviews. The results were analysed qualitatively using thematic analysis. The results are presented in Tables 1 and 2, with extracts from the interviewees shown. All participants were adults above the age of 18 years who consented to take part in the study with full knowledge of the purpose of the study results.

Table 1 Responses from focus groups

Group	Cyclone Idai	COVID-19	Themes emerging
1-(5 ppl)	– Nation caught unaware – No immediate response to save lives – Lack of Communication	– Nation caught unaware – Could not handle the pandemic – The first victims died due to lack of equipment and preparedness	– State of preparedness and readiness still need to be improved. The country is in a reactive than proactive mode
2-(5ppl)	– Delay in reaction leading to loss of lives – Items meant for the poor stolen – Poor communication channels – No taskforce set	– Responsible ministry was overwhelmed – Lack of protective equipment for health staff – Nation reacted by setting up SIs to guide people. Task force set – More communication channels set up for information sharing on media	– An improvement noted communication and information sharing as compared to Cyclone Idai period. Responsible Ministries did their part although resources were lacking
3-(5ppl)	– Victims remained in tents up to now – Misuse of donations by officials	– Misuse of Government resources by leaders – Exclusion of the poor majority from life-saving machinery. Only those known by leaders benefited from life machinery	– Disasters and pandemics still used as opportunities to grab resources meant for the poor – Recovery and rehabilitation of victims still poor

Source Authors (August 2022)

Table 2 Statutory instruments for COVID-19 2020

Date	Instrument	Description
23 March 2020	SI 76/2020	Declaration of state of disaster. COVID-19
24 March 2020	SI 77/2020	COVID-19 prevention, containment and treatment regulations
27 March	SI 82/2020	Further COVID-19 measures
28 March	SI 83/2020	National lockdown order
29 March	SI 86/2020	National lockdown amendment No 1 ‘essential services’
14 April	SI 88/2020	Rebate of customs duty on goods imported to fight COVID-19
19 April	SI 92/2020	Mandatory Standards for COVID-19 PPE and Disinfectants
21 April	SI 94/2020	National Lockdown Measure Amendment No 4, Cross Border Goods Vehicles

Source Facts adapted by Authors (2022) from Ndlovu ([17]: 396–398)

5 Results

Responses from Table 1 show that Zimbabwe has remained in a reactive than proactive state in terms of disaster management in the period between Cyclone Idai and COVID-19.

There was a general increase in response from the government as evidenced by the setting up of a COVID-19 taskforce that is giving periodic reports on the state of affairs to the public on national media. To give each update on the state of affairs of the disease in the country, a move highly commended and in line with disaster management. Observations made by the study during the data gathering period reveal that many meetings and report backs were being done by the task force and other key stakeholders. However, such structures should always be in existence if one follows the disaster management cycle where disaster management is a cyclic on-going process that should never be broken.

There was an increase in communication channels on print and non-print media as of July 2020. Radio, television, social media, newspapers, billboards, private organisations, posters and so on had taken an active critical role in raising levels of awareness and sensitisation better than was done during the Cyclone Idai pandemic.

Statutory Instruments (SIs) were established to deal with the pandemic at hand, an indication that the Government had a better response. Table 2 shows some of the SIs established to curb the spread of COVID-19 pandemic in Zimbabwe.

The instruments shown in Table 2 are all efforts by the Government of Zimbabwe to curb the spread of COVID-19 through legislation and this shows much commitment, concern and care for the citizens during life threatening pandemics.

Interviews with participants revealed that the state of disaster preparedness, response and readiness is still problematic in the country. One key informant asked to comment on Zimbabwe's readiness and preparedness to disasters and pandemics had this to say:

> There is a lot of corruption in Zimbabwe and the authorities have to very strict on offenders. It was pathetic to notice that some rich people even stole groceries meant for the people affected by Cyclone Idai. The same thing happened during COVID-19 when well renowned officials could fake travel documents and steal public funds meant to benefit the poor and vulnerable.

This response is an indication that there was no improvement or stringent measures to curb corruption, abuse of funds and resources meant for the less privileged and this is very unfortunate. During Cyclone Idai, Government officials and law enforcement agents were reported as stealing cash and material resources meant for the cyclone victims (newzimbabawe.com.10 April 2019, The Herald 9April 2019, Daily News August 1 2019, Newsday 22 May 2019) and the same reports are appearing during the COVID-19 pandemic while the frontline staff and the vulnerable people are suffering and ailing in poverty. Some politicians and well-connected people are profiteering by supplying COVID-19 Protective clothing and Equipment at exorbitant prices so that they can be paid the national aid money meant to help more vulnerable people. While all this resource grabbing and abuse was reportedly taking place, the health practitioners from doctors to nurse aids were also showing their cry out for protective clothing, medication and equipment for COVID-19 patients and the already suffering patients from other diseases. This shows a lack of transparency, accountability, monitoring and evaluation in the whole Disaster Management system in Zimbabwe.

Another participant in the interviews lamented the way poor people remain the most affected in both scenarios without much attention to help them as vulnerable groups.

> We expected the Government to help poor people with food humpers, cash or other basic needs but nothing came during COVID-19. During Cyclone Idai, what was collected for the needy ended up in the hands of rich thieves, which is an indication that no cares about the poor

Although the COVID-19 pandemic affected everyone across the world without discrimination, the issue of endemic poverty remains outstanding and a discriminating factor, just like in the Cyclone Idai disaster. The poor, who included women, children, the disabled, the elderly and other vulnerable communities with unsustainable livelihoods specifically in the rural areas and the informal sector in urban areas suffered more because they did not have savings or fall-back plans for food security during the business and movement restrictions period from March 30 2020.

6 Discussion

The state of disaster preparedness and response in Zimbabwe still leaves a lot to be desired. The period between Cyclone Idai (March 2019) and COVID-19 outbreak (March 2020) have seen no increase in the state of preparedness. The observations made by the article and responses from key informants to the article concur with Mavhura [13] who posits that the country's state of disasters is reactive than proactive. Institutions only start to be seen after a disaster strikes. The Government, The Ministry of local Government Rural and Urban Planning, the Ministry of health and other line ministries with crucial roles in disaster management in the country were caught unprepared as evidenced by lack of resources at Wilkins Hospital, a referral hospital for infectious disease. The hospital could not respond and treat its first cases of COVID-19-a great lack of preparedness-leading to the first death in the country, of case number 2 in the same hospital on March 23 2020.

Most of the institutions responsible for disaster mitigation and preparedness were not ready, but the Government improved through setting up of taskforces, communication channels as compared to the level they were during the Cyclone Idai disaster, an indication that no learning and improvement is taking place in the country in terms of disaster preparedness. This shows that Disaster Management is not a priority in the development policies of the nation. The disaster management cycle still has to be adhered to with the attention and priority it deserves so that improvements can be seen through the passage of time. The period between disasters/pandemics should be used by any country to increase education, training, awareness and sensitisation campaigns so that communities, institutions and individuals are ready to respond whenever an epidemic or disaster occurs.

The responsible ministries should be given full mandate of disaster management so that centralisation of all disaster management action is reduced. In as much as a multi-stakeholder approach is called for when a nation faces disasters, the responsible institutions should be given full mandate to exercise full power of all disaster preparedness activities. The problem in the Disaster management system of Zimbabwe is that there is generally lack of resources committed to disasters, thus, institutions find it hard to plan and implement preventive schemes all the time. This has cost the country in terms of loss of life and infrastructure, as can be measured by the state of affairs in the period between Cyclone Idai and COVID-19 pandemic. However, there has been a significant increase in levels of communication on national media and social media and this is highly recommended. All critical channels of communication including traditional systems in rural areas have all been equipped with information on the COVID-19 pandemic that has been shared continuously since March 2020 when the disease had its first case in the country. All concerned citizens of Zimbabwe knew the times when such critical updates from the taskforces and advertisements were being run on national television. This made some people aware and ready to learn and receive information concerning the pandemic. However, the challenge remains that only a few languages are used in most of these communication channels, while the constitution of Zimbabwe (Chapter 1: 6(1) states that Zimbabwe has 16 officially recognised languages, thus other minority groups of people are being excluded. It is also sad to note that it's not everyone in Zimbabwe with access to information, thus, those in marginalised areas are still excluded from the vital lifesaving information, a situation that has to be addressed to save lives and infrastructure from destruction.

The setting up of a COVID-19 taskforce at national level that has been cascaded to lower- level structures have increased interconnectedness in dealing with the disease that is very effective. One can only hope that such systems stay in place even after COVID-19 so that training meetings, sensitisation workshops and campaigns, including resource mobilisation, can be implemented to improve the state of preparedness and response in case of future disasters. In addition, the nation made the response to COVID-19 mandatory by establishing so many Statutory Instruments, a move that helped the nation in compliance with the rules and regulations.

When something is made law, people are more likely to comply than when a section or a department is giving out information without the support of the government. People may take it for granted, thus, increasing hazards and losses of life and infrastructure. Such a move should have been done during Cyclone Idai so that people respond to the call to move out of danger within the specified time. Sometimes issues to do with disasters should not be left to the affected people to make own decisions at their own pace and time. An example can be sited of the period before Cyclone Idai when people in the Chimanamani area in Manicaland province were just encouraged to move out of the way of the cyclone, without actually being forced to comply by the Government.

Since the Government is the custodian of the rights of all the people, establishing laws and regulations should be matched with enforcing them within the specified period so that the country does not continue to lose lives due to situations that can be averted. As stated earlier, disasters cannot be prevented or stopped but nations can reduce their impacts through proper preparedness and response. It takes much time and effort to reach a good state of preparedness at national level, more so at community level where different beliefs and attitudes may hinder acceptance of information on early warning signs and systems as explained in the Weberian theories of disaster management. This means much continued effort in communication, education, sensitisation and awareness should be the order of the day for disaster management institutions without waiting to appear only when a disaster strikes, that is the norm in Zimbabwe.

Corruption has always been found to be the worst enemy of development worldwide. In Africa and Zimbabwe, in particular, it remains pathetic to find out that high profiled people continue to be found wanting when dealing with aid money and resources. During both disasters, some people have always been found on the wrong side of the law, an indication of selfishness, shamelessness, and lack of prudence, professionalism, accountability, transparency, monitoring and evaluation. It is a shame for the country to be found on the list of countries with the most corrupt officials in terms of handling aid money every time there is a disaster. One may be tempted to think that such corrupt officials may be praying for such disastrous events to occur more often so that they find an opportunity to enrich themselves with resources meant for the most vulnerable members of society. Tolerating such behaviours and allowing such people to continue serving the people in public offices is a scandal in itself. Some of these things may be caused by the failure to separate politics from humanitarian activities and development. With proper separation of power and mandate, politicians may not find an opportunity to hijack humanitarian activities from the ministry or department given mandate to implement disaster management issues. If one has never done any meaningful development activity in their constituency before, why should they be given an opportunity to misuse aid money to gain political mileage at the expense of the poor people? Even the international community may never trust such a nation, thus, there is great need for Zimbabwe to increase security systems and accountability in the way public funds are spent so that confidence can be instilled in the people at all levels.

The state of preparedness and levels of response to disasters in Zimbabwe continue be affected by the rate of endemic poverty in the country. As indicated by the Marxist perspective of disaster management, poor people are always the worst affected in any disaster. During the Cyclone Idai disaster, the poor farmers in the rural areas were the most affected and they are still struggling to recover from the impacts of that disaster more than a year later. If people were better placed in terms of livelihoods and savings, they could have recovered from the impacts of Cyclone Idai a long time ago. OCHA [18] and Reliefweb [21] show that 198 families displaced by Cyclone Idai are still in tents, making them vulnerable to cold, unhygienic conditions and food shortages during the COVID-19 period. This shows lack of recovery and reconstruction efforts and abilities in the country.4.3 million people are food insecure in rural Zimbabwe while 1.1 million women and children need nutrition assistance. At the same time 1.2 million school age children are failing to access education because they do not have access to internet or at least the phones that access social media, and the data to join other more privileged children accessing online classes. These facts compound the already existing inequality and, like the Marxist point of view in disaster management, the poor are the hardest hit, always side-lined and left to suffer without anyone caring for them. This shows that Zimbabwe is still struggling to cope with disasters and pandemics, a situation which can be improved through a multistakeholder approach because the Government alone cannot manage disaster preparedness, readiness and response.

This is a critical area where the leaders should give priority to reducing poverty and hunger as indicated by the SDG 1 and 2. More programmes and policies targeted for the poor should be increased at household and community level so that the people do not rely on aid but become sustainable in terms of food security. Even during the COVID-19 pandemic, the poor people were affected more across the country due to lack of savings and fall-back plans during such disasters. When the disease remained statistical game for other people who have more resources, the poor, the elderly, and the disabled experienced nightmares in terms of feeding their families during the national lockdown periods experienced in the country from March 30 2020. Informal traders have insecure livelihoods and they leave from hand to mouth. This meant that they could not continue in lockdown without food and most people from such backgrounds were in need of food aid. This was evidenced by the way informal traders in Mbare Musika overcrowded the open spaces with or without face masks and taking very little heed of the social distancing requirements stated by the Ministry of Health of the Statutory Instruments concerning prevention, containment of COVID-19.

Disaster management can therefore be best practised if priority is given to eliminating poverty and creating an enabling environment for job security, food security and sustainable livelihoods for people in rural and urban areas. Such communities will be found resilient during and after disasters. The nation should take a leading hand in ensuring that the majority for the people are able to respond to disasters at household, community and national levels giving priority to grassroots structures. The fact that some of the victims of Cyclone Idai are still in tents is an indication of stagnation and failure to follow the disaster management cycle that emphasises reconstruction, recovery and healing after the occurrence of a disaster. This means a second disaster (COVID-19) is catching up with them while they are still suffering from the impacts of another one that occurred a year ago. Sustainability, improvement and empowerment are never in the lives of such people and a great violation of their human rights.

7 Conclusions and Recommendations

This article concludes that the state of disaster management in Zimbabwe is still in its infancy. An analysis of the period between Cyclone Idai and COVID-19 outbreak in the country shows that no significant improvement occurred in terms of disaster preparedness, mitigation or response levels, thus, past events were never used as learning points in the disaster management life of the nation. The institutions concerned with disaster management are incapacitated due to lack of finances committed to disaster management thus, research, education planning and implementation of policies is not always possible. High levels of poverty in the nation expose the majority of people to bear the worst effects of any disaster and taking more time to recover as compared to richer communities and families. While a rise has been seen in the setting up of task forces, SI instruments and communication channels during the COVID-19 era, such trends should be maintained (free from corruption) and remain continuous with guidelines from the disaster management cycle and other international disaster management guidelines line the Hyogo Framework of action (2015–2030) and the 17 sustainable development goals (SGDs). The article therefore recommends that:

- Women and girls, all youths and university students (who are idle now) should be given the lead in disaster management so that they learn and practice what they are being taught at school, in a mentoring style with close monitoring and evaluation. This should be done in a bid to separate politics and disaster management and opening up forums for a multi stakeholder approach.
- Continuous research should be prioritised in national budgets and be given to all private and public institutions of higher learning. When students research and make recommendations, they should be given a chance to implement their ideas.
- The nation should take lessons from other countries regionally and internationally to improve on the state of disaster management. Policies should be more proactive than reactive so that disaster management is treated as a continuous process not an event.
- Stiff penalties should be given to corrupt officials misusing disaster response funds.
- Local Disaster Management regimes (IKS) should be strengthened financially and materially not ignored since the local people are the first to render assistance to victims and they are the ones who actually save lives.

References

1. Alfredo D (2017) Disaster management theories. Available online: https://bizfluent.com/13364656/disaster-management-theories. Accessed on 27 May 2020
2. Betera L (2011) Overview of disaster risk management and vulnerability. Available online: https://www.itu.int/en/ITU-D/Emergency-Telecommunications/Documents/Zambabwe_2011/Overview%20of%20disaster%20management%20and%20vulnerability%20-%20Civil%20Protection%20Unit.pdf. Accessed on 25 May 2020

3. Bhavnani R, Owor M, Vordzorgbe S, Bousaret F (2008) Report on the status of disaster risk reduction in the Sub-Saharan African Region. Available online: http://www.gfdrr.org/sites/gfdrr/files/publication/AFR.pdf. Accessed on 23 Aug 2020
4. Chanza N, Siyangwana PO, Williams-Bruinders L, Gundu-Jakarasi V, Mudavanhu C, Sithole VB, Manyani A (2020) Closing the gaps in disaster management and response: drawing on local experiences with cyclone Idai Chimanimani, Zimbabwe. Int J Disaster Risk Sci 31:37–44
5. Chatiza K (2019) Cyclone Idai in Zimbabwe: an analysis of policy implications for post disaster institutional development to strengthen disaster risk management. Available online. https://policy-practice.oxfam.org/resources/cyclone-idai-in-zimbabwe-an-analysis-of-policy-620892/. Accessed on 18 Jan 2021
6. Gonye N (2014) The civil protection unit's preparedness and its role in disaster management: the case of Nyanga District. Available online.ir.msu.ac.zw.>xmlui>handle. https://www.semanticscholar.org/paper/The-civil-protection-unit's-preparedness-and-its-in-Gonye/99e21d0bcc96691812c3af0943e66dacf0075d22. Accessed on 18 Jan 2021
7. Government of Liberia (2012) National disaster management policy. UNDRR. PreventionWeb.Preventionweb.net/English/policies. Available online. https://www.preventionweb.net/english/policies/index.php?o=ent_title&o2=ASC&cid=98 Accessed on 26 May 2020
8. Government of Zimbabwe and United Nations (2017) Capacity assessment of the disaster risk management system in Zimbabwe. Cadri.net/sites.files.report-may-2017. Available online. https://www.cadri.net/sites/default/files/Zimbabwe-Report-May-2017.pdf. Accessed 30 May 2020
9. Holgren TE, Moe AK (2012) Disaster risk reduction in Liberia. University of Stravanger. Available on line https://www.gdnonline.org/resources/Holmgren%20T%20E%20%20%20Moe%20A%20K%20%20(1).pdf. Accessed on 23 July 2020
10. Khan H, Vasilescu LG, Khan A (2008) Disaster management cycle-a theoretical approach. J Manage Market 6(1):43–50
11. Manyena SB, Mavhura E, Muzenda C, Mabaso E (2013) Disaster risk reduction legislations: Is there a move from events to processes? Glob Environ Chang 23(6):1786–1794
12. Mhlanga C, Muzingili T, Mpambela M (2019) Natural disasters in Zimbabwe: the primer for social work intervention. Afr J Soc Work 9(1):46–54
13. Mavhura E (2016) Disaster legislation: a critical review of the civil protection act of Zimbabwe. Nat Hazards 80(1):605–621
14. Mavhura E (2017) Chapter 24: disaster risk reduction policy and management in Zimbabwe. Handbook of Disaster Risk Reduction & Management. Available online. https://www.worldscientific.com/doi/https://doi.org/10.1142/9789813207950_0024. Accessed on 19 June 2020
15. Mlevu S (2020) Women hard hit by COVID-19. Available online https://www.cnrgzim.org/women-hard-hit-by-covid-19. Accessed on 10 Jan 2021
16. Munsaka E, Mudavanhu C, Sakala L, Manjeru P (2021) When disaster risk management systems fail: the case of cyclone Idai in Chimanimani Disaster Chimani mani Zimbabwe. Int J Disaster Risk Sci. Springer Link. Available online. https://www.researchgate.net/publication/355194985
17. Ndlovu L (2020) Government must account for Resources donated to fight Coronavirus' Available online. https://cite.org.zw/govt-must-account-for-resources-donated-to-fight-coronavirus. Accessed on 10 Jan 2021
18. OCHA (2020) Zimbabwe situation analysis report. Available online. https://reports.unocha.org/en/country/zimbabwe/. Accessed 18 Jan 2021
19. Ravoala R (2014) Improving disaster preparedness in madagascar. Available online. https://www.google.com/search?newwindow=1&sxsrf=ALeKk00amoQ6EnjTRstShJ_Pqd3Q8Npghw:1610964187911&q=Rapala,+R.+(2014)+Improving+Disaster+Preparedness+in+Madagascar.+Available+online.+Reliefweb+Int%3Ereport/improving-disaster-preparedness-Madagascar.&spell=1&sa=X&ved=2ahUKEwiX-qmhnaXuAhURohQKHbwIDToQBSgAegQIBBAt. Accessed 20 May 2020
20. Reliefweb (2019) Zimbabwe Cyclone Idai Response: Situation Report#2.IOM.unOCHA Services. Available online. https://reliefweb.int/report/zimbabwe/zimbabwe-cyclone-idai-response-situation-report-27-24-30-april-2019www.unocha.org. Accessed 10 May 2020

21. Reliefweb (2020) COVID-19 Preparedness and response: stranded and vulnerable. Zimbabwean migrants returning from within South African Region. Issue 02. Available online. https://reliefweb.int/report/zimbabwe/covid-19-preparedness-and-response-stranded-and-vulnerable-zimbabwean-migrants. Accessed on 30 June 2020
22. Shaban AR (2020) COVID Organics. Madagascar launces Africa's first cure for virus. Africanews.com/2020. Available online https://www.africanews.com/2020/04/22/covid-organics-madagascar-launches-africa-s-first-cure-for-virus//. Accessed on 15 May 2020
23. Shamim AB (2016) Theoretical approach to disaster management. Research gate. Net/ publication. Available online. https://www.google.com/search?newwindow=1&sxsrf=ALeKk03QdahIUqPub-ooVXf9CA2FBqlmkw:1610961896620&source=univ&tbm=isch&q=Shamim,+A.+B.+(2016)+Theoretical+Approach+to+Disaster+Management.+research+gate.+net/publication.&sa=X&ved=2ahUKEwjhxODclKXuAhV-ZxUIHdphDWEQjJkEegQIAxAB. Accessed on 18 Jan 2021
24. World Bank (2019) Restoring Zimbabwe's livelihoods, infrastructure after Cyclone Idai. World Bank. Available online https://www.worldbank.org/en/news/feature/2019/09/16/restoring-zimbabwes-livelihoods-infrastructure-after-cyclone-idai. Accessed on 18 Jan 2021

COVID-19 Insights on Geriatric Care Systems' Vulnerabilities and Utility

Felistas R. Zimano

Abstract The study investigates the usefulness of existing systems for looking after the elderly people in different societies. The two systems identified are the care home and the home care system. Wisdom shared in this chapter is mainly informed by understandings as exposed during the COVID-19 pandemic. The study digs into literature and follows a comparative approach with the predominantly Western world care-home system on one side whilst the home-care system as widely used in African societies on the other hand. Findings show care-home systems as more vulnerable to the effects of contagious diseases like COVID-19. However, the care-home systems can also work well if well quarantined as opposed to the home care-systems.

Keywords Care-homes · Home-care systems · COVID-19 · Geriatric care

1 Introduction

The Coronavirus of 2019 (COVID-19) emerged as a highly contagious respiratory infection with a high mortality rate. The World Health Organisation (WHO) declared it a world pandemic due to the rate at which it was spreading and the challenges obtaining in containing it. This prompted some questions giving impetus to this chapter over the possible challenges or prospects encountered in the caring for the aged people in various parts of the world. As such, two main geriatric care systems from developed and developing countries are unpacked herein. These are care-homes and home-care systems to geriatric care. Care-homes for the elderly and vulnerable are common in developed countries. Care-homes are also dotted in developing countries but still they are mostly used by white Europeans and Americans. Whilst the arrangement of old age 'homes' is a standard part of life in the developed countries, this is not the same in most parts of Africa [13]. The utility of various geriatric care systems comes under spotlight due to the mortality patterns brought to the fore by COVID-19. The COVID-19 pandemic has exposed gaps in geriatric care systems

F. R. Zimano (✉)
Great Zimbabwe University, Masvingo, Zimbabwe
e-mail: zimanof941@gmail.com

R. Kasimba and I. Chirisa (eds.), *Disaster Risks and Hazards*, Advances in 21st Century Human Settlements, https://doi.org/10.1007/978-981-96-4577-0_11

[16]. 'The virus affects everyone indistinctly, but its lethality and gravity take a very significant logarithmic curve if people are aged and suffer multiple pathologies' [30: 1]. True to the word, there have been high case-fatalities among older adults in care-homes during the COVID-19 pandemic. This exposed some inadequacies of this geriatric care method. A question to ponder is: Is it wise to congest people vulnerable to the wrath of contagious pandemics in one place? Is it wrong to conclude that care-homes are potential drivers of contagion? Some car-homes lost more than fifty percent of residents to the pandemic. Care-home workers were equally ravaged. This warrants the need to interrogate the institution of care-homes alongside alternative geriatric care systems used in other parts of the world.

Of concern and imperative for comparative reasons, is the scant utilisation of care-homes by people of African origin and the subsequent less severity of COVID-19 in the African elderly. In Zimbabwe, for example the first COVID-19 death was on the 23rd March 2020. In the subsequent month up to the 23rd of April 2020 the total number of deaths increased to four. According to the Zimbabwe Ministry of Health and Childcare, two of these were old aged individuals: the first—79-year-old male succumbed on the 3rd of April in Bulawayo. The fourth COVID-19 death on the 22nd of April was another elderly—82-year-old female from rural Mashonaland central province case [17, 18]. The two individuals were clinically identified as cases number 11 and 27 respectively. These two cases separated by nineteen days in far removed places paint a different picture in COVID-19 progression in geriatric population between developed and developing countries. Zimbabwe's case 11 was a care-home resident whilst case 27 lived in traditional family home-care systems. In developed countries the virus progressed very fast in the elderly members of the society. In one weekend, Stanley Park Care-Home in Italy lost thirteen residents, in Milan another care-home lost 190 residents whilst a care-home in Brooklyn in New York lost 55 residents [1, 3, 9, 30]. These stark differences between geriatric COVID-19 deaths in developed and developing countries expose some strengths and weaknesses in the different care systems employed in various parts of the world.

To show geriatric care systems strengths and deficiencies, the chapter also unpacks the underlying reasons for low utilisation of care-homes by people of African descent. Key among the reasons are economic and socio-cultural considerations. As statistics above show, the failure to embrace the care-home system seem to have worked for the good of African geriatrics. The spread of COVID-19 among African elderly was evidently not as fast, spontaneous and wrecking as happened to their European and American counterparts in care homes. This validates the proposition that the African indigenous systems of keeping the vulnerable in family home-care reduces exposing a particular group in society to the severity of a pandemic. Family home-care arrangement thinly spreads out the vulnerable population. The COVID-19 pandemic exposed care-homes as possible drivers of and weak points during contagious pandemics [1, 3, 9]. This is likely to leave a lasting dent on the stature of care-homes that instead of being places to be for convenience, they may recede into places to be only due to the absence of options. Even people in the developed countries who are typically in favour of care-home systems were sickened by the mortality patterns. Some sections of this society even called for class action against care-homes calling them with unforgiving names ranging from 'massacre camps', 'ghost-towns', 'hell', 'heart-breaking' to 'surreal' [1, 9, 30]. In making geriatric care systems' comparison, this chapter responds to calls for interventions that can be implemented to save humanity.

The chapter will show the economic and socio-cultural factors behind the low movement of African aged into care-homes. The chapter also unpacks the sociodemographic changes in developed countries that make care systems highly untenable. On the downside, the chapter appreciates that the continued stay of the aged in family home-care set-up come with its burdens. However, with hindsight from the COVID-19 pandemic, the chapter argues that the home-care model works for the good as Africa's senior citizens were relatively spared whilst their counterparts from developed world were ruthlessly ravaged. In terms of perception, the effects of COVID-19 on care-homes is not mere 'passing wind' but 'a game changer'. Facilities may never be revered as much as before. Care-homes' future is likely to fall from glory into a place only to be out of desperation.

2 Care-Home Set-Up

Care-homes are gated communities providing a range of services. Admission to such facilities is voluntary or by consent from parent or guardian. Care-homes are sometimes referred to as residential homes or nursing homes depending on the services on offer. Residential homes normally offer 24 h personal support [11]. In addition, nursing homes offer nursing services (ibid.). A person of any age can stay in a care home depending on needs [4]. However, this chapter is only committed to those care-homes dealing with older people.

3 The Convenience of Care-Homes

Care-homes are institutions that provide services for the care of vulnerable people due to their old age, chronic ailments or a combination of both [20]. They have a number of variations and prestige. A common phenomenon for all is that one leaves their family home and moves into an institutionalised care system. In terms of definition for old age, according to the World Health Organisation (online), developed countries seem to have accepted 65 years as the mark in line with onset of receiving pension benefits whilst for developing countries it is not really cast in stone.

To these elderly citizens, their families and society, care-homes offer much convenience. As people age, their health care requirements become more pertinent and frequent. Most people in their advanced ages have multiple chronic conditions that require more specialised health care [23, 27]. This may be burdensome for those who have to look after the elderly whilst having to go on with their other occupational and ordinary responsibilities. In addition, costs especially in terms of time to transport the elderly to seek health care from time to time can become strenuous. These challenges in accessing specialist health care as and when necessary is a major barrier in health care [14]. This results in most people in family home-care set-up risking foregoing health check-ups more often. Care-homes come with an advantage in managing the movement of older people as most have arrangements for health checks within the homes. This becomes useful because most elderly suffer from mobility issues. The value of having medical checks without having to wait for long cannot be overstated [12, 14]. Good as they seem, do these advantages outweigh the disadvantages of care homes?

4 Care-Homes—The Other Side

The COVID-19 pandemic illuminated the place of care-homes across the world where thousands of elderly residents, who are most vulnerable to the disease, are believed to have died after contracting it [1]. Even long before the COVID-19 outbreak, there are a number of people who held reservations over care-home systems. Some of their reasons are informed by economic, socio-cultural, and medical factors. Although care-homes are very popular in most developed countries, studies show that some people only move into homes out of desperation. Others do not want the thought of moving into a care-home at all. This is because taking up life in a care-home comes with a raft of changes in one's life. That is why for a good number of the elderly, the transition from their private lives into care facilities is a challenge [15]. The combination of staying in a care-home with its associated restrictions and aging alter the residents' identities [22]. Restrictions associated with care-homes are likely to jeopardize individual autonomy and push individuals into forced redefinition of life routines (ibid.). Such things as leisure time, experience with nature, family and issues of spirituality are some of the important life meaning-giving areas that get distorted by the movement into care-homes (ibid.). Matters of spirituality give a person the inner sense of being in communion with God [21]. As such, anything that affects one's spirituality may affects one's mental stability.

Studies have also shown that it is rare for care-homes' residents to be taken out of the homes for enjoyment whilst issues of neglect have been reported as frequent [7]. Cases of neglect have been widely reported in media. In some care-homes, patients who are sick and those who not are mixed up, share rooms and almost always, share dining facilities [9]. In times of communicable diseases pandemics like the COVID-19 such neglect exposes all residents to infection.

Another challenge common in care-homes is insomnia. World Health Organisation (online) defines insomnia as 'taking longer than 30 min to fall asleep, awakenings during the night totalling more than 30 min of wakefulness, and less than 6.5 h of total sleep'. Life in care-homes can contribute to increased risk of sleep disorders [8]. This issue of sleep disorder is worrisome because sleep is essential for the general health of a person due to is physiological functions (ibid.) Sleep disorders can be a direct consequence of high depression, limited occupational activities and minimal or nil exposure to sunlight [24].

All these and more care-home challenges can account to rife depression and suicidal tendencies in care-homes. The incident of depression and suicidal depression in care-homes points to other underlying challenges. There are wide ranging medical and social interventions to deal with depression. However, the problem of depression remains rife in care-homes despite the uptake of antidepressants and the development of intervention programs [6]. The reported cases of depression and suicidal tendencies within care-home systems are higher as opposed to those of the same groups in community setup [10]. Accordingly, calls have been made for more spiritual and psychosocial help for people in institutionalised care systems in order to improve their feelings of security [15].

There are several characteristics that influence one's settling into any environment that also affect individuals as they move into care-homes:

Psychological factors

- *Sense of Acknowledgement*
- *The preservation of one's habits and values*
- *Autonomy*
- *Control, and*
- *Coping*

Social factors

- *Interaction and relationship with staff, residents' family and friends, and pets.*

The built environment

- *Private and (quasi-) public space*
- *Personal belongings*
- *Technology*
- *Look and feel*
- *the outdoors and location*

Adapted from: [25]

The above summarised factors influence one's sense of being home. Challenges in any of the above normally result in one failing to settle in smoothly. That is the scenario that most aged find themselves in as they move into care-homes which may account for most care-home residents having psychological disorders (ibid.). The psychological state of care-home residents seems to have compromised the safety and precautions adopted during COVID-19 outbreak. In some care-homes, care-home staff were not allowed to wear face masks as this was deemed as inappropriate. Face masks were discouraged as they would cause fear and unrest within care-home residents [9]. This speaks of a psychologically volatile group of people that demands extra caution to avoid slightest triggers. However, the caution adopted to avoid disturbing the elderly in care-homes worked against the most basic first-line measures in containing the spread of COVID-19—the use of face masks.

Usually, the nature of care-homes also have the challenge of striking a balance between indoor air quality and patients' warmth and security. This has always been a disadvantage of care-homes and it worsens the COVID-19 situation. According to [31: 1].

> Respiratory infections can be transmitted through droplets of different sizes... Droplet transmission occurs when a person is in close contact (within 1 m) with someone who has respiratory symptoms (e.g., coughing or sneezing) and is therefore at risk of having his/her mucosae (mouth and nose) or conjunctiva (eyes) exposed to potentially infective respiratory droplets. Transmission may also occur through fomites in the immediate environment around the infected person. Therefore, transmission of the COVID-19 virus can occur by direct contact with infected people and indirect contact with surfaces in the immediate environment or with objects used on the infected person (e.g., stethoscope or thermometer).

As such, whilst trying to deliver on a residents' warmth and security, care-homes may unconsciously overlook the need to prioritise air quality within their holding facilities. '[Elderly home residents] are incredibly vulnerable because they are within a situation that we would call a super spreading environment: Being kept inside an age care facility, often sharing rooms and having common dining rooms' [1: 1]. Air humidity levels are usually high in congested homes giving a fertile breeding environment for bacteria. Poor air circulation also means fresh air quickly gets used up and people end up inhaling poor quality air. This deprives the body of sufficient oxygen supply. In some countries, the limited availability of beds at hospitals resulted in them opting to continue caring for elderly COVID-19 patients within care-homes. Coupled with challenges of air quality, this most likely fuelled the spread of the virus in care-homes. 'Hosting COVID-19 patients in nursing homes was like lighting a match in a hay stack' [30: 1]. This arrangement was reportedly implemented in Italy.

Having this array of disadvantages, why do care-homes remain popular in European and American communities whilst alien to Africa? In South Africa for example, a study with the LGBTI community revealed that most indicated that they would opt to die rather than move into a care-home [13]. The answer maybe in the socialisation and economic systems of the African society as will be exposed forthwith.

5 The African Geriatrics' Care Systems

The Afrocentric style of taking care of the aged in family home-care set-up is informed by several reasons. Africans revolve around Ubuntu (*I am because we are*). Ubuntu is the unseen part of the Africans inner persons. In African society, the task of caring for the vulnerable in the society rests on the society, which is not typically the case in European and American systems. 'Through its emphasis on humanity, compassion, and social responsibility' African systems form a bond between the old and the young [26: 2]. Both groups feel indebted to take up certain responsibilities on behalf of the other. The relationship and responsibility allocation is cemented by the principle of reciprocity. According to [5] older people are not as irrelevant as sometimes assumed as they play a crucial role in knowledge transfer that underpins the healthy functioning of the society. This healthy intergenerational knowledge transfer keeps both generations up to date with the past and the present.

The elders also play a crucial role in the socialisation process coming in as the bridge for the young, old and ancestors by passing on a society's norms and values [29]. There are several African rituals and genealogy wisdom that is not readily documented. Such wisdom is passed from generation to generation through oral tradition [32]. Usually societies jealously guard such wisdom as sacred inheritance that is only shared within family and societal confines. This also makes it worthwhile for the elderly, as the custodians of such wisdom, to remain in the family home-care setup. Within a family system, the elderly remains accessible whenever need arises. It is prudent to also appreciate that in African tradition, there are some conversations that cannot be held through modern communication devices. Face to face dialogue is still held dear especially between the young and the elderly. In addition to giving advice and direction, the older people in African societies also play a crucial role in rituals that connects the living to the ancestral realm (ibid.). They also mediate in conflicts between the living and conflicts between the living and the dead. In essence, the elderly are pacifiers. 'In African societies this notion of intermediaries attributed to living and dead ancestors prevails. This is the stance where the socio-religious functionality of elderly people is manifested' [2: 148].

In Africa also, 'grandparents are such an important source of child-care and knowledge' [5: 1]. Most people in Africa will testify having stayed under the care of their grandparents, older aunts and/uncles. 'Within the African family historically, the elders provided care to the children who in turn provided care to them in their old age, hence vernacular Zimbabwean sayings, "*umcede, uzakunceda kusasa,*[1] *karere kagokurerawo*[2]" (Look after it and it will look after you)' [29: 1]. This is unlike the scenario in most developed societies where socio-demographic changes have displaced families. In Africa, it is normally during the early years that a child stays with their grandparents that lasting bonds and memories are created. In rural areas, the extended family system is still intact with several people from various generations staying in one place and helping in caring for the family's disabled and elderly members [28]. It is these bonds that make the act of looking after the elderly bearable. Whilst outsiders to this system can view care giving as burdensome, those who grew up in this system will submit otherwise. This is why it is common for young adults in Africa to swear that they would rather hunger themselves than have their grandparents go without. The memories of the sacrifices made by their elders as they grew up informs this thinking.

In the African culture and value system, there is also a place for proverbs that people need to appreciate in understanding African approaches to life. There are proverbs for almost every aspect of life in African wisdom. The richness of proverbs is such that they come as codified messages that takes some maturity, experience, teaching and wisdom to interpret their meanings. The following are few of the countless proverbs that shed insights into the role and status of the elderly in Africa.

---

[1] Ndebele.

[2] Shona.

6 African Sayings on the Elderly

- *The mouth of an elderly man is without teeth, but never without words of wisdom.*
- *Those who respect the elderly pave their own road toward success.*
- *A village without the elderly is like a well without water.*
- *The youths walk faster but the elderly knows the road*
- *The elders of a community are the voice of God.*
- *The words of the elders do not lock all the doors, they leave the right door open.*
- *The strength of the elderly is in the ears and on the lips.*

This chapter will not attempt to interpret each of these proverbs to avoid adulterating them. It is up to each reader to attach meaning based on own interpretation. These are some of the proverbs that African children grow up hearing from elementary stages. In proverbs, children are taught the value of wisdom from their elders; the value of elders in paving the young one's future; the value of the elderly in giving a community dignity; the value of the elderly in guiding the youth's ways, the omnipotent role of the elderly; the power of the elderly's advice; the contractual relationship between the young and the old and so forth. These teaching etch some values in African children's minds. Like the notching of a sculptor's chisel on a stone, proverbs shape the African children. This in turn influences them to hold dear their elderly such that they do not readily subscribe to care systems that take the elderly out of their families. To them, care-homes will take away not only the elderly from them but custodians of family wisdom, their family shields and mediators.

More can be said about the socio-cultural factors, but for the purpose of this chapter, those listed above should suffice. In addition, there is also the economic angle to the care systems that need to be unpacked. Unlike developed countries, most developing countries do not have robust state-driven welfare systems for the elderly. In developed countries, National Health Service (NHS) is available for the maintenance of the elderly people's health [2]. There are funding systems at the state level committed to the geriatrics. In Africa however, the state is yet to set aside funds for the care of the elderly. As such, even if people were to disregard the socio-cultural points in the foregoing paragraphs, the economic set-up in Africa gives no option but to care for the elderly in the family home-care set-up. This is a home-care set-up without a litany of luxuries like routine access to care givers and mobility aids.

7 Discussion: Both Systems Must Stay but…

The two competing care systems presented above seem inferior to people who hold different opinions to life. The COVID-19 pandemic exposed the care-home systems' vulnerability to contagious pandemics. The Afrocentric approach of caring for the aged within family home-care systems reduce the possibility of congesting vulnerable people in one place. However, both systems remain viable options in geriatric care. Since there are a myriad of sociodemographic changes that have eroded traditional family system in developed countries, there is no way anyone can recommend the return to the status quo of home-care. In Africa also, countries are still lagging behind in terms of economic development that expecting robust NHS in the immediate future is fallacy. As such, there is a need for stakeholders to take time to appreciate the positives of both systems and find ways to improve both systems. A proper care system should be one that guarantees the multidimensional Quality of Life (QOL). There are various aspects of life like one's health status, their psychological disposition and general upkeep that feed into this QOL measure [19]. This chapter submits that it is possible to adjust both geriatric care systems and reach QOL levels. It will be good for humanity if both systems are improved to better the life of both the elderly and their caretakers.

First, there is a need to work on ways to improve air quality in care-homes. Having seen the high fatalities in these facilities [16] called for robust approaches to be put in place for geriatric care. Care-homes also need to be decongested. More stringent measures need to be put in place to ensure continued surveillance of all people who get into contact with the elderly in care-homes to avoid incidents of outsiders bringing viruses into care-homes or taking viruses from care-homes into communities. COVID-19 has proved that once a contagious disease finds its way inside, it is likely to spread like veld fire as most residents will be vulnerable. It is an already known fact that 'a range of infections can spread easily in vulnerable populations who live in nursing or residential care, for example norovirus, scabies, influenza, and here, the novel coronavirus.' [3: 1]. Once the COVID-19 pandemic has gone, it is likely to leave a trail of low confidence in care-home systems. As such, it will take a raft of measures to rebuild people's confidence in this geriatric care system. Until then, the care-home systems status will be deemed a place to be out of desperation than for convenience.

In light of the COVID-19 pandemic, the Afrocentric home-care system triumph as the surest way to avoid congesting vulnerable people in one place. For most in developing countries, decisions against care-homes are not directly related to the possibility of outbreaks. However, whatever reasons, those elderly in family home-care systems were less exposed to the effects of COVID-19 than their counterparts in care-homes. Most Africans prefer staying in family home-care rather than in care-homes as they age [13]. Whilst the socio-cultural reasons behind the African approach to geriatric care are justified and in need of preservation, there are economic and progressive factors that need to be addressed. There is a need to work on improving the availability of equipment and aids for the elderly like wheelchairs to reduce the burden on the family carers.

8 Conclusion

Both care-home and home-care geriatric systems with their challenges, remain valuable. They both need improvements if they are to deliver optimum care. In light of communicable diseases, care-homes emerge as more vulnerable. There is a need to work on congestion and air quality levels in care-homes to reduce the transmissibility rates in case of communicable disease pandemics. These changes, although bringing safeguards in case of communicable diseases, they also come with their disadvantages like the furtherance of isolation that is already a leading factor in geriatrics depression and insomnia. There is a need for more robust studies to ensure that as measures recommended herein are taken on board, measures to curb their drawbacks are also activated. The family home-care system's triumph in saving older persons from communicable diseases cannot be over emphasised. However, as earlier highlighted, it comes with its challenges. There is a need to bring in several other measures to alleviate care burden on the family. Without proper management, home care systems can result in depression, not in the older persons, but the younger persons caring for the older persons as they will have to sacrifice most of their youthful time to providing these unpaid care services.

References

1. Aljazeera (2020) Coronavirus: spotlight on hard-hit elderly homes as toll emerges Authorities in US, Canada, Italy, France and Spain launch investigations into response of several nursing homes. Available at: https://www.aljazeera.com/news/2020/04/coronavirus-spotlight-hard-hit-elderly-homes-toll-emerges-200418083848929.html. Accessed 25/04/2020
2. Ayete-Nyampong S (2014) A study of pastoral care of the elderly in Africa. An interdisciplinary approach with a focus on Ghana. Bloomington, Author house
3. Booth R (2020) Half of coronavirus deaths happen in care homes, data from EU suggests. The Guardian, 13 April. Available: https://www.theguardian.com/world/2020/apr/13/half-of-coronavirus-deaths-happen-in-care-homes-data-from-eu-suggests. Accessed 20 April 2020
4. Bradshaw SA, Playford ED, Riazi A (2012) Living well in care homes: a systematic review of qualitative studies. Age Ageing 41:429–440
5. Brown RL (2016) In dramatically young Africa, elderly are a valuable resource. The Christian Science Monitor. Available: https://www.csmonitor.com/World/Africa/2016/1229/In-dramatically-young-Africa-elderly-are-a-valuable-resource. Accessed 26 April 2020
6. Chau R, Kissane DW, Davison TE (2019) Risk factors for depression in long-term care: a prospective observational cohort study. Clin Gerontol. https://doi.org/10.1080/07317115.2019.1635548
7. Cooper C, Marston L, Barber J, Livingston D, Rapaport P, Higgs P et al (2018) Do care homes deliver person-centred care? A cross-sectional survey of staff-reported abusive and positive behaviours towards residents from the MARQUE (Managing Agitation and Raising Quality of Life) English national care home survey. PLoS ONE 13(3):e0193399
8. Da Silva RM, Afonso P, Fonseca M, Teodoro T (2019) Comparing sleep quality in institutionalized and non-institutionalized elderly individuals. Aging Mental Health 17(1):1–7
9. Giuffrida A (2020) Italian police broaden inquiry into care home where 190 have reportedly died. The Guardian, 16 April. Available: https://www.theguardian.com/world/2020/apr/16/italian-police-broaden-care-home-coronavirus-milan. Accessed 20 April 2020
10. Gleeson H, Hafford-Letchfield T, Quaife M, Collins DA, Flynn A (2019) Preventing and responding to depression, self-harm, and suicide in older people living in long term care settings: a systematic review. Aging Ment Health 23(11):1467–1477
11. Goodman C, Dening T, Gordon AL, Davies SL, Meyer J, Martin FC, Gladman JRF, Bowman C, Victor C, Handley M, Gage H, Iliffe S, Zubair M (2016) Effective health care for older people living and dying in care homes: a realist review. BMC Health Services Res 16:269

12. Grant RW, Ashburner JM, Hong CS, Chang Y, Barry MJ, Atlas SJ (2011) Defining patient complexity from the primary care physician's perspective: A Cohort study. Ann Intern Med 155:797–804
13. Henderson N, Khan J (2020) "I will die if I have to go into an old age home": afrocentric options for care of older LGBT people in South Africa. Agenda. Available: https://doi.org/10.1080/10130950.2020.1728064. Accessed 10 April 2020
14. Liddy C, Drosinis P, Joschko J, Keely E (2016) Improving access to specialist care for an aging population. Gerontol Geriatric Med. Available: https://doi.org/10.1177/2333721416677195. Accessed 10 April 2020
15. Man-Ging CI, Uslucan JO, Frick E, Büssing A, Fegg MJ (2019) Meaning in life of older adults living in residential and nursing homes. J Religion, Spirit Aging 31(3):305–322
16. Mazumder H, Hossain M, Das A (2020) Geriatric care during public health emergencies: lessons learned from novel corona virus disease (COVID-19) pandemic. J Gerontol Social Work, Available https://doi.org/10.1080/01634372.2020.1746723. Accessed 24 April 2020
17. MOHCC (2020a) Coronavirus COVID-19 update: 22 April 2020. Available: http://www.mohcc.gov.zw/index.php?option=com_phocadownload&view=category&id=15&Itemid=741. Accessed 28 April 2020
18. MOHCC (2020b) Zimbabwe Ministry of health and childcare. Available: http://www.mohcc.gov.zw/. Accessed 1 May 2020
19. Mudey A, Ambekar S, Goyal RC, Agarekar S, Wagh VV (2011) Assessment of quality of life among rural and urban elderly population of Wardha district, Maharashtra, India. Stud Ethno-Med 5(2):89–93
20. Noam E (1975) Homes for the aged, supervision and standards: a report on the legal situation in European countries. National Clearinghouse on Aging, Washington
21. Norberg C, Fischer RS, Isaksson U, Lämås K (2019) Something wonderful and incomprehensible: expressions and experiences of spirituality among very old people. J Study Spirit 9(2):125–137
22. Paddock K, Wilson CB, Walshe C, Todd C (2019) Care home life and identity: a qualitative case study. The Gerontol 59(4):655–664
23. Ramage-Morin, P. L. 2008. Chronic pain in Canadian seniors. Ottawa, Ontario: Health Reports, Statistics Canada
24. Raquel R (2018) Urban warfare: housing under the empire of finance. Blackwell
25. Rijnaard MD, van Hoof J, Jansen BM, Verbeek H, Pocornie W, Eijkelenboom A, Beerens HC, Molony SL, Wouters JM (2016) The factors influencing the sense of home in nursing homes. J Aging Res 2016:1–16
26. Sambala EZ, Cooper S, Manderson L (2020) Ubuntu as a framework for ethical decision making in Africa: responding to epidemics. Ethics Behav 30(1):1–13
27. Schoen C, Osborn R, How SK, Doty MM, Peugh J (2009) In chronic condition: experiences of patients with complex health care needs, in eight countries, 2008. Health Aff 28:w1–w16
28. Tarricone R, Tsouros AD (2008) The solid facts: home care in Europe. World Health Organisation. Available: http://www.euro.who.int/__data/assets/pdf_file/0005/96467/E91884.pdf?ua=1. Accessed 11 April 2020
29. Tavengwa M, Nhongo TM (2002) The changing role of older people in African households and the impact of ageing on African family structures. Conference proceedings: the ageing in Africa conference, Johannesburg. Available: http://archive.kubatana.net/docs/hivaid/helpage_impact_of_ageing_0408.pdf. Accessed 12 April 2020
30. Tavernini A, Di Rienzo A (2020) The 'massacre' of Italy's elderly nursing home residents. TRT World magazine. Available: https://www.trtworld.com/magazine/the-massacre-of-italy-s-elderly-nursing-home-residents-35575. Accessed 30 April 2020
31. WHO (2020) Modes of transmission of virus causing COVID-19: implications for IPC precaution recommendations. Scientific brief, 29 March 2020. Available: https://www.who.int/publications-detail/modes-of-transmission-of-virus-causing-COVID-19-implications-for-ipc-precaution-recommendations. Accessed 30 April 2020

32. Zimano FR, Keresencia M, Chilunjika A (2018) 'Clutch-the-ear' and get enrolled: the antagonistic intrusion of indigenous knowledge systems to the detriment of contemporary educational developments. South African J Childhood Educ 8(1):a557

Felistas R. Zimano holds a Ph.D. in Management and Public Administration (University of KwaZulu Natal) and M.Sc. International Relations (University of Zimbabwe) among other qualifications. Her areas of research interest are regional and global governance taking a multidisciplinary approach. She has researched and published in research methodologies, trade facilitation, classical political theories, and conflict resolution. Occasionally, she researches inclusive education issues and law.

Harnessing Local Knowledge for Resilience Building: The Case Study of Lower Muzarabani

Vincent Chenzi and Mathew Douglas Chihobo

Abstract This chapter analyses the harnessing of home-grown knowledge towards resilience building in Lower Muzarabani—a natural disaster-prone area from 1980 to 2021. Natural disasters and diseases are largely triggered by the region's physical landscape. Floods and droughts are the most common natural disasters whilst malaria and cholera are the region's endemic diseases. These hazards have resulted in the destruction of infrastructure, famine, poverty, and loss of life. Nevertheless, the inhabitants continue to thrive in the area through the utilisation of indigenous knowledge. Findings from the study identified several home-grown resilience mechanisms which included, disaster prediction, rearing of drought tolerant livestock and crops, flood plain cultivation, harvesting of water and wild fruits to mention but a few. External stakeholders especially the Government of Zimbabwe (GoZ) and several Non-Governmental Organisations (NGOs) have also contributed towards resilience building by bankrolling various community empowerment projects including provision of drugs, food, and agricultural inputs, among others. Despite the existence of numerous challenges affecting resilience building, the chapter concluded that the harnessing of local knowledge by the Lower Muzarabani community has been more effective than the role of external stakeholders in enabling the inhabitants to survive in the region for generations.

Keywords Lower Muzarabani · Resilience building · Zimbabwe · Natural disasters · Indigenous knowledge

---

V. Chenzi (✉)
Department of Peace, Security and Society (DPSS), University of Zimbabwe, Harare, Zimbabwe
e-mail: chenziv05@gmail.com

M. D. Chihobo
History Department, University of Zimbabwe, Harare, Zimbabwe

R. Kasimba and I. Chirisa (eds.), *Disaster Risks and Hazards*, Advances in 21st Century Human Settlements, https://doi.org/10.1007/978-981-96-4577-0_12

1 Introduction

This study examined the role played by local knowledge towards resilience building in disaster prone communities with special reference to Lower Muzarabani. The term '*Muzarabani*' is a Shona word which directly translates to the phrase 'filling the plain' referring to the flood waters that cover much of the region during the rainy season [46]. Local or indigenous knowledge generally refers to information shared by local inhabitants about surviving, predicting, preventing and preparing for disasters which is passed down for generations originating from the interaction between members of the community and the environment in which they live [17]. The physical landscape of Lower Muzarabani can be classified as semi-arid and it is an incubator of natural disasters that afflict the inhabitants of this region. Droughts, floods, and malaria are the chief natural disasters discussed in the study [34]. Droughts in this region are primarily driven by erratic rainfall and high temperatures that impede crop growth leading to hunger and starvation of the inhabitants [42]. Floods are caused by the back-flow of several rivers in the region especially Musengezi, Kadzurure, Hoya, Nzoumvunda and Musingwa [35]. Malaria on the other hand is caused by high temperatures coupled with stagnant water sources which act a breeding spaces for malaria mosquitoes. This situation is further compounded by the fact that Lower Muzarabani is situated within a flood plain located inside a valley with few trees due to deforestation resulting in rapid flooding throughout the region. These natural disasters have resulted in devastating effects on the lives and livelihoods of the inhabitants. For example, between the 1st of January to the 3rd of May 2020 Zimbabwe reported 262,968 malaria cases and 246 deaths. The bulk of these cases were reported in Mashonaland Central which encompasses Lower Muzarabani [28]. Also, in 1997, 10 hectares of crops were uprooted whilst 2500 hectares of arable land was smitten by water logging in Chidodo. Again in 2007, floods destroyed 3000 hectares of maize, 6000 hectares of cotton and 3000 hectares of sorghum through uprooting and water logging [43].

Despite these catastrophic natural disasters, the inhabitants of Lower Muzarabani are attached to their physical landscape. This was evidenced by the testimony of an elderly woman at the aftermath of Cyclone Japhet in 2003 that:

> We are so much attached to this ancestral land. No one wants to leave, and no one is prepared to go elsewhere. We are used to surviving under these circumstances here. I would have high blood pressure if I were to be moved from this land. Our relatives' graves are here and we cannot abandon them just like that. Whoever feels that we have to be moved must be prepared to rebury our relatives wherever we go. [16]

There are several reasons why denizens of Lower Muzarabani prefer to stay in the hostile environment apart from staying connected to their 'ancestral graves'. These reasons include fertile soil in the valley, access to land, and social considerations of calling a place home where inhabitants have attachments to family, relatives and friends [41]. It is against this background of a hostile environment that the Lower Muzarabani community resorted to harnessing indigenous knowledge towards resilience building. Constructing community resilience is a process which a society adopts to reduce the effects of disasters [20]. In Lower Muzarabani, these resilient building mechanisms have ranged from traditional early warning systems for disasters, water harvesting, flood plain cultivation, among others. However, external stakeholders including the Go), World Health Organisation (WHO) and Zimbabwe Red Cross Society (ZRCS) have also participated in the building of community resilience in the area by constructing Blair toilets, malaria treatment, key hole gardening among other activities. This research embraces the values enshrined in the Sendai framework for disaster risk reduction especially priorities three and four which focus on investing in disaster risk reduction for resilience and enhancing disaster preparedness for effective response in recovery, rehabilitation and reconstruction respectively [1]. It examines how these values have been successfully inculcated in the fight against disasters in Zimbabwe's Lower Muzarabani region. The chapter is divided into four segments apart from the introduction. Section 2 is the methodology of the research followed by a combined section containing the literature review and theoretical framework. Section 4 comprises of the study's main findings whilst the Sect. 5 is made up of the conclusion and recommendations section.

2 Methodology

This study was largely qualitative and descriptive in nature [31]. A qualitative approach was the most appropriate since the study sought to shed light on how exactly the inhabitants of Lower Muzarabani utilised their local knowledge towards building resilience during disasters including the effectiveness of these measures. Lower Muzarabani is made up of different communities including Chadereka, Dambakurima, Chimoio, Hoya, Chiwenga, Kapembere, and Gunduza among others. This study mainly relied on face-to-face unstructured oral interviews with both key and non-key informants. The key informants included Muzarabani's District Administrator (DA) who is also the chairperson of Muzarabani Civil Protection Committee (MCPU), the district field officer of the Zimbabwe Red Cross Society (ZRCS), Health personnel at Muzarabani clinic and World Vision field officers. These key informants were selected via snowball sampling whereby the researcher(s) initially embarked on the study with the above mentioned three key informants who were able to introduce the researcher(s) to the other key and non-key informants [11]. The informants provided valuable information ranging from knowledge about the district pertaining to natural disasters and interventions by the GoZ and other NGOs. These interview session with the key informants lasted between 30 min and 1 h. Unstructured interviews or non directive interviews with the key participants were relevant since they gave room for the interviewees to provide ample information on the topic [12].

The study also utilised structured interviews with the inhabitants. A total of twenty-three interviews were conducted in different parts of Lower Muzarabani. However, only twelve interviewees were recorded to prevent data redundancy. Although a sample of twenty three respondents may appear small, however huge sample sizes are unnecessary for qualitative studies [13]. Hence the researcher(s) were able to draw adequate information from the sample whilst avoiding data redundancy [31]. Each of these structured interviews lasted between ten to fifteen minutes per session and the issues that were discussed included: the history and nature of natural disasters specific to certain areas of Lower Muzarabani; resilient strategies used by the inhabitants; nature and type of external interventions and areas of attention to improve on disaster risk reduction endeavours. All the informants in the study agreed to participate on the basis that the information they provided was to be used for the sole purpose of this research. Furthermore, all the participants were able to read and write in English and they were aged between 24 and 72 years. The data collection process was conducted between March and July 2021.

The major limitations for the study was the issue of information accuracy and reliability. Since the research mainly relied on unstructured oral interviews on a period of over forty years the accuracy of the oral information was affected by exaggeration, mix-up of events, and data omission. Some informants failed to provide accurate data especially in terms of dates and chronology of events. Furthermore, some participants were not aware of the precise years of the events they were narrating. In order to overcome these challenges the researcher(s) utilised Critical Discourse Analysis (CDA) which is an analytical tool that is concerned with studying and analysing both written and spoken texts to reveal the discursive sources of power, dominance, inequality and bias [33]. Added to that, the oral data was triangulated with information drawn from alternative sources especially primary data from Muzarabani district, ZRCS archives, and newspaper reports about the events that transpired. The Muzarabani district administration and the ZRCS offices had archived documents spanning for more than two decades. This information included accurate records on natural disasters that affected the district, state and non-state intervention efforts, casualties incurred, damage recorded among other pertinent data. Several Newspapers archived in the Herald House were used to identify natural disasters in Lower Muzarabani from 1980 to the present (2021). These alternative data sources were analysed through summative content analysis which involves counting and comparisons, usually of keywords or content, followed by the interpretation of the underlying context [27].

3 Literature Review and Theoretical Framework

This study adopted the community resilience theory [25]. This is a theory of resilience that incorporates contemporary understandings of stress, wellness, adaptation, and resource dynamics. It is based on the belief that communities are capable of effectively adapting in the face of disasters. The theory emphasises that communities need to reduce risk and resource inequities, create organizational linkages, involve local residents in mitigation, boost and protect social support systems, as well as plan for the possibility that there will not be a plan. This requires flexibility, decision-making skills, and trusted sources of information that will work in the face of unknowns [25]. The theory also posits that resilient communities must be capable of adapting to changing conditions that are determined by interconnected resources, such as social capital, economic development, information, communication, and community aptitude [2]. Community resilience is therefore a crucial link between resilience and local knowledge since communities share values, experiences, knowledge and information which is distributed among members and is vital when confronted by disasters. Accordingly, Mayer [23: 167] postulated that, "Community disaster resilience offers much promise as a guiding paradigm for the promotion of disaster risk reduction and the enabling of disaster recovery through attention to, and investment in, community capacities for adaptation to a changing and uncertain environment." Therefore, this theory is appropriate for this research because the various local disaster resilience measures discussed were drawn from members of Lower Muzarabani community. This theory is thus critical in understanding the harnessing of local knowledge for building disaster resilience as it encompasses the agency of both individuals and the entire community.

There exists a growing body of academic literature on the subject of resilience, indigenous knowledge and disasters. Regarding literature on local knowledge, a study which was conducted by Jigyasu [18] revealed that local knowledge possessed by disaster prone communities is very vital towards predicting the location, intensity, occurrence, manifestation, frequency and duration of disasters prevalent in those communities. This knowledge is collected by reading various environmental signs through trial and error for generations. It is passed on to future generations becoming part of that community's collective intangible heritage. The nomadic Moken tribe from the southern seas of Thailand and Myanmar and the indigenous tribes of the Andam Islands utilise their knowledge of the sea to predict the occurrence of tsunamis which has saved the lives of both their tribesmen and tourists [18]. On a similar note, indigenous farmers from the Kutch region of Gujarat utilise their local knowledge based on the direction of the wind to predict droughts. This helps them take precautionary measures. Apart from predicting disasters, local knowledge is also used in resilience building. The *Bhunga* structures constructed by the Banni tribe in India's Kutch region survived the 2001 Gujarat earthquake whilst modern buildings were completely destroyed. In addition, Jigyasu [18] stated that local knowledge is a vital social asset especially in providing localised social, economic and institutional coping skills and capacities during and after a disaster situation. The aftermath of the 2015 Nepal earthquake brought about a collective voluntary communal initiative as community members worked together towards repairing damaged houses, shrines and community buildings in order to restore community life.

Ariando's [5] study also examined how traditional knowledge helped indigenous groups adapt to climate change. The Orang Suku Laut are indigenous Malay people who have been sea nomads since the sixteenth century. According to Ariando [5] 80.18% of the Orang Suku Laut were affected by climate change evidenced by changing seasonal patterns, water scarcity, and disease outbreaks. In order to overcome these climate change challenges, the Orang Suku Laut applied ancestral cultural practices including adopting a nomadic lifestyle to avoid the impact of disasters in the region, weather forecasting through observing natural signs, such as the atmosphere, temperature, wind patterns, fish or bird behaviour and ocean currents [5].

On a similar note, Dasgupta et al.'s [10] study examined how indigenous knowledge practices of shifting cultivation or *Jhum* practised by the farmers from India's Zunheboto district of Nagaland contributed towards disaster risk reduction, forestry and biodiversity conservation. Although shifting cultivation is generally associated with land degradation and disaster risk, Dasgupta et al. [10] argued that *Jhum* can contribute towards the attainment of Sustainable Development Goals (SDGs), especially, SDG-1(No poverty), SDG-2 (Zero hunger), and SDG-15 (Life on land) if all the *Jhum* traditional protocols are followed through. Contrary to mainstream research on local knowledge as an asset towards disaster risk reduction, Muga et al.'s [24] article argued that local knowledge practices by the Somali nomadic pastoralists was undermining North Eastern Kenya's public health system by spreading the Rift Valley Fever (RVF). They mentioned that indigenous traditional practices such as animal ritual sacrifices and drinking raw animal blood and milk were responsible for spreading the RVF in the region. The study's findings demonstrated that the nomadic pastoralist were aware of all the public health practices to safeguard against the RSV but ignored them because of the entrenched socio-cultural and religious practices associated with the pastoralist production system [24].

On the African continent, Iloka's [17] article examined the role of indigenous knowledge for disaster risk reduction in Africa. The study contends that although a lot has been written about indigenous knowledge, the area of disaster risk reduction and climate change in Africa has received very limited scholarly attention. Colonialism and lack of proper documentation of indigenous knowledge on disasters were singled out as the main reasons behind the failure of in-cooperating disaster risk reduction indigenous knowledge to institutional policies. As a result, policymakers continue to depend on Western scientific mitigation measures whilst overlooking locally accessible and inexpensive indigenous knowledge in their disaster risk reduction efforts [17]. This oversight has resulted in failed disaster mitigation efforts. Apprehending how indigenous African communities have adapted to myriad disasters for generations may hold the key to ushering sustainable policies to mitigate future disaster related challenges in the continent. In addition, Manyena's [21] study investigated the challenges faced by African countries in adopting resolutions passed by the Sendai framework for disaster risk reduction (2015–2030). Despite the fact that most African countries subscribed to the Sendai Framework resolutions, [21: 41] argued that the African continent still remains the most vulnerable region to disasters and is the least resilient which is worsened by poverty, structural transformation, climate change and rapid urbanisation. In order to address this anomaly, Manyena [21] argued that since disasters are caused by both internal and external factors, there is a need to strengthen the region's political commitment to addressing disasters, 'Africanizing' the Sendai goals and promoting domestic solutions and local investment.

In Zimbabwe, one of the key studies on resilience is Mavhura and Manyena's [22] research which examined the spatial quantification of resilience in relation to collective hazards, with special reference to Muzarabani District. Through the use of the principal component analysis technique and the Geographical Information Systems (GIS) tool, they were able to design a community resilience spatial variation model for effective policy intervention in disasters. However, their findings indicated that there was a geographical variation in community resilience with most communities bordering between low to below low levels of overall resilience [22]. Lunga and Musarurwa [19] explored how Indigenous Knowledge Systems (IKS) were utilised by disaster prone regions in responding to disasters especially in Zimbabwe's Matabeleland and Mashonaland provinces between 2012 and 2013. Their primary focus was on how inhabitants of these two regions made use of local knowledge in mitigating the severe impact of natural disasters. The IKS responses encompassed the usage of water conservation techniques, mushroom eating, treatment of epidemics, and protection against wild animals among others [19]. The study concluded that there is need to increase the effectiveness of disaster response effort by integrating scientific methods on disaster mitigation with local knowledge. Lunga and Musasrurwa's [19] study was very critical for this study especially in appreciating the diverse nature of indigenous knowledge towards resilience building in Zimbabwe's disaster-prone regions other than in Lower Muzarabani. However, unlike Lunga and Musarurwa [19] this study investigated how the various local resilience building measures are utilised towards early warning, preparation, prevention and recovery in Lower Muzarabani and how this contributes towards fulfilling the 3rd and 4th priorities of the Sendai framework for disaster risk reduction [1].

Chenzi and Ndamba's [9] study examined the resilience mechanisms that were adopted by Harare's informal sector workers (vendors, foreign currency dealers, shebeen operators, cross-border traders and private taxi operators) in response to the country's strict Covid-19 lockdown measures between March 2020 and May 2021. They argued that in the wake of the Covid-19 pandemic, the Zimbabwean government adopted western-based lockdown protocols which prohibited all informal activities without any meaningful government assistance which resulted in rebellion from members of the informal sector. Most African states are largely very poor and depended on the informal economy. Hence, adopting western lockdown measures which were designed for high income formal economies created serious administrative challenges for the African governments as citizens flaunted them citing hunger and starvation. Consequently, this situation undermined the Covid-19 resilient efforts by the GoZ [9]. Also, Chenzi's [8] study examined the negative consequences of spreading false and misleading information via social media platforms (Face book, WhatsApp and Twitter) across Zimbabwe's urban population during the Covid-19 pandemic. The fake news was mainly in the form of false preventive measures, miraculous remedies, myths about the African race and rumours about Covid-19 vaccines [8]. The study concluded that the fake news greatly undermined the government's Covid-19 resilience building endeavors resulting in the rapid spread of the pandemic and the failure to attain head-immunity due to low vaccination turn-out.

On a similar note, Bwerinofa et al.'s [7] research examined the community resilience mechanisms adopted by Zimbabwe's rural community in responding to the Covid-19 pandemic. The study's findings showed that local knowledge, social networks, flexible adaptation and communication were key factors in responding to the pandemic situation. They stated that Zimbabwe's perennial harsh economic environment was a driving factor behind the citizen's resilient response to the pandemic in the absence of state support. Nevertheless, they argued that although there was evidence of adaptation, innovation and shared learning capabilities, the nature of the resilience was spontaneous, passive, and resigned instead of being a positive, empowering, and transformational [7]. They thus propounded that instead of viewing community resilience as the remedy for disaster preparedness and response in the context of pandemics, variations of community resilience practices should be viewed as vital components of a wider, plural health system.

The above literature shows that different populations from various regions have generated a repository of indigenous knowledge which they draw upon to develop resilience against disasters unique to their regions. In Africa in general and Zimbabwe in particular, there is evidence of such initiative. However, these studies also revealed challenges affecting the effective usage of local disaster resilience knowledge and they include failure to inculcate local knowledge to government policy, over-dependence on expensive western scientific methods by African governments when addressing local disasters, and lack of available documented local resilience data on disaster risk reduction. Drawing from these insights, this chapter aims at examining the various indigenous knowledge measures adopted by the Lower Muzarabani community and how they contributed towards resilience building against disasters; the challenges that the communities face in building resilience towards disasters; role of external stake holders; and provide possible recommendations in enhancing local knowledge for resilience building during and after disasters.

4 Discussion of Findings

This section explores various locally developed knowledge, skills and activities including their effectiveness in building resilience against natural disasters in Lower Muzarabani. The section also examines the role of external stakeholders in strengthening community resilience against disasters. It concludes by identifying several challenges that affect community resilience against natural disasters. Resilience building has become one of the major concepts in disaster management studies. Traditional/Local/Indigenous Knowledge Systems are the backbone of resilience building in Lower Muzarabani. Some of these locally developed skills and approaches for resilience building peculiar to Lower Muzarabani include changing settlement patterns, flood plain cultivation, growing drought tolerant crops, water harvesting, gathering of wild fruits, domesticating livestock, digging wells inside dried rivers, relocating to safer areas during the flooding season among other activities [46]. Findings from the study indicated that these indigenous knowledge mechanisms developed in Lower Muzarabani were designed to deal with natural disasters especially floods and droughts. Some of these resilience strategies date back to the pre-colonial era and still continue to be utilised.

On the matter of floods, there were several public awareness campaigns spearheaded by senior local inhabitants who shared and exchanged their knowledge pertaining to the traditional early warning systems for floods. In 2012, there was a community exchange exercise between inhabitants of Chadereka and Dambakurima on the use of traditional warning systems against floods during the rainy season [29: 2–11]. Some of these signs include, the presence of foam and weeds in the river. This indicates that the river is about to flood. Moreover, certain positions of the moon are another indigenous warning system which is used to predict the amount of rainfall for the season. Likewise, certain types of clouds for example *Hanga*[1] clouds are a sign of heavy rainfall and flooding [46]. Furthermore, the observation of wind strength and direction is a locally developed early warning system for floods. However, these environmental signs are not only used to predict floods but they are also used in identifying the nature of the rains in the forthcoming season. For example, strong winds or *Mbambara* wind blowing from north to the south are indicative of a good rainfall season. In addition, observation of indigenous vegetation especially trees locally referred to as the *Pfumvudza*—when it sheds its leaves earlier than usual means there will be a good rainy season [35].

Another local resilience strategy which is widely used by inhabitants of Lower Muzarabani especially during the flooding season is relocating to safer areas. Floods always leave these communities with a myriad of challenges. One of these problems being the destruction of infrastructure especially houses and bridges. Therefore, residents of Lower Muzarabani sometimes relocate themselves to safer zones in the valley [44]. Examples of these refuge zones include Keche, Chadereka, and *Magarakata*[2] villages which were specifically established by the community to provide a safe haven from floods [46]. In the aftermath of the flooding season, inhabitants return from the refuge villages to rebuild their destroyed homes on the same location. This is one of the reasons why some of the inhabitants of the valley continue to be affected by floods. It is a sort of maladaptation of the denizens to their physical landscape. Building foot bridges using locally sourced materials especially wood, stones and sand to traverse over flooded rivers is also another resilient building strategy adopted by the community to ensure mobility and accessibility to social services. Consequently, there are several foot bridges criss-crossing Nzoumvunda, Kadzurure and Hoya rivers [35].

Although floods are generally considered a threat to livelihoods the community also consider floods as an annual blessing by capitalising on flood plain debris which is left at the close of Zimbabwe's rainy season which occurs between October and February. After the rainy season, there is a lot of moisture and debris that is left in the Muzarabani flood plain. The flood debris contains fertile alluvial soils that can sustain crops for several months without rainfall. Hence, they plant their crops (maize, sugar beans, and pumpkins) in March and harvest them in June which will be in the middle of the winter season throughout the Southern Hemisphere [16]. Flood plain cultivation is locally referred to as *Mudzetse* or *Matimba* farming and it is a crucial source of food and income for the community [26: 12]. Flood plain cultivation is one of the most effective resilient strategies used by the inhabitants of Lower Muzarabani and it is similar to how Egyptians rely upon the Nile floods since

[1] Cumulus clouds.

[2] A Shona word meaning a high place.

ancient times [6]. Nevertheless, floodplain cultivation also threatens the ecosystem through siltation [45].

Apart from floods, droughts are also another dominant natural disaster in Lower Muzarabani. The reason why droughts are prevalent in Lower Muzarabani despite the heavy flooding is largely because the area itself receives very little rainfall and floods are generally caused by the back-flow of several rivers in the area which are flooded with rainfall received from various parts of the country. Therefore, one of the popular drought resilient strategies has been the cultivation of drought resistant crops especially rapoko, millet and sorghum [30]. The growing of drought tolerant crops has been a very effective drought resilience measure since the 1981/82 and 1984 droughts such that small grains have been the staple diet of the community [45]. Livestock production is another drought resilience measure and one of the mainstays of Lower Muzarabani's economy. They practice mixed farming by growing crops and keeping livestock. Animal husbandry is very lucrative in the area due to the availability of good quality pastures, high temperatures, and fertile soils with high quality salts consumed by cattle. Animals raised in the area include cattle, sheep and goats [45]. Small livestock kept in the region also comprise of chickens, guinea fowls and ducks. These animals are essential for dietary needs of the inhabitants. During periods of drought, livestock is used for both sustenance and barter exchange [46]. Cattle and goats are usually exchanged for grains especially maize, millet and sorghum.

Water harvesting is also locally practised as a drought mitigation measure because of the scarcity of safe drinking water. During the rainy season, Lower Muzarabani residents harvest river water and during the summer period when most of the rivers such as Hoya, Nzoumvunda and Kadzurure are dried up, they dig wells along these dried rivers. The water harvested from the rivers is used for both domestic and agricultural purposes. Other than sourcing river water, the community also practices traditional water harvesting methods by identifying aquifers or water bearing rocks [46]. Aquifers are traditionally identified by the type of vegetation that grow close to them. After identifying the vegetation, they then dig deep wells, which contain enough water to support an entire village. Consequently, some of the villages in the community are constructed in the vicinity of aquifers. For example, Keche village is situated close to a deep well on an aquifer that serves the entire village [39]. Thus, wells have greatly assisted in the provision of water in the community. Therefore,, local knowledge for identifying water in the community has resulted in the digging of dependable wells, that ensure the availability of water in the community.

Another local method adopted by inhabitants of Lower Muzarabani for building food security in the drought prone community is the gathering of wild fruits. Wild fruits are an essential livelihood aspect in this region and they constitute part of the local diet especially during drought periods. The most important fruits in Lower

Muzarabani are *Mauyu*[3] and *Masau*[4] [36]. During drought periods, some families survive on these wild fruits for several months. These fruits produce plentifully in Lower Muzarabani compared to other regions. Hence, local inhabitants gather these wild fruits and store them. They either eat them in their raw state, cook porridge or brew *Maheu.*[5] These fruits proved invaluable during the 1992, 1997 and 2008 drought periods as most families survived by subsisting on them [15]. Other than being used as sustenance, these wild fruits are also used for barter trade especially with people from outside the valley. These fruits are normally exchanged for money and grains. For a 5-kg bag of maize, one gets a 20-kg bag of *Masau* [38]. Barter trade is an ancient system of economic exchange that pre-dates the colonial era and it still continues to be practised in the community as a resilience building strategy to reduce the effects of droughts and economic hardships. In order to safeguard the continued nourishment from these wild fruits, the community leaders made it illegal to cut down any of these wild fruit bearing trees throughout Lower Muzarabani.

Although the inhabitants of Lower Muzarabani mainly depend on their home-grown resilience measures, the GoZ and other NGOs have also been involved in the building and strengthening of local resilience in the community. One of the first interventions in post-colonial Zimbabwe by the GoZ was the drilling of boreholes to ensure that the drought-prone community had access to clean and safe drinking water [14]. However, this initiative though helpful did not adequately address the community's water needs since only a few boreholes were drilled in the area due to financial constrains. In order to complement the GoZ's efforts to provide safe drinking water NGO's especially the ZRCS and World Health Organization (WHO) assisted by providing aquatic tablets for water treatment before consumption. This measure was an effective drought resilience measure since the tablets were used to treat water harvested from wells and other open water bodies which supplemented the inadequate borehole water [45].

Another GoZ initiative to address the drought challenges of Lower Muzarabani was the growing of cotton. Cotton is a drought tolerant crop that is grown in the region after the GoZ encouraged farmers from the community to venture into cotton farming as a cash crop to supplement their financial needs [45]. Cotton, requires very little rainfall, and it flourishes under high temperatures. However, the cotton market was gravely affected by international price fluctuations since 2008 which led to a reduction in cotton cultivation in favour of other drought resistant food crops especially small grains so as to mitigate the effects of droughts in the area.

The GoZ in collaboration with the WHO also embarked on an initiative to control the spread of malaria by providing mosquito nets and indoor spraying services [37]. Mosquito nets contribute towards preventing mosquito bites, one of the main ways that malaria is spread. In-door house spraying is done in the area at least once annually.

[3] Baobab fruits.

[4] A fruit grown on trees, *Masau* (Ziziphus Mauritania) has many names in English, including "Indian Jujube," "Chinese Apple" and "Indian Plum." It originated in Central Asia and spread to North Africa, where it became distinctly Zimbabwean.

[5] A local traditional brew made of powdered fruits mixed with water and left to ferment.

Added to that, mosquito repellents such as lotions and incense are also donated to the community since it was observed that though the inhabitants used nets and in-door spraying, they still remained prone to mosquitoe bites especially when working in their fields and gardens. The authorities also deployed health experts in the community to test malaria and provide treatment to infected people. This initiative has been effective as evidenced by the reduction in malaria cases and fatalities thereby reducing the patient load on health centres (Ibid).

Apart from strengthening the community's drought and malaria resilience, external stakeholders especially the ZRCS launched several projects in 2012 which focused on afforestation, key hole gardening, construction of Blair toilets and pioneered the construction of Rocket Lorena mud cooking stoves [3: 1]. The aim of the ZRCS was to shift from their previous short term role of reacting in the aftermath of disasters and becoming proactive through focusing on building sustainable community resilience by strengthening the community and environment. This initiative dovetails into the community resilience theory since the community is the first to be affected whenever disasters occur, so it is imperative to equip the community to withstand the effects of natural disasters [25]. They borrowed this idea from the 2005 Hyogo framework. At the Hyogo conference, it was agreed that communities need to be resilient, in order to mitigate the loss associated with disasters [3: 1].

The Rocket Lorena Cook stove is an energy saving innovation originally designed in Uganda. It was brought to Zimbabwe in 2014 through a Zimbabwe and Uganda Red Cross exchange learning agreement [32: 1–2]. These stoves assist users to save firewood by 50–60% compared to the traditional open fire stoves. Furthermore, the stoves have a good heat retention capacity which keeps food warm for longer periods thereby indirectly reducing the spread of food related diseases. In addition, the firewood that can be used for one day under a traditional open fire stove could be used for 4–5 days in a Rocket Lorena stove (Ibid). The merits of using the stoves in these communities is that they are cheap and local material is used. Clay and anthill soil are used for the main structure. In Lower Muzarabani, these stoves are popularly known as *Tsotso stoves*[6] meaning that small amounts of firewood are required for cooking. Which helps in reducing deforestation thereby enhancing environmental resilience against disasters. This initiative is further complemented by tree planting (afforestation). Trees are essential since they act as windbreaks. The ZRCS spearheaded the tree planting initiative especially the planting of eucalyptus trees around schools and clinics. An example was at Chadereka clinic. This is meant to avert the destruction of infrastructure by hailstorms like what transpired during the 2009 and 2014 rainy seasons [40]. The trees also provide twigs which lessens the burden of women who in this patriarchal society are responsible for fetching firewood which usually involves the carrying of heavy loads by head over long distances resulting in health challenges especially back aches [32: 1].

Another ZRCS project is the building of Blair toilets as an initiative to end open defecation in the wild. Lower Muzarabani is a remote area, most people do not have Blair toilets, hence open defecation is usually the norm. This practise is a health

[6] Tsotso is the Shona venecular term for twigs.

hazard because whenever floods occur, water contamination usually ensues resulting in diarrhoeal diseases. However, this initiative has been met with resistance in the community. Most of the inhabitants are reluctant to cooperate, arguing that they do not have Blair toilets [46]. Key hole gardening is also another resilience building initiative which was introduced by the ZRCS. The programme was introduced in 2016, to ensure good health practices through improved nutrition and food security [4]. A Key hole garden refers to a raised bed with a wedge cut out for easy access to the compost basket which is at the middle. An assortment of solid material with different granular sizes are placed in the compost basket (Ibid). These materials along with a grass wall filter the waste water before it goes into the plant bed. Biodegradable household waste is thrown in the compost basket and starts decomposing. When watering the compost, water and compost nutrients filter into the plant bed. Water is extremely scarce which is worsened by climate change induced prolonged dry seasons. Boreholes and other water sources usually dry up before the rainy season. This restricts households' interest and possibility to grow vegetable gardening and that affects nutritional diversity. Key hole gardens are more successful when established within the homestead for ease of access when disposing waste water and other domestic waste material. It fosters the idea of community resilience since it uses local materials (Ibid). The adoption rate has been remarkable in Lower Muzarabani, as families embraced the garden as a practical solution to their nutritional challenges.

5 Conclusion and Recommendations

The study's findings demonstrated that the inhabitants of Lower Muzarabani greatly rely on the local knowledge about their environment to overcome disasters that affect them especially droughts and floods. However, these local knowledge practises are insufficient to address all disasters in the region particularly sickness related disasters such malaria and diarrhoeal epidemics. This has resulted in external stakeholders including the GoZ, ZRCS, and the WHO intervening to strengthen local resilience against diseases. However, from a comparative stance. It appears that the Lower Muzarabani community has contributed more through their local knowledge in building resilience against disasters compared to the external stakeholders. Several challenges were identified which have affected the harnessing of knowledge for strengthening resilience. These problems include lack of funding for projects, socio-cultural norms, bad habits, community resistance, and failure to corroborate local knowledge with scientific resilience building mechanisms.

There are several recommendations that need to be inculcated towards the effective harnessing and utilisation of knowledge for resilience building in Lower Muzarabani. One of the most crucial steps towards this goal is the need to combine local knowledge with external/scientific methods in order to strike a balance so as to maximise on opportunities presented by both. This challenge is not only unique to Lower Muzarabani but is predominant in most disaster prone communities [19]. This is compounded by community resistance to external/scientific interventions. For example, the fight against the spread of malaria in the community has been greatly undermined by resistance from the local inhabitants. Some of the inhabitants do not cooperate with health and government officials especially during malaria treatment and prevention campaigns. They resist using mosquito nets whilst others refuse indoor spraying on account of the smell of the malaria drug dichloro-diphenyl-trichloroethane (DDT) [37]. Health authorities reported that this resistance against complying with health regulations is also responsible for the continued spread of maria. This challenge of resistance by locals to apply scientific measures was also raised by Muga et al.'s [24] study on the spread of the RVF in North Eastern Kenya where they identified indigenous traditional practices as the primary driver for resisting scientific prevention and treatment methods.

The problem by external stakeholders of heavily depending on scientific resilience building measures is a challenge not only limited to Lower Muzarabani. Iloka's [17] research argued that this problem is widespread in the African continent where evidence of local disaster resilience practices is available but there is lack of documentation and formalisation of these practices by policy makers and NGOs due to over-reliance on scientific western measures. In Lower Muzarabani, no effort has been made by the major NGOs which operate in the region including to properly document and formalise the local resilience knowledge [35]. What these organisations have done is to introduce foreign scientific measures whilst ignoring locally developed practices which results in community resistance in adopting foreign approaches regardless of their effectiveness as highlighted above. Furthermore, the imposition of external measures results in the emergence of a dependency syndrome within the community which ends up abandoning its local resilience measures in favour of expensive foreign scientific measures.

The other solution is to address bad habits of open deification which results in diarrhoeal disease outbreaks especially during floods. A possible solution to this challenges can be the introduction of fines to households without Blair toilets. This can speed up the construction of Blair toilets in the community and ultimately end the culture of open defecation. Building of schools inside villages to avoid crossing of rivers though constrained with resources is a sustainable measure. It ensures safety of school children from drowning and reduces absenteeism during the rainy season. Last but not least, there is need for the GoZ to fund projects in the area which can enhance long-term community resilience against droughts. For example, the construction of a lake in Mavhuradonha mountain range has been suspended due to lack of funding for the project. Such an initiative can address the long term water challenges affecting the region. Furthermore, this initiative will also enhance the agricultural practices thus providing ample water for irrigation purposes.

References

1. Aitsi-Selmi A, Egawa S, Sasaki H, Wannous C, Murray V (2015) The Sendai framework for disaster risk reduction: renewing the global commitment to people's resilience, health, and well-being. Int J Disaster Risk Sci 6(2):164–176
2. Aldunce P, Beilin R, Handmer J, Howden M (2014) Framing disaster resilience: the implications of the diverse conceptualisations of "bouncing back". Disaster Prevent Manage
3. A paper on building community resilience in Lower Muzarabani by the Zimbabwe Red Cross Society, Muzarabani District Administrator's Office, April 2018
4. A paper on Key Hole Gardening in Lower Muzarabani, Muzarabani Red Cross Offices, 2014
5. Ariando W (2020) Traditional knowledge helps indigenous people adapt to climate crisis, research shows. https://www.preventionweb.net/news/traditional-knowledge-helps-indigenous-people-adapt-climate-crisis-research-shows
6. Blokland B (2022) Volcanic eruptions, resilience and vulnerability: the impact of Nile flood variability on Ptolemaic Egypt (261–30 BC). Master's thesis
7. Bwerinofa IJ, Mahenehene J, Manaka M, Mulotshwa B, Murimbarimba F, Mutoko M, Sarayi V, Scoones I (2022) What is 'community resilience'? Responding to COVID-19 in rural Zimbabwe. BMJ Glob Health 7(9):e009528. https://doi.org/10.1136/bmjgh-2022-009528
8. Chenzi V (2023) Fake news, social media and the COVID-19 pandemic: the Zimbabwean experience. In: Chapungu L, Chikonzi D, Kaitano D (eds) COVID-19 in Zimbabwe: trends, dynamics and implications in the agricultural, environmental and water sectors. Springer, Cham, pp 263–280
9. Chenzi V, Ndamba A (2023) Surviving the COVID-19 lockdown: the case of Zimbabwe's informal sector 2020–2021. Rev Afr Polit Econ
10. Dasgupta R, Dhyani S, Basu M, Kadaverugu R, Hashimoto S, Kumar P, Johnson BA, Takahashi Y, Mitra BK, Avtar R, Mitra P (2021) Exploring indigenous and local knowledge and practices (ILKPs) in traditional Jhum cultivation for localizing sustainable development goals (SDGs): a case study from Zunheboto district of Nagaland, India. Environ Manage. https://doi.org/10.1007/s00267-021-01514-6
11. Felix-Medina MH, Thompson SK (2004) Combining link-tracing sampling and cluster sampling to estimate the size of hidden populations. J Off Statistics-Stockholm 20(1):19–38
12. George T (2022) Unstructured interview: definition, guide & examples. Scribbr. Retrieved February 14, 2023, from https://www.scribbr.com/methodology/unstructured-interview/
13. Heale R, Twycross A (2015) Validity and reliability in quantitative studies. Evidence-Based Nurs 18(3):66–67
14. Herald Reporter, 9 October, 1984. Several schools are threatened with closure in the Lower Zambezi valley. The Herald.
15. Herald Reporter, 11 May, 1992. Dish full of nuts only meal for a day. The Herald.
16. Herald Reporter, 2 April, 2005. Villagers attached to the area despite disasters. The Saturday Herald
17. Iloka NG (2016) Indigenous knowledge for disaster risk reduction: an African perspective. Jamba. Potchefstroom, South Africa 8(1):272. https://doi.org/10.4102/jamba.v8i1.272
18. Jigyasu R (2020) Harnessing the knowledge of indigenous communities for disaster risk reduction. https://www.preventionweb.net/blog/harnessing-knowledge-indigenous-communities-drr
19. Lunga W, Musarurwa C (2016) Exploiting indigenous knowledge commonwealth to mitigate disasters: from the archives of vulnerable communities in Zimbabwe
20. Manyena SB (2006) Rural local authorities and disaster resilience in Zimbabwe. Disaster Prevent Manage: Int J
21. Manyena B (2016) After Sendai: is Africa bouncing back or bouncing forward from disasters? Int J Disaster Risk Sci 7(1):41–53
22. Mavhura E, Manyena B (2018) Spatial quantification of community resilience in contexts where quantitative data are scarce: the case of Muzarabani district in Zimbabwe. Geo: Geogr Environ 5(2):e00065

23. Mayer B (2019) A review of the literature on community resilience and disaster recovery. Curr Environ Health Reports 6(3):167–173. https://doi.org/10.1007/s40572-019-00239-3
24. Muga GO, Onyango-Ouma W, Sang R, Affognon H (2021) Indigenous knowledge of Rift Valley Fever among Somali nomadic pastoralists and its implications on public health delivery approaches in Ijara sub-County, North Eastern Kenya. PLoS Negl Trop Dis 15(2):e0009166. https://doi.org/10.1371/journal.pntd.0009166
25. Norris FH, Stevens SP, Pfefferbaum B, Wyche KF, Pfefferbaum RL (2008) Community resilience as a metaphor, theory, set of capacities, and strategy for disaster readiness. Am J Community Psychol 41(1–2):127–150. https://doi.org/10.1007/s10464-007-9156-6. PMID: 18157631
26. Proceedings of the disaster risk management training workshop for Muzarabani, Muzarabani District Administrator's Office, 2015
27. Rapport F (2010) Summative analysis: a qualitative method for social science and health research. Int J Qual Methods 9(3):270–290
28. Relief Web (2020) Zimbabwe: malaria outbreak—Mar 2020. https://reliefweb.int/disaster/ep-2020-000127-zwe
29. Report on the public awareness campaign on hazards associated with the rainfall season; Mashonaland central province 23 to 28 September 2012, Muzarabani District Administrator's Office, 2012
30. Sunday Mail Reporter (1987) Grow drought resistant crops, Gutu folks urged. The Sunday Mail, 1 March 1987
31. Tetnowski J (2015) Qualitative case study research design. Perspect Fluency Fluency Disorders 25(1):39–45
32. The Zimbabwe Red Cross Society booklet on Lorena Stoves, construction manual for household rocket Lorena cook stoves, Muzarabani Red Cross Offices, August 2017
33. Van Dijk TA (1993) Principles of critical discourse analysis. Discourse Soc 4(2):249–283

Interviews

34. Chadereka Nyapewa, Villager, Chadereka Shopping Centre, 03/06/2021
35. Chihobo John (Mr), Muzarabani District Administrator, Altenna Farm, 01/07/2021
36. Chidemba Etifare, Villager, Gomo Village, Female, 17/06/21
37. District Health Personnel, Muzarabani Growth Point, Muzarabani, 18/03/2021
38. Eveness Kamuteyo, Gatsi village, 06/07/2021
39. Kamupango (Mr), Keche Village, 15/03/2021
40. Karise (Mr), Red Cross Field Officer, Muzarabani Growth Point, 28/03/2021
41. Keche (Mrs), Keche Village, 13/06/2021.
42. Luck Champion, Villager, Keche Village, 30/06/2021
43. Matsveru (Mr), World Vision Field Officer for WASH Programme, Muzarabani Growth Point, 06/06/2021
44. Matadzei Tauzeni, Gomo Village, 12/04/2021
45. Munyoro (Mr), Agronomist for Muzarabani District, 24/04/2021
46. Ruza Keche, Keche Village Headman, Keche Village, 30/06/2021

Vincent Chenzi is a Lecturer of Conflict, Peace, Strategic Studies and Disaster Management at the University of Zimbabwe's Department of Peace, Security and Society (DPSS). He is also an external examiner at Allan Gray Cape Town University's Graduate School of Business. He is a holder of a B.A. Hons. and M.A. in War and Strategic Studies (UZ). Currently he is enrolled as a D.Phil. Student at the University of Zimbabwe's DPSS. His research interests include:

Disaster Management, digital media, African politics, conflict resolution and nonviolence. He has published widely in refereed academic journals including Conflict Studies Quarterly, African Security Review, Journal of Contemporary African Studies, African Identities among others.

Mathew Douglas Chihobo is an academic, humanitarian worker base in Lower Muzarabani district. He is also a former Graduate Teaching Assistant at the University of Zimbabwe's former History Department. He is a holder of a BA. Hons. and MA. in African History from the University of Zimbabwe.

The Impact of Tropical Cyclones on Human Security: A Case of Mozambique and Zimbabwe (2000–2019)

Samuel Banda and Lucky Bassie Bangidza

Abstract The paper set out to analyse the effects of tropical cyclones on human security with regards to the economic security of rural households in Mozambique and Zimbabwe. The three key objectives linked to human security are to a ascertain the levels of awareness on tropical cyclones of the rural households, assess the threats of tropical cyclones on rural household socio-economic security and examine the link between tropical cyclones and rural household social vulnerabilities. Empowerment Theory (ET) and the Social Vulnerability Theory (SVT) are the two theoretical frameworks which helped the researchers to conceptualise and broaden the analysis of the effects of tropical cyclones on human security with regards to rural household socio-economic security. The findings of the investigation confirmed that the governments of Mozambique and Zimbabwe were not doing enough to raise the levels of awareness of rural communities regarding the threats of tropical cyclones on human security. The critical recommendation of this investigation was that the governments of both Mozambique and Zimbabwe, through their respective line ministries within grassroots disaster risk reduction and management initiatives systems approach to reduce the risks imposed by tropical cyclones on human security.

Keywords Tropical cyclones · Human security · Rural households and vulnerabilities

S. Banda · L. B. Bangidza (✉)
Harare, Zimbabwe
e-mail: bangidzaluckybassie@gmail.com

S. Banda
Zimbabwe National Defence University, Harare, Zimbabwe

R. Kasimba and I. Chirisa (eds.), *Disaster Risks and Hazards*, Advances in 21st Century Human Settlements, https://doi.org/10.1007/978-981-96-4577-0_13

1 Introduction

This article focuses on effects of climate change based on tropical cyclones effects on human security with regards to the economic security of rural households in Mozambique and Zimbabwe. Internationally, climate change is real and it is manifested through the ever-increasing extreme weather conditions linked with destructive winds, storm surges, droughts, flash floods, the destruction of plants and animals, the depletion of household economic sources of income, the destruction of the food sources and most critically, an increase in the number of international climate refugees [19, 23].

Given this critical background, [2: 95], stresses that, "over the past two centuries, tropical cyclones have accounted for the deaths of about 1.9 million people worldwide, whereas more than 10,000 people are expected to perish each year due to tropical cyclone disasters." Shultz et al. [16], define a tropical cyclone as a rapidly turning storm system characterised by heavy rain and windstorms. The United Nations University 2013 Annual Report [27], noted that during the tropical cyclone induced rainy season of 2000, devastating floods in Mozambique affected approximately 4.5 million people and destroyed vast tracts of agricultural land, buildings and other infrastructure.

Secondary hazards linked to tropical cyclones are caused by primary hazards, for example, floods and fires caused by lightning. Critically, tropical cyclone hazards continue to inflict both short and long-term socio-economic insecurities within communities by causing human deaths, injuries and massive human displacements including the destruction of household property and community infrastructure and the destruction of agricultural land and staple crops. This threat can be attributed to the successive effects of Tropical Cyclone Idai and Kenneth that made strike within the same tropical cyclone season in Mozambique. The catastrophic effects of tropical cyclones Idai are on the increase in Zimbabwe as well.

2 Statement of the Problem

The frequency and severity of harmful tropical cyclones are socio-economic concerns with serious human security threats especially in developing countries. Although the two neighbouring countries Zimbabwe and Mozambique reacted to the harmful effects of tropical cyclones by declaring states of emergency, the devastating effects of tropical cyclones continue to escalate in terms of human and physical impact. The reaction to tropical cyclones effects on rural household socio-economic security and rural household vulnerabilities was not defensive but reactive in approach.

3 Conceptual Framework

The two key variables are namely tropical cyclones and human security with a deliberate focus on the sub-variable of rural household socio-economic security. Tropical cyclone is the independent variable (The actor- subject) and human security is the dependent variable (The acted upon-object) and its sub-dependent variable (Rural household socio-economic security).

Household socio-economic security is closely linked to all other variables of human security including food, political, personal, community and environmental security in that it is concerned in shaping how rural communities and individuals top-gain the resources needed to meet their basic human needs. Rural households should be able to meet their immediate food, income and social needs if they are to remain free from socio-economically threats.

3.1 Research Objectives

- Ascertain the levels of awareness on the tropical cyclone threats on human security.
- Assess the effects of tropical cyclones on rural household socio-economic security.
- Examine the effects of tropical cyclones on rural household vulnerabilities.

3.2 Research Questions

- To what extent do tropical cyclones threaten human security?
- How do tropical cyclones affect rural household socio-economic security?
- How are rural households vulnerable to the effects of tropical cyclones?

3.3 Significance of the Investigation

The investigation analysed the effects of tropical cyclones on human security with focus on the economic security for rural households in both Mozambique and Zimbabwe. Accordingly, the investigation adds value to the existing body of knowledge on the effects of tropical cyclones and mitigating the effects of tropical cyclones. The investigation focused on the rural village of Sussundenga that is located in the Manica Province, whereas in Zimbabwe the investigation focused on the rural village of Chimanimani in the Manicaland province.

The investigation focused on the effects of tropical cyclones on household socio-economic security of two rural household settlements of Sussundenga and Chimanimani. Accordingly, this investigation targeted the affected households from the two rural settlements including the District Development Coordinators, Civil

Society Organisations (CSOs), District Civil Protection Units (DCPUs) and traditional leaders from Sussundenga and Chimanimani who were judged to be part of the investigation population and under COVID -19 pandemics. The researchers engaged the Mozambican Embassy and also used emailed questionnaires to strategically positioned officers to gather information from the government officials in Mozambique. The assumption is that, tropical cyclones may continue to affect human security in particular socio-economic security in most rural household settlements in Mozambique and Zimbabwe.

4 Literature Review

The investigator reviewed literature related to the effects of tropical cyclones (the independent variable) and human security (the dependent variable). Since the year 2000, an average of nine (9) intense tropical cyclones have been recorded in this basin during the months of November to April and this period is now commonly referred to as tropical cyclone season [2] (Table 1).

According to the Saffir–Simpson Hurricane Wind Scale [29], shown above, Category 1 tropical cyclone is the lowest class that is characterised by less dangerous wind but can cause isolated and scattered damages. Any tropical cyclone above category 2 range is linked to natural disasters, such as storm surge, strong wind, heavy rain, flash floods, mudslides and landslides.

It is also critical to mention the fact that tropical cyclones do not observe political boundaries, since they can potentially affect quite a number of countries per each tropical cyclone season. Furthermore, the uniqueness of a tropical cyclone disaster is that it continues to inflict damages and destruction well after the tropical cyclone event has passed. In this regard, [7], observed that tropical cyclones can produce widespread, torrential rain that often cause devastating floods. Tertiary hazards are the last category of the effects of tropical cyclones.

Table 1 Categories of tropical cyclones

Category	Wind speeds (for 1-min maximum sustained winds)			
	m/s	knots (kn)	mph	km/h
5	≥ 70	≥ 137	≥ 157	≥ 252
4	58–70	113–136	130–156	209–251
3	50–58	96–112	111–129	178–208
2	43–49	83–95	96–110	154–177
1	33–42	64–82	74–95	119–153

Source Saffir–Simpson Hurricane Wind Scale [29]

5 Tropical Cyclone Eline, Favio, Idai and Kenneth

This described scenario best illustrates the fact that tropical cyclone hazards definitely affect human security especially the household socio-economic issues. For instance, the rural settlements of Sussundenga in Mozambique and Chimanimani in Zimbabwe are always at the mercy of tropical cyclone risks since these rural settlements cannot withstand the effects of tropical cyclones. This scenario is aggravated by the fact that it is easier for a natural disaster to escalate in rural areas since most of the rural communities are settled in poverty-stricken zones that cannot counter the immediate risks of tropical cyclones. Despite the massive human displacements, the rural communities in Mozambique and Zimbabwe were further exposed to widespread hunger and starvation prompted by a combination of primary, secondary and tertiary hazards of Tropical Cyclone Eline.

In 2007, Tropical Cyclone Favio confirmed that Mozambique was exposed to tropical cyclone threats. It is also important to note that Tropical Cyclone Favio mainly affected people with disabilities, older people and women by exposing them to risks of injuries, death, diseases, poverty, displacements and the general loss of access to essential social and economic services.

Tropical Cyclone Idai was an intense Category 3 storm that made landfall on the night of 14 to 15 March 2019 near Beira in Mozambique.

The Post-Cyclone Idai Recovery Programme [21], estimated that Tropical Cyclone Idai flooded 3000 m^2 of land and caused approximately 1297 deaths. Accordingly, the United Nations Office for the Coordination of Humanitarian Affairs [26] estimated that over 750,000 people needed urgent humanitarian aid in Mozambique following the destructive effects of Tropical Cyclone Idai. Tropical Cyclone Idai caused widespread flash flooding claiming human life and wildlife, displacing communities and caused extensive damage to property and infrastructure in Mozambique and Zimbabwe.

Mozambique experienced two deadly cyclones in 2019, namely Tropical Cyclone Idai and Kenneth. According to UNISDR [19, 23], Cyclone Kenneth with wind blasts of up to 220 km/h, was recorded as the strongest cyclone to ever hit the African continent, surpassing the effects of Cyclone Idai in Mozambique.

6 Summary of Tropical Cyclone Trends and Threats on Human Security

The historical review of tropical cyclones reinforces the notion that threats of tropical cyclones on human security are real and are increasing. Tropical cyclones continue to threaten human security by causing loss of human life, destruction of homes and property including the destruction of agricultural systems (Table 2).

Table 2 Summary of tropical cyclone trends and effects

Serial No.	Trend	Cyclone Eline	Cyclone Favio	Cyclone Idai	Cyclone Kenneth
(a)	(b)	(c)	(d)	(e)	(f)
1	Duration	4 weeks	11 days	3 weeks	8 days
2	Affected areas	Mozambique and Zimbabwe	Mozambique, Zimbabwe and Malawi	Madagascar, Malawi, Mozambique and Zimbabwe	Mozambique and Zimbabwe
3	Human fatalities	722 people killed	70 people killed	1303 people killed Many are still missing	52 people killed
4	Financial losses	USD 309 m	USD 71 million	USD 2.2 billion	USD 100 m
5	Socio-Economic Impact	(a) Destroyed 20,000 head of livestock (b) Damaged 60,000 homes (c) Damaged 230 dams (d) Destruction of crops and livestock	(a) Destroyed 130,000 homes (b) Damaged 130 schools (c) Destroyed buildings and infrastructure (d) Destroyed 277,000 ha of crops	(a) Massive damage on socio-economic infrastructure Destroyed banana plantation and several market gardens (c) Destroyed 20,000 houses in Mozambique (d) People lost food, clothing and their source of income Massive environmental damages	(a) Destroyed 60 to 80% of staple crop in Mozambique (b) Damaged 2500 homes in Mozambique (c) Environmental damages
6	Displaced people	329,000 people	33,000 people	128,941 people were displaced in Chimanimani alone	30,000 people were evacuated in Mozambique

Source Relief Web (2019)

7 Human Security and Tropical Cyclone Threats

The concept of Human Security and the relationship between tropical cyclones and socio-economic security is discussed. In this investigation, human security is the dependent variable that is acted upon by the effects of tropical cyclones. The UNDP Human , Development Report [17, 22], made the specific claim that individuals should be the referent of security instead of the states since the state's security was now less vulnerable while the security of individuals was fast becoming more vulnerable especially from internal and external threats. In general terms, human security combines the overarching tenets of both household socio-economic security and human development concepts in addressing the security concerns of humans as part of the community. At its basics, security relates to being free from internal and external threats. Accordingly, the United Nations Commission on Human Security [24: 23], defines human security, "as the means of protecting people from critical and universal threats and situations including building on their strengths and human aspirations." In this description, the concept of human security stresses the need to empower households by creating functional systems that provide individuals the basic building blocks for basic survival, dignity and descent livelihood. It is however important to note that socio-economic security has a domino effect since its insecurities can trigger a chain reaction of all other security components of human security. The United Nations Security Handbook [11], identifies five vital principles of human security, all in all, these human security principles can be implemented with a deliberate emphasis on reducing the harmful effects of tropical cyclones on human security giving focus on the vulnerable rural communities.

8 Tropical Cyclones and Socio-economic Security

This viewpoint of human security therefore resonates well with the focus of this investigation that is meant to analyse the effects of tropical cyclones on human security focusing on two rural households (Sussundenga village in Mozambique and Chimanimani rural village in Zimbabwe). In line with this outlook, household socio-economic security entails the need to ring-fence all the basic human needs, such as food, water, shelter, health care, basic incomes, education and the provision of essential social services.

FAO, IFAD and WFP [20], acknowledges that the smallholder farmers and less-privileged rural households are more vulnerable to socio-economic insecurities since they mainly settle on areas prone to tropical cyclone hazards. Authorities should consider relocating all vulnerable households settled along preferred tropical cyclones, to safer zones besides prioritising the construction of tropical cyclone resilient buildings and infrastructure.

On the other hand, economic security directly contributes to household socio-economic insecurity. To this end, the loss of household economic income sources coupled with the destruction of household property and belongings, the destruction of household-based income-generating projects and the loss of agricultural produce are all indicators of household socio-economic insecurity.

With regards to the two rural villages of Sussundenga and Chimanimani, household economic security was severely affected when the succeeding tropical cyclones destroyed economic facilities, such as bottle stores, grocery shops, grinding mills, banana plantations, market gardens, irrigation projects, or worse still the destruction of the entire business centres like Kopa Growth Point in Chimanimani. The existing gap established the need for the governments of Mozambique and Zimbabwe to seriously consider tropical cyclone resilient buildings and infrastructure owing to the increase in climate change-induced tropical cyclones.

Beyond causing production and socio-economic losses, recurrent tropical cyclones also cause widespread unemployment.

9 Tropical Cyclones and Household Food Security

Besides threatening the rural household economic sources of livelihood, climate change-induced tropical cyclones also threaten household food security. Subsequently, the World Food Summit (1996), concluded that food security exists when all people, have physical and economic access to adequate, safe and nutritious food to meet their food choices. The United Nations Development Programme [22], views food insecurity as the inadequate or lack of access at all times of food needed by a household to lead a healthy life. Category 4 and 5 tropical cyclones often damage agricultural and food systems, food reserves and they can also destroy livestock including livelihood-related infrastructure thereby posing a serious threat to household food security. On the other hand, transitory food insecurity is intermediate in nature and will require the mobilisation of external food resources in order to restore food security.

Environmental security also plays a critical role in promoting household socio-economic security. The United Nations Human Security Handbook [11], views environmental security as related to the harm inflicted on the living and health of humans or other lives by climate change induced hazards, such as storms, floods, droughts, rising sea waters and pollution. In this description, environmental security therefore covers such critical environmental issues, such as water pollution, land degradation and environmental degradation that can potentially cause socio-economic challenges within communities. Similarly, mudslides are responsible for massive human casualties, including the destruction of homes, buildings, dams, bridges, schools, clinics and roads that threatens community security.

10 Empirical Review of the Investigation

The researcher conducted an empirical review of similar studies in order to fully understand the link between the two variables of the investigation (tropical cyclones and the human security) funnelling it to household socio-economic security. The first empirical investigation focused on the effects of natural disasters in Africa whereas the second empirical investigation assessed the risks of cyclone-induced storm surge and sea level rise in Mozambique.

The primary objective of the impact of natural disasters in Africa was to assess the social, economic and political insecurities caused by the natural disaster hazards in the different regions of Africa in order to recommend sustainable disaster management policies and strategies. The gap left by this case investigation therefore needs to be exploited. The methodology used by Lukamba [13], was based on a statistical survey that covered all the five (5) major regions of Africa namely East Africa, West Africa, North Africa, South Africa and Central Africa. In addition, Africa has not developed human resource, social and economic capacity to adapt to the impacts of natural disasters. The major conclusion of this investigation was that natural disasters were increasing in Africa, thus, affecting the social, economic and political stability.

11 Cyclone-Induced Storm Surge and Sea Level Rise in Mozambique

The key findings of this investigation were that owing to its geographical location, Mozambique was a preferred path for deadly tropical cyclones that makes it one of the most vulnerable countries in Africa to the effects of tropical cyclone disasters and the ever-rising sea levels. The investigation further concluded that the frequency of storm surges emanating from tropical cyclones are subsequently causing severe economic damages whereas flooding is creating numerous humanitarian consequences as many people are killed, injured and displaced. Since natural disasters have no regard for political boundaries, this particular investigation recommended that African countries should effectively communicate so as to share up to date weather forecasts and storm tracking messages that will enable contingent planning and the implementation of evacuation procedures once a natural disaster strike.

12 Theories Guiding the Investigation

In view of this assertion, the researcher examined the Empowerment Theory (ET) and the Social Vulnerability Theory (SVT) in order to fully understand the effects of tropical cyclones on human security. According to Rappaport [15], the ET advocates the need to develop capacity that enables the vulnerable households within a community set-up to increase their own awareness, local resources and self-esteem including the need for the locals to build their own capacity to act correctly in the face of natural disasters. According to the UNODRR [28], community involvement enables rural households to literally own the imminent disaster since it is the very same community that suffers most when a tropical cyclone strikes.

One unique method of empowering rural households as part of the community is to uphold and promote indigenous knowledge systems (IKS) that is still intact in many African countries. The best way to make use of the IKS is to involve the households within a community set-up to solve problems affecting their socio-economic well-being. IKS is a concept that is useful if it is properly understood and conceptualised into practice in mitigating the effects of tropical cyclones. According to Blaike [1: 9], social vulnerability is a, "set of characteristics of a group or individual in terms of their capacity to predict, deal with, resist and recover from the threats of a natural hazard." In basic terms, social vulnerability entails being susceptible to multiple stressors and social shocks, including natural hazards, like tropical cyclones. Guided by the statement of the problem, research objectives, the Researcher discussed the link between tropical cyclones and human security including its sub-variable, socio-economic security.

13 Research Methodology

This research methodology focused on analysing the effects of tropical cyclones on human security in Mozambique and Zimbabwe in order to appreciate what is true and what is held to be true. In this investigation, a research philosophy was mainly concerned with the manner in which knowledge was acquired about the effects of tropical cyclones. By design, the major shortcoming of the positivism research approach is that it is not a particularly good approach to be undertaken when trying to explain the effects of tropical cyclones on human security, hence the need to combine this approach with other available research philosophies. The need to compensate the weaknesses of both the positivism and the interpretivism motivated the researcher to adopt the pragmatism research philosophy.

The pragmatism philosophy is a hybrid philosophy that combines both the positivism and interpretivism approaches, that is the triangulation of various research philosophies. On this basis, the pragmatism research philosophy helped the researcher to focus on what works as the truth basing on the research objectives.

It is a prerequisite for an investigation to adopt an appropriate research design that enables researcher to plan and structure the investigation in a manner that eventually addresses the research objectives and questions. In the same vein, Polit and Hungle [9: 50], define a research design as, "an overall plan for obtaining answers to the questions under investigation and for handling some of the difficulties encountered during research." Simply stated, a research design is an investigative method employed to gather evidence that is needed to answer research questions. As a result, the overall research design was descriptive in nature and focused on creating a holistic and

Table 3 Research design matrix

Research objectives	Research questions	Key variable indicators	Research instruments	Target groups	Assumptions
(a)	(b)	(c)	(d)	(e)	(f)
To assess awareness levels on tropical cyclones (TCs)	Are rural households aware of TC hazards?	Awareness of TC hazards and attitude towards TCs	Interviews and review of related literature	Traditional Leaders and affected families	Households have limited knowledge on TC hazards
To ascertain TC threats on rural household socio-economic security	How are TCs threatening rural household socio-economic security?	What are the indicators of socio-economic insecurities?	Questionnaire, personal interviews and FGDs	Traditional leaders, DCPUs, DDCs, CSOs and displaced households	TCs threaten the socio-economic security of rural households
To examine the effects of TC induced rural household vulnerabilities	What are the effects of the TC induced rural household vulnerabilities?	The effects of TCs on rural households	Questionnaire, interviews, FGDs and review of literature	Disaster experts, DCPUs, DDCs, CSOs and displaced households	TC escalates social vulnerabilities among rural households

Source Author's own construct (2012)

meaningful picture relating to the effects of tropical cyclones on human security with a specific focus on socio-economic security of two rural settlements in Mozambique and Zimbabwe. Choguil [12: 4], stresses that "the research design matrix is a system of rows and columns into which the components of a research project fit, including the aim, objectives, hypotheses, variables, methods of analysis and likely conclusions." In this way, the research design matrix that is shown at Table 3, summarises what the researcher intends to do in the investigation from the beginning right to the end of the completion of the investigation.

14 Research Approach

Three possible research approaches that were available to the researcher are namely the qualitative, quantitative and mixed research approaches. On the other hand, the quantitative approach gives emphasis to a systematic investigation of the noticeable phenomena through a statistical or computational method [3].

Having considered the characteristics, strengths and weaknesses of these two research approaches, the researcher selected the mixed research approach. The choice of the mixed research approach was determined by the fact that the approach permits a more complete, systematic and logical utilisation of data since it integrates both quantitative and qualitative research approaches.

15 Data Collection Methods

Data collection methods available to the researcher included the survey, experiment, observation, content analysis, research interviews and the Focus Group Discussion (FGD). In tandem with the research objectives, the researcher adopted the survey research method as the primary data collection method.

A survey research method is the procedure in which an investigator administers an investigation to a sample of the population in order to describe their attitudes and characteristics [10]. The researcher found out that the survey method was very flexible and convenient as it enabled the researcher to engage the research respondents through a variety of ways, like using the telephone, electronic mail (e-mail) and the use of hard copies (the questionnaire method).

Section A dealt with demographic information while Section B assessed the levels of awareness on tropical cyclone threats in line with the first research objective. Finally, Section D focused on the effects of tropical cyclones on human security with regards to rural household vulnerabilities in line with the third research objective.

Subsequently, the researcher also interviewed District Development Coordinators, District Civil Protection Members and Disaster Management Experts. According to Nachmias and Nachmias [6], an interview guide is critical when collecting information about personal feelings and opinions as it allows more questions to be asked and most importantly, it captures the respondents' own words thereby clarifying any emerging doubts. Personal interviews also enabled the researcher to further probe the respondents in order to obtain additional information and most importantly enabled the researcher to reduce costs through the conduct of telephone interviews. The researcher further conducted an FGD so as to triangulate the data collection methods. Morse [5], observed that an FGD method allows for an interactive discussion between pre-selected members led by a moderator and focuses on a specific topic. In order to achieve promptness and to keep the discussion focused to the research objectives, the researcher developed an FGD guide.

Basically, sources of data in a research investigation fall under two distinct categories. These are namely primary and secondary sources of data. This particular investigation relied on both the primary and secondary sources of data (Table 4).

The sample as a sub-set of the population used by the researcher is composed of thirty (30) members. Additionally, a convenient sampling method was used to select two (02) DDCs, four (04) CSOs, two (02) DCPUs, two (02) Disaster Management Experts and four (04) traditional leaders. The approach to determine the sample size was based on the assumption presented by Kothari [4], that the sample size should be at least 30 respondents or 1% of the population. The researcher was mainly worried with how well the data collection instruments measured what they purported to measure. Therefore, to achieve validity, the researcher triangulated the data collection

Table 4 Sample frame

Serial	Sample frame	Population	Sample size
1	Affected rural households	1144	16
2	District development coordinators (DDCs)	02	02
3	Civil society organisations (CSOs)	10	04
4	District civil protection units (DCPUs)	02	02
5	Disaster experts/academics	06	02
6	Traditional leaders	12	04
7		1176	30

Source Author (2019)

instruments through the use of the questionnaire, interviews and the FGD. To achieve reliability the researcher considered participant error, participant bias, researcher error and bias.

16 Ethical Considerations

Ethics was an important consideration since the investigation involved human feelings, emotions and beliefs. Ethical considerations enabled the researcher to gain maximum cooperation from the respondents. In this investigation, the researcher made sure that the respondents had a thorough understanding of the research and all other relevant risks. Furthermore, the researcher issued Letters of Consent (Appendix 4) to all the research participants. The researcher also ensured that the research participants were given sufficient detail about the objectives and purpose of the research and potential risks and benefits. As part of the ethical guidelines the researcher valued the confidentiality of information obtained from the respondents as well their anonymity. Accordingly, the researcher emphasized to the respondents that information gathered was exclusively used for academic purposes and names were not published during the process.

17 Data Analysis Procedure, Presentation and Analysis

In this section, the researcher describes the general plan used for data analysis. Data was analysed on a descriptive and interpretative manner because most of the data was collected through questionnaires, the interview guide and a FGD guide. The research methodology by describing the research philosophy, research design, research approach, data collection methods alongside data collection instruments. The chapter further covered the sources of data, the population and sample and data analysis methods.

The presentation and thematic analysis of data helped the researcher to methodically address the problem statement and research objectives. The second section presents thematic analysis of data based on the three key research investigation objectives. The first thematic analysis was based on the first research objective that sought to assess the levels of awareness on the threats of tropical cyclones on human security. The second thematic analysis focused on the effects of tropical cyclones on rural household socio-economic security whereas the third thematic analysis concentrated on the effects of tropical cyclones on human security with regards to rural household social vulnerabilities.

18 Response Rate and Demographic Attributes

A total of thirty (30) questionnaires were distributed to population sample composed of the affected rural households, District Development Coordinators, CSOs, DCPUs, traditional leaders and Disaster Management Experts. Whereas there is no general agreement on what is a good response rate, Fryrear (2015), argued that it is preferable to get a high response rate of 80% or higher to guarantee the reliability of the research results (Table 5) below.

Figure 2 shows that the respondents were made up of 16 females (57% of the respondents) and 12 males (43% of the respondents). The obtaining research results shows that the investigation involved both females and males although the females were on the majority heading rural households in Mozambique and Zimbabwe (Fig. 1).

Table 5 Questionnaire response rate

	Target sector	Issued	Returned	Variance	Response rate (%)
(a)	(b)	(c)	(d)	(e)	(f)
1	Affected rural households	16	14	− 2	87.5
2	District development coordinators	02	02	0	100
3	Civil society organisations	04	04	0	100
4	District civil protection units	02	02	0	100
5	Disaster management experts	02	02	0	100
6	Traditional leaders	04	04	0	100
7	Total	30	28	2	93

Source Author (2019)

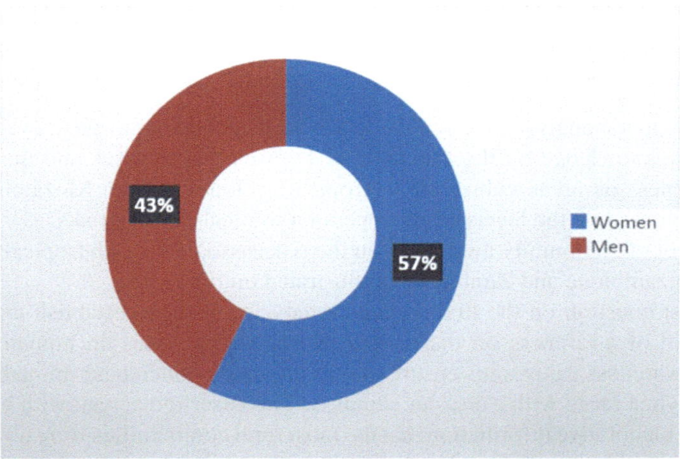

Fig. 1 Gender distribution. *Source* Author (2020)

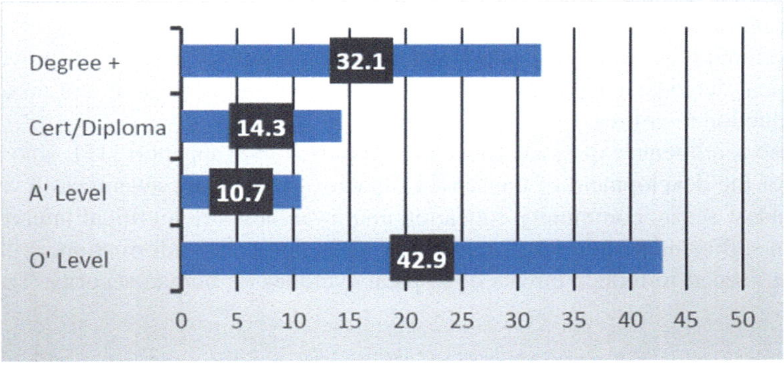

Fig. 2 Summary of educational qualifications. *Source* Author (2020)

19 Educational Qualifications

Figure 2 shows the educational qualifications of the respondents.

According to Fig. 2, 42.9% of the respondents held 'O' level qualifications, 10.7% of the respondents held 'A' level qualifications while 14.3% of the respondents had either certificates or diplomas. It can be deduced from these literacy statistics that the respondents fully understood the research questions and in turn responded objectively basing on their higher levels of education.

20 Thematic Analysis 1: Awareness on Tropical Cyclone Threats on Human Security

The first thematic analysis covered the levels of awareness on tropical cyclones, the community knowledge on the functions of Disaster Committees and the community awareness on areas vulnerable to tropical cyclones in both Mozambique and Zimbabwe based on the levels of awareness on tropical cyclone risks.

The level of community awareness on the risks associated with tropical cyclones in both Mozambique and Zimbabwe is illustrated on Fig. 3.

The first question on the first thematic analysis sought to establish the respondents' level of awareness on the threats of tropical cyclones on human security. Further, awareness campaigns ensure that family units understand immediate steps to follow when faced with a disaster situation. This assertion agreed with interviews that raised elaborative information that the most rural communities were aware of the threats imposed by tropical cyclones. Tracking the movement of cyclones empower households to take part in reducing tropical cyclone risks before, during and after a disaster. These findings are collaborated by the Focus Group Discussion (FGD that reiterated the fact that government authorities in Mozambique and Zimbabwe were not putting in place aggressive measures needed to raise the awareness levels of the rural communities vulnerable to tropical cyclones. Similarly, tracing the movement of tropical cyclones enables households to adopt proper corrective and protective risk reduction measures.

These sentiments expressed above are supported by Rappaport [15], who advocates for the development of household capacities to increase awareness levels. In its broadest sense, community education and awareness are a critical intervention that can sufficiently equip rural households with appropriate information, skills and attitude needed to reduce threats of tropical cyclones on human security. Tropical

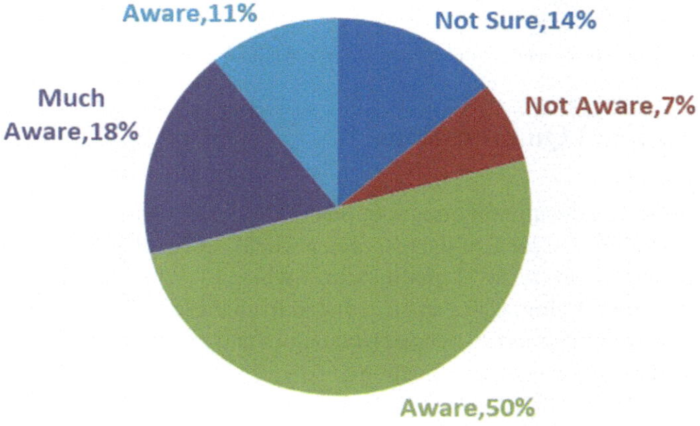

Fig. 3 Tropical cyclone risks. *Source* Author (2020)

cyclone education and awareness campaigns in Mozambique and Zimbabwe must be intensified to enable rural households to reduce the effects of tropical cyclone disasters.

21 Threats of Tropical Cyclone on Human Security

Images captured on Fig. 4 portrays the infrastructural destruction and the havoc that was wrecked by the deadly Tropical Cyclone Idai (2019), in Chimanimani, Zimbabwe.

The night of 15 March 2019 remains vivid in the minds of most rural households in Chimanimani almost a year when strong winds averaging 200 m/hr accompanied by heavy rains slashed through the Chimanimani Valley, bursting river banks, while landslides washed away numerous homes including food sources, medication and other critical sources of livelihood.

The threats of tropical cyclones on human security in both Mozambique and Zimbabwe are illustrated below.

As shown on Fig. 5, 21% of the respondents were not sure while 18% of the respondents were not aware of the threats posed by tropical cyclones on human security in both Mozambique and Zimbabwe. On the main, the majority 61% were quite aware of the threats of tropical cyclones on human security in both Mozambique and Zimbabwe.

The critical issue raised from the questionnaire and interviews was that most rural communities were either not sure or not aware of the threats of tropical cyclones on human security simply because the structure of disaster planning and management at community level was inadequate. The rural communities therefore require relevant information that will enable them to fully prepare to reduce all the insecurities occasioned by tropical cyclones.

These sentiments and opinions are in resonance with the United Nations Commission on Human Security Report [24] that called for human security to be viewed as protecting people from universal threats and situations including building their strengths and human aspirations.

22 Functions of Disaster Committees

The sub-question sought to assess the community knowledge regarding the functions of Disaster Committees in both Mozambique and Zimbabwe. Figure 6 shows that 21% of the respondents were not sure while 43% were not aware of the functions of the grassroots disaster committees. 18% of the respondents were aware of the functions of disaster committees while 11% of the respondents were much aware and 7% were very much aware of the functions of local disaster committees.

Fig. 4 Tropical cyclone Idai damages in Chimanimani. *Source* Herald 20 March 2019

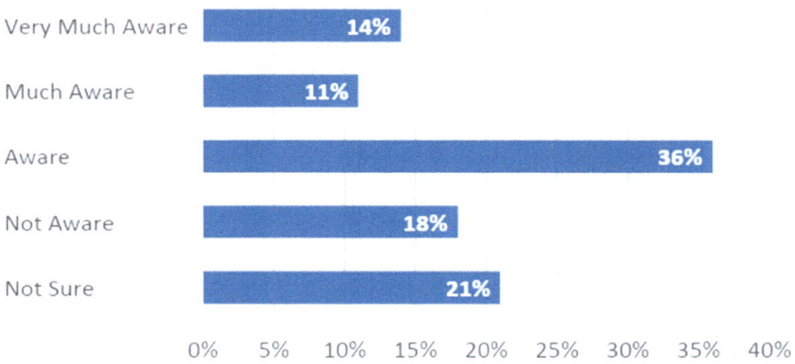

Fig. 5 Threats of tropical cyclone on human security

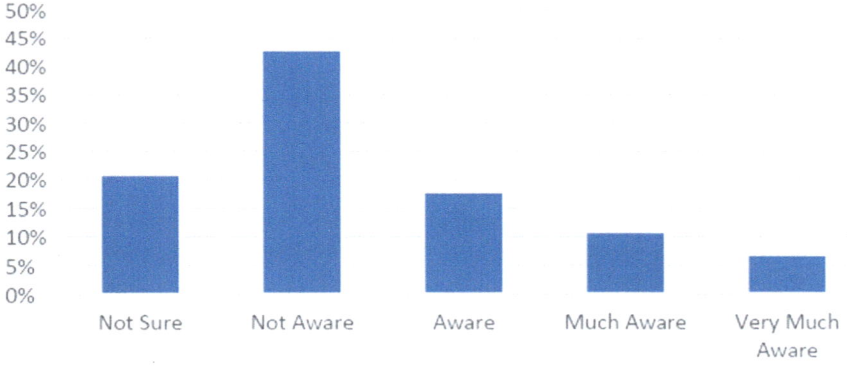

Fig. 6 Functions of disaster committee. *Source* Author (2020)

Interviews suggested that the fact that tropical cyclones continue to cause human loss and damages to property, buildings and infrastructure, it can be concluded that rural communities are not capacitated to mitigate the effects of tropical cyclones. Of concern is the fact that these DCPU are either not active on the ground and they remain reactive in nature thereby creating gaps within the grassroots level disaster structures, that exposes the rural households to frequent tropical cyclone hazards.

Interviews and the FGD revealed that rural communities can use IKS and local resources to reduce the threats of tropical cyclone on human security. This viewpoint is supported by [8], who maintained the need to acknowledge the fact that rural communities possess new insights and IKS that may help to identify, manage and resolve natural disasters, such as tropical cyclones. If properly utilised and put into practice, IKS can help to mitigate the effects of tropical cyclones and are very useful with regards to crop production, food preservation, environmental conservation, health care, disaster early warning systems and disaster management.

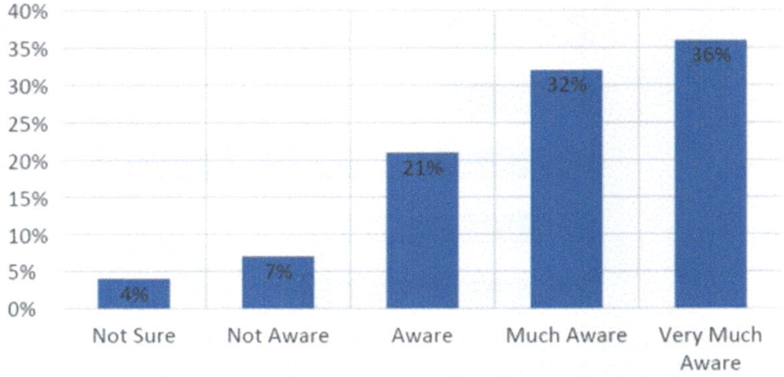

Fig. 7 Awareness on areas vulnerable to tropical cyclones. *Source* Author (2020)

23 Tropical Cyclone Vulnerable Areas

The sub-question sought to assess the levels of awareness on areas mostly susceptible to tropical cyclone hazards in Mozambique and Zimbabwe. 21% of the respondents were aware while 32% were much aware and 36% were very much aware of areas most vulnerable to tropical cyclone hazards (Fig. 7).

Information obtained from questionnaires was that most rural areas in Mozambique and Zimbabwe were vulnerable to tropical cyclone hazards mainly due to higher levels of poverty and lower adaptive capacities. Interviews and the FGD indicated that households settled in rural areas of both Mozambique and Zimbabwe possessed little financial resources and struggled the most in the aftermath of tropical cyclones. This situation is compounded by the fact that most of the rural households in Mozambique and Zimbabwe are settled along river lines and within low-lying areas that are prone to tropical cyclone hazards.

24 Thematic Analysis: Effects of Tropical Cyclones on Socio-economic Security

The second thematic analysis covered the effects of tropical cyclones on rural household survival, the relationship between tropical cyclones and environmental change and the threats of tropical cyclones on rural household food security in both Mozambique and Zimbabwe based on the effects of tropical cyclones on rural, household economic survival. The question sought to establish the respondents' views on the effects of tropical cyclones on the sources of economic survival for rural communities in both Mozambique and Zimbabwe.

Figure 8, shows that there were no respondents who were either not sure or not aware that tropical cyclones affects the sources for economic survival of the rural

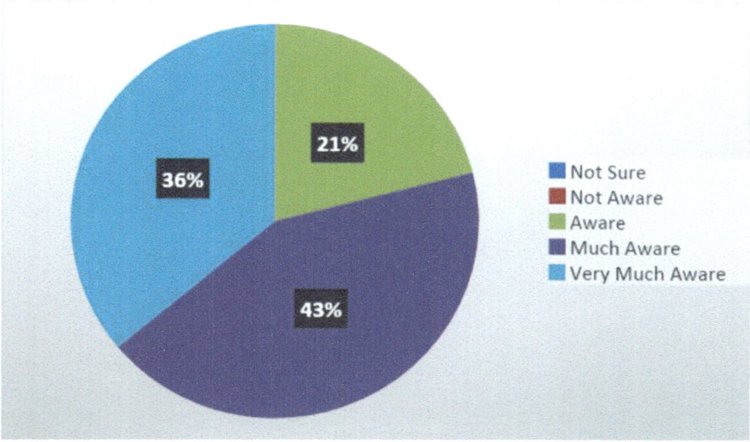

Fig. 8 Tropical cyclone effects on rural household economic survival. *Source* Author (2020)

communities. However, the majority of the respondents (21% were aware, 43% were much aware whereas 36% were very much aware) that tropical cyclones are increasingly affecting all the sources for economic survival in most rural communities in both Mozambique and Zimbabwe. To make matters worse, the rural household survival options are further affected by the sudden increase in tropical cyclone calamities that tend to disturb their basic social and economic means of survival.

Interviews reinforced the statement that the national economies of both Mozambique and Zimbabwe are largely based on agricultural activities and livestock products. Therefore, livestock possession act as strategic household investment since cattle, goats, sheep and chickens can easily be disposed for cash when the need arises especially within the rural communities.

The FGD further proved that the disturbance of agricultural systems occasioned by frequent tropical cyclones events is severely affecting the basic socio-economic sources for survival of most rural communities in the two countries under investigation.

24.1 Threats of Environmental Change on Rural Livelihood

Information gathered from questionnaires as shown on Fig. 9, indicates that there were no respondents who were not sure that environmental changes threatens rural livelihood in Mozambique and Zimbabwe. 43% were aware, 39% were much aware and 14% were very much aware that tropical cyclone-induced environmental changes were progressively threatening the livelihood of rural communities in Mozambique and Zimbabwe.

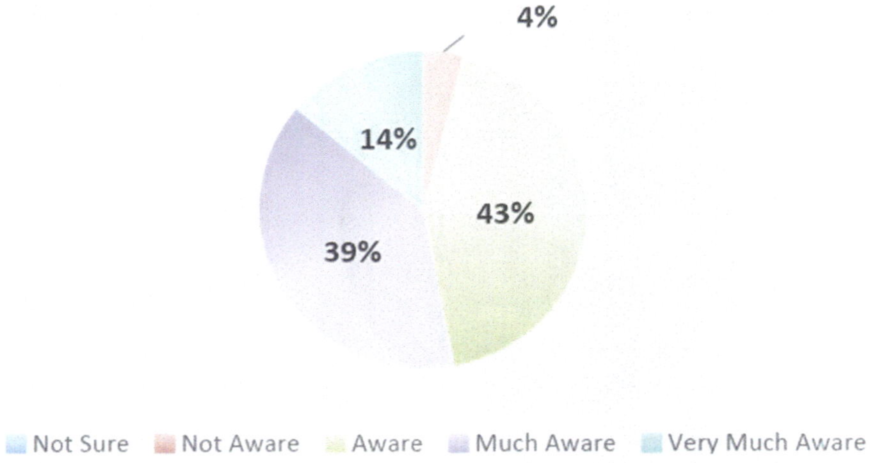

Fig. 9 Threats of environmental change on rural livelihood. *Source* Author (2020)

The thematic analysis from interviews revealed that most rural households rely on climate-sensitive natural systems, such as water supplies, subsistence agricultural practices, livestock husbandry, natural resources, wild fruits and herbs. Subsequently, this often limits the survival options of the rural households since they over-rely on natural resources and systems for both basic survival and economic survival.

Findings from interviews and the FGD are consistent with views expressed by the Intergovernmental Panel on Climate Change (2007), that environmental change has immediate and direct effects on the socio-economic well-being of millions of rural households who depend on natural resources for their basic survival. According to Moffett (1994), environmental changes imposed by ravaging tropical cyclones affect the quality of soil that is a critical concern to all rural households who mostly rely on traditional agricultural practices.

24.2 Threats of Tropical Cyclones on Rural Household Food Security

Figure 10 illustrate the threats of tropical cyclones on rural household food security in both Mozambique and Zimbabwe.

Information from interviews reinforced the fact that the increase of tropical cyclone hazards is contributing to the destruction of arable land that is mainly used for subsistence farming by mostly peasant rural communities thereby upsetting rural household food security. Given this scenario, there is need to equip rural communities with the requisite knowledge and skills that empowers them to be more cyclone

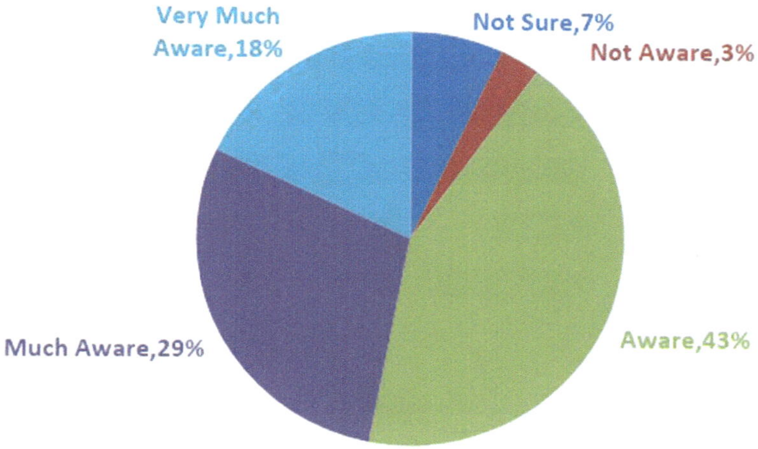

Fig. 10 Tropical cyclone threats on household food security. *Source* Author (2020)

resilient and that they can be more actively involved in reducing food insecurities that threaten their livelihoods.

These views are consistent with the FAO, IFAD and QFP Report [20] that outlined most rural communities as exposed to chronic, seasonal and transitory food insecurities. Seasonal food insecurity and is linked to cyclical floods and winds, that temporarily upsets food systems and food reserves. The last category is the transitory food insecurity that constrains rural households to meet their minimum food requirements especially after a catastrophic disaster.

25 Thematic Analysis: Tropical Cyclones and Household Social Vulnerability

This thematic analysis covered the relationship between tropical cyclones and rural household displacements, the effects of tropical cyclones on socio-cultural systems and the effects of tropical cyclones on rural development in both Mozambique and Zimbabwe based on Tropical Cyclones Contributing to Rural Household Displacements.

Figure 11 illustrate the relationship between tropical cyclones and rural household displacements in both Mozambique and Zimbabwe.

Accordingly, Fig. 11 shows that 11% of the respondents were not sure that tropical cyclones contribute to the displacement of the rural population in Mozambique and Zimbabwe. 7% of the respondents were not aware that tropical cyclones contribute to the displacements of the rural communities. The majority of the respondents, 36% were aware, 23% were much aware and 21% were very much aware that tropical

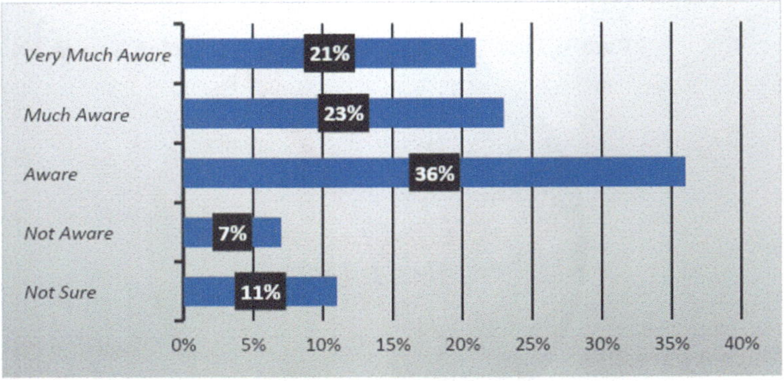

Fig. 11 Tropical cyclones displacing rural households. *Source* Author (2020)

cyclones are contributing to the displacement of the rural population in Mozambique and Zimbabwe. In the rural areas of Mozambique and Zimbabwe, rural communities are forced to relocate following the destruction of their homes, personal belonging, crop fields and livestock.

Interviews further confirmed that tropical cyclone-induced disasters are on the increase hence the need to strengthen community resilience by improving the food security of the rural communities and by ensuring that rural households are better equipped to cope with all tropical cyclone risks. Above all else, the vulnerable rural communities must be engaged and involved in the disaster planning process so that their unique and specific needs are factored into the area-based disaster risk reduction initiatives.

Similar sentiments were expressed through the FGD that stressed the importance of community-based disaster management initiatives. Communities must also be aware of the local resources at their disposal and how to use them to reduce the risks of natural disasters that frequently threaten their community well-being. These above sentiments are consistent with the views expressed by [1], that several factors increase the susceptibility of a community to the risks of cyclones, which forces them to relocate. A combination of these factors can force the rural communities to settle in disaster-prone rural areas.

25.1 Effects of Tropical Cyclone on Social and Cultural Systems

Figure 12 illustrate the threats of tropical cyclones on rural household food security in both Mozambique and Zimbabwe.

Figure 12 shows that 25% of the respondents were not sure while 3% of the respondents were not aware that cyclones were affecting social and cultural systems

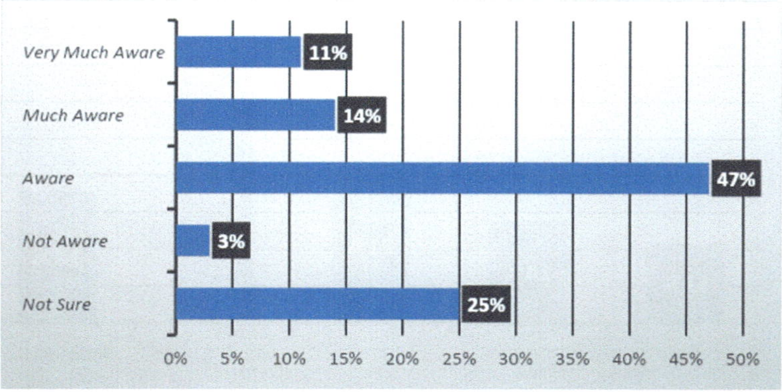

Fig. 12 Tropical cyclones affecting social and cultural systems. *Source* Author (2020)

in the rural areas of Mozambique and Zimbabwe. However, the majority 47% of the respondents were aware, 11% of the respondents were much aware while 14% were very much aware that tropical cyclones were threatening social and cultural systems in the two countries. The major theme emerging from interviews was that tropical cyclones are causing untold social and cultural insecurities.

In the same vein, the FGD confirmed that cyclones are increasingly destroying sacred places, cultural heritage sites and cultural centres. There is need for the governments of Mozambique and Zimbabwe to put in place measures necessary to reduce the risks of cyclones on social and cultural systems in order to safeguard human security.

25.2 *Tropical Cyclones Affecting Rural Development*

Figure 13 illustrate the effects of tropical cyclones on rural development in both Mozambique and Zimbabwe.

According to Fig. 13, 11% of the respondents were not sure that tropical cyclones retard rural development in Mozambique and Zimbabwe. 14% of the respondents were not aware that tropical cyclones retard rural development. Alternatively, 21% of the respondents were aware that tropical cyclones retard rural development, 36% of the respondents was much aware whereas 18% of the responds was very much aware that tropical cyclones are retarding rural development within rural communities in Mozambique and Zimbabwe. Accordingly, the diversion of national resources meant for rural development to cater for the effects of tropical cyclone disasters continues to retard rural development in both Mozambique and Zimbabwe. Critically, most developing countries lacked the economic capacity to withstand the effects of natural disasters hence the continuous retardation of rural development.

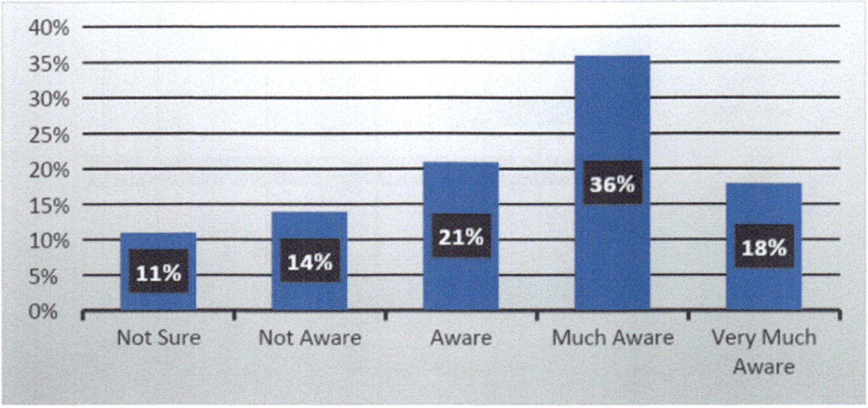

Fig. 13 Tropical cyclones affecting rural development. *Source* Author (2020)

26 Critical Analysis of the Submission

Tropical cyclones continue to threaten human security in both Mozambique and Zimbabwe yet awareness and sensitisation campaigns are not enabling the affected communities to appreciate their roles before, during and after a disaster. The major findings of the investigation will assist the researcher to address the three objectives of the investigation and the problem astatement. The next chapter presents the major findings and conclusions of the investigation based on the research objectives and questions.

27 Findings, Conclusions and Recommendations

A summary of the critical findings, conclusions and recommendations of the investigation in line with the research objectives and the problem statement are outlined hereunder. The purpose of this research investigation was to analyse the effects of tropical cyclones on human security with regards to the socio-economic security of rural households assess the levels of awareness of tropical cyclones on rural households, ascertaining the threats of tropical cyclones on rural household socio-economic security and examining the link between tropical cyclones and rural household social vulnerabilities. In order to fully appreciate and understand the threats and effects of tropical cyclones on human security with regards to rural socio-economic security, the researcher undertook a comprehensive literature review on factual works by scholars who have written on the link between climate change-induced tropical cyclones and human security.

28 Summary of Critical Findings

In line with the investigation objectives, the following critical findings are made:

- The threats posed by tropical cyclone hazards are real and their impact on national security should be taken seriously.
- Most rural households in both Mozambique and Zimbabwe are settled along river lines and in low-lying areas, exposing them to severe physical vulnerability.
- The effects of tropical cyclone hazards on human security are forcing rural households in Mozambique and Zimbabwe to disintegrate and abandon their homes thereby exposing them to extreme emotional aftershocks, tensions among social groups including undesirable socio-economic insecurities.
- Tropical cyclone disasters continue to destroy rural infrastructure, such as schools, clinics, roads, bridges and dams, that hampers rural development since the central government of Mozambique and Zimbabwe is forced to concentrate on rebuilding and reconstructing the damaged or destroyed infrastructure.

29 Discussions and Conclusions

From the findings, the investigation concluded that the threats of tropical cyclones are real and greatly affects human security with regards to the socio-economic security of rural households in Mozambique and Zimbabwe during the period 2000 to 2019.This shortcoming is attributed to the fact that the governments of Mozambique and Zimbabwe are lacking in raising the levels of awareness of the rural communities on the human security threats imposed by the ever-increasing tropical cyclones episodes. Most rural communities in Mozambique and Zimbabwe are settled on the preferred path of tropical cyclones, yet they have the lowest levels of awareness on tropical cyclones, that further exposes them to disaster risks. In conclusion, the research investigation was restricted to assessing the levels of awareness on tropical cyclones, ascertaining the effects of tropical cyclones on rural household socio-economic security and examining the effects of tropical cyclones on human security with regards to rural household vulnerabilities with recommendations.

30 Recommendations

In line with the research objectives, the research findings and conclusions, this investigation recommends the following practical solutions to address the effects of tropical cyclones on human security with regards to rural socio-economic security in Mozambique and Zimbabwe:

- Both the governments of Mozambique and Zimbabwe must ensure that community-based education and awareness campaigns regarding tropical cyclone hazards should be viewed as a top national priority in order to safeguard human security.

The Meteorological department, Community leaders, traditional leaders, churches, schools, medical centres, the business community, CSOs, the media and all other community-based groups should all be involved in these community-wide grass roots awareness campaigns

- The governments of Mozambique and Zimbabwe, through their respective line ministries must ensure insure grassroots disaster risk reduction and management initiatives are contributing significantly towards safeguarding human life, property and infrastructure from the devastating effects of tropical cyclones.

References

Books

1. Blaikie P (1994) Natural hazards, people's vulnerability and disasters. Routledge, London
2. Christie F, Halon J (2001) Mozambique and the great flood of 2000. Indianan University Press, Bloomington
3. Given LM (2008) The Sage Encyclopedia of qualitative research methods. Sage Publications, Los Angeles
4. Kothari C (2004) Research methodology: methods and techniques. New Age International Publishers, New York
5. Morse JM (2003) A handbook of mixed methods in social and behavioural research. Sage Publications, New Delhi
6. Nachmias C, Nachmias D (1985) Research methods in social sciences. Edward Arnold Publications, London
7. Neumann CJ, Jarvinen BR, McAdie CJ, Elms JD (1993) Tropical cyclones of the north Atlantic ocean: historical climatology. Department of Commerce, Washington DC
8. Noyoo N (2007) Indigenous knowledge systems and sustainable development: relevance for Africa, 1st edn. University of Cape Town, Cape Town
9. Polit D, Hungler B (1999) Nursing research: principles and methods. Lippincott, Philadelphia
10. Shaughnessy J, Zechmeister E, Jeanne Z (2011) Research methods in psychology, 9th ed. McGraw Hill, New York
11. United Nations Human Security Handbook (2016) Human Security handbook An integrated approach for the realization of the sustainable development goals and the priority areas of the international community and the United Nations system

Journals

12. Choguil (2005:4) The research design matrix: a tool for development planning research studies. Habitat Int 29(4):615–626

13. Lukamba MT (2010) Natural disasters in African countries. J Trans-Discip Res Southern Africa 2(6):17–19
14. Neumann JEK, Sai AE, Lindsay R, Caroleen V (2013) Assessing the risk of cyclone-induced storm surge and sea level rise in Mozambique. UNI-WIDER 1(3):45–56
15. Rappaport J (2008) Studies in empowerment; introduction to the issue. Prevent Human Serv J 2(3):1–7
16. Shultz JM, Russell J, Espinel Z (2005) Epidemiology of tropical cyclones: the dynamics of disaster. Dis Dev 1(27):21–35

On Line Journals

17. UNDP (1994) Human development report 1994: new dimensions of human security. http://www.hdr.undp.org/en/content/human-development-report. Accessed Oct 2019
18. United Nations Office for Disaster Risk Reduction (Online). 2018 Annual report. Available at:https://reliefweb.int/sites/reliefweb.int/files/resources/64454. Accessed 20 Oct 2019

Reports

19. United Nations Office for Disaster Risk Reduction (UNISDR) (2019)—Report
20. FAO, IFAD and WFP (2015) The state of food insecurity in the world 2015: meeting the 2015 international hunger targets: taking stock of uneven progress. Rome.
21. The Post-Cyclone Idai Recovery Programme (2019) Multinational—Programme post cyclone Idai and Kenneth emergency recovery and resilience programme for Mozambique, Malawi and Zimbabwe—(PCIREP)—Appraisal report
22. UNDP (2014) Human Development Report 2014: sustaining human progress—reducing vulnerabilities and building resilience. New York. http://hdr.undp.org/en/content/human-development-report-2014
23. United Nations Office for Disaster Risk Reduction (UNISDR) (2019) ISBN/ISSN/978-92-1-004180-5
24. United Nations Commission on Human Security UNCHS) (2003), New York
25. United Nations High Commissioner for. Refugees (UNHCR) (2003) Human security now. New York
26. United Nations Office for the Coordination of Humanitarian Affairs (UNOCHA) (2019) Annual report. New York
27. United Nations University (UNA) 2013 Annual report
28. United Nations Office for Disaster Risk Reduction (UNDRR) (1999) Report

Wind Scale

29. Saffir-Simpson Hurricane Wind Scale (SSHWS) (1971)

Natural Resource Governance and Disaster Risk Management in Zimbabwe: Lessons from Past Experiences

Sifiso Mpofu, Gay Tapiwa Gweshe, and Barbra Maiwasha

Abstract The rising trends of conflicts and disaster occurrences around the extractive industry in Zimbabwe has led to extensive efforts by the government in putting into place essential frameworks of good governance and disaster risk reduction. The essential frameworks of natural resource governance are consistent with the UNDP's Sustainable Development Goals and the United Nations' Sendai Framework for Disaster Risk Reduction (2015–2030) whose thrust is to prioritise strengthening of disaster risk reduction governance and economic disaster preparedness. Given that resource governance is a prime function of the State, often characterised by hierarchical involvement of several stakeholders, it is pertinent to interrogate the concept and frameworks of natural resource governance and how it is of strategic value to both disaster risk reduction and national development in Zimbabwe. Whilst governance is often cast as encompassing the legislative frameworks, institutional and structural engagement; natural resource governance is even a more critical aspect of governance warranting scholarly attention. Using qualitative research method, and drawing lessons from past experiences, this chapter is poised to examine resource governance within the context of disaster risk management in Zimbabwe. The need to prioritise strengthening natural resource governance policies, frameworks and institutions in order to enable greater community beneficiation from its resource wealth is glaring.

Keywords Zimbabwe · Governance · Good governance · Resource governance · Sustainable development · Disaster risk management

S. Mpofu (✉) · G. T. Gweshe · B. Maiwasha
University of Zimbabwe, Harare, Zimbabwe
e-mail: mpofuskbm@gmail.com

R. Kasimba and I. Chirisa (eds.), *Disaster Risks and Hazards*, Advances in 21st Century Human Settlements, https://doi.org/10.1007/978-981-96-4577-0_14

1 Introduction

Zimbabwe is a land locked country located in Southern Africa (Reference for this factual statement required!). Economically, Zimbabwe is classified as a poor 'third world' country yet this country is endowed with varied natural resources. Zimbabwe's diverse mineral wealth and its economic performance does not reflect qualitative sustainable development premised on the efficacy of its natural resources thereby making the country one of the poorest countries in the world since it is ranked at number 27th out of 194 countries in terms of the World Bank poverty ranking index (International Monetary Fund, World Economic Outlook April 2021). This rating is based on the country's Gross Domestic Product (GDP) that is built on purchasing-power parity (PPP) per capita.

Besides its varied mineral resources, Zimbabwe hosts rare wildlife in its national parks and enjoys very good climatic conditions in addition to fertile agricultural land—these natural resources make Zimbabwe an ideal rich country, but it is not since less than **20%** of children between 6 and 23 months receive a minimal adequate diet thus compromising their quality of life (*2020 Global Report on Food Crises*). Zimbabwe is one of the many heavily indebted poor countries despite being "endowed with enormous natural wealth that includes Marange diamond fields, thought to be the biggest diamond find in over 100 years, and the second largest platinum reserves in the world" (www.centrefornaturalresourcegovernance). Such realities for a country blessed with natural resources, spells a crisis of resource governance hence good governance and disaster risk management interventions are essential.

In this chapter, we discuss and interrogate the need for clear and inclusive frameworks for resource governance for economic development within the context of Zimbabwe's laws that govern its natural resources. The express goal of this chapter is to promote qualitative resource frameworks and stakeholder participation in Zimbabwe's governance of her natural resources. The thesis of the chapter is premised on the theoretical assumption that in order to address disaster risk, not only do the underlying drivers of disaster risk need to be addressed to avoid risk generation [29], but the governance of natural resources must be reinforced. There is need for the country to prioritise strengthening natural resource governance frameworks and institutions in order to enable greater community beneficiation from its resource wealth.

2 Context

In this study, we are mindful that "governance of natural resources is complex, with a myriad of competing rights, users and governance structures, involving a wide range of actors at multiple levels and across a broad range of temporal and geographical scales" [11]. In our analysis of the Zimbabwe context, we observe that the existing state institutions have severe challenges in executing their mandate when it comes to resource governance resulting in increased vulnerability of the natural resources hence illicit deals that undermine the beneficiation policies when it comes to the benefits that must be derived from the same. On paper, Zimbabwe has fairly good policy frameworks but these are not supported by good governance practices on the ground. Observably, there is need to involve independent and civic society stakeholders in governance structures for the greater good of the country.

In our view, the state must promulgate smart governance policies and frameworks that mitigate potential disaster in terms of the extraction and usage of natural resources by capitalist players to no benefit for the country. Lessons from the Gulf countries (Middle East) that are rich in crude oil and have strong resource governance policies and structures have resulted in these arid countries having a high gross domestic product (GDP) complimented by its purchasing-power parity (PPP) per capita. This is true of Saudi Arabia that is ranked the 24th richest country in the world (www.indextofpoorcountries). Clearly, effective resource governance is a critical element of sustainable development. In fact, it is an integral part of good governance and sustainable development resource governance is an important aspect of accountability since all facets of public governance benefits directly from efficient resource governance hence the need for effective resource governance and management policies and structures towards greater state beneficiation from its God given resources. We concur with Amaratunga et al.; that the state must be magnanimous in involving more stakeholders when it comes to resource governance for transparency and accountability purposes. In the case of Zimbabwe, where mineral resources are in abundance, we believe that effective mining laws and contracts must spell out clearly the terms of what the country must gain from its extractive resources.

For effective resource governance, strong regulatory frameworks and competent institutions are the bedrock towards achieving sustainable development from natural resources. This calls for "flexible and adaptive forms of governance, with systems that can respond quickly to change and are firmly anchored in the local political economy" [11]. The fundamental principles of resource governance include respect for "human rights, political openness, participation, tolerance, administrative and bureaucratic capacity and efficiency. The rule of law, transparency, equity, consensus-orientated accountability and strategic vision are other essential elements of good governance" [8]. Such critical regulatory frameworks make resource governance an integral and essential aspect in the realm of good governance and risk management.

3 Methodology and Theoretical Framework

Our research is premised on the qualitative method. Denzin and Lincoln understand qualitative methodology as an "interpretive naturalistic approach to the world" [12]. The nature of this work involves examining socio-economic perceptions and experiences within the context of natural resource governance in a social setting hence issues discussed here are premised on a qualitative research methodology discourse. Mason [20] argues that qualitative research engages the researcher(s) with issues that matter and ways that matter in exploring the dynamics of our economic and social world as well as our experiences and feelings. Through the qualitative paradigm, we are fully engaged in the exploration of best resource governance models that result in a win–win-solution for both the state and the local communities. Our methodology is fully engaged in collecting and analysing opinions and experiences; meaning and perspectives in the exploration of natural resource towards promoting good governance and mitigating avoidable disaster risk reduction.

The study draws its theoretical basis from the collaborative governance theory advanced by Ansell and Gash [6]. We define collaborative governance as the processes, systems, structures and arrangements in modern public management in which people across the boundaries of public agencies, levels of government, private and civic spheres deliberately and constructively engage to solve public problems or satisfy a common good [6, 13; Sun 2017). This theory asserts, a co-design, co-production, co-assessment of policies from community stakeholders with the goal of pursuing community resilience and sustainable socio-economic development [7]. Drawing on collaborative governance theory, the study contends that governance and disaster risk management cannot be tackled within single defined boundaries.

As a consequence, to address the frequent challenges besetting effective disaster management in Zimbabwe, collaborative approaches to natural resource governance are pertinent to create public value, and ensure that transparency and accountability are better achieved. Our theory implies that better natural resource governance will generate foreign currency, strategically positioning the country for sustainable development. Guided by Ansell and Gash [6], the study is not blind to the fact that certain critical conditions would determine whether or not this mode of governance produces anticipated results in disaster risk management. Such factors include stakeholders' participation, community commitment, power and resources imbalances, leadership pragmatism, collaboration and dialogue, trust building, shared understanding and smart institutional frameworks.

4 Literature Review: Governance and Good Governance

In our analysis, we understand governance to imply the interaction between formal and informal structures, processes, rules, traditions, laws, regulations, discursive debates, and negotiations. We further define governance to mean decision-making processes that affect a country's wellbeing in terms of social, political, and 'economic activities and its relationship with other economies'. For the benefit of doubt, governance is a "multi-level and multi-actor phenomenon having spread the role and power of governing away from the central nation state to many different actors involved in different aspects of society" [17: 47]. Our view of governance has major implications for equity, poverty and quality of life [8: 7].

In sustainable development circles, governance must be measured hence in modern management circles we talk of good governance. Governance without the component of 'good' does not always deliver sustainable development. In this context, it is critical that we explore the concept and meaning of good governance. But one may ask, what constitutes good governance? Good governance relates to how a society is governed and how it achieves its developmental goals (social, political and economic) for the greater benefit of citizens [26]. From a socio-political perspective, good governance is a "public good that citizens should be entitled to" [25]. More specifically, we understand good governance as qualitative leadership and decision making that is informed by transparency, accountability, fairness, efficiency, participation and decency towards enhancing economic and sustainable development.

Good Governance influences the way in which national and sub-national actors (including governments, parliamentarians, public servants, the media, the private sector, and civil society organizations) are willing and able to coordinate their actions to manage and reduce disaster-related risk [28]. The definition of governance varies in the literature resulting in terminological confusion. In order to confiscate this conceptual confusion, we adopt the following definition:

> Governance, as defined by UNDP, is the exercise of political, economic and administrative authority in the management of a country's affairs at all levels. It comprises mechanisms, processes and institutions through that citizens and groups articulate their interests, exercise their legal rights, meet their obligations and mediate their differences. Governance encompasses, but also transcends, government. It encompasses all relevant groups, including the private sector and civil society organizations. [28: 1]

We note that resource governance is a ubiquitous concept. And it is common knowledge that where there are too many cooks, the soup will be spoilt. Within the disaster risk management context, the term is increasingly becoming catchy and increasingly elusive. Despite the challenges associated with the term, governance is the umbrella under that disaster risk reduction takes place [28: 1].

The fundamental principles of good governance include "respect for human rights, political openness, participation, tolerance, administrative and bureaucratic capacity and efficiency. The rule of law, transparency, equity, consensus-orientated accountability and strategic vision are other essential elements of good governance" [8]. Under good governance, policies and regulatory frameworks are "stabilised and turned into institutions that enjoy the confidence of citizens" [25]. While good governance is mainly associated with the principles that are unique in any state/country towards ensuring the smooth flow of operations in dealing with the national agenda, the capacity of a country to sustainably manage natural resources is largely depended on critical institutional frameworks and deliberate stakeholder inclusion complimented with good governance policies.

5 Natural Resource Governance

In our discourse, natural resource governance refers to "the norms, institutions and processes that determine how power and responsibilities over natural resources are exercised, how decisions are taken and how citizens—men, women, indigenous people and local communities—participate in and benefit from the management of natural resources" (International Union for the Conservation of Nature, IUCN website). Resource Governance is, therefore, an "authoritative tool to help decision-makers, and those who hold them accountable, to ensure that extractive industries are responsibly managed as countries navigate the challenges ahead" [8: 8]. As a measure of good stewardship of the natural resources dotted across the world, the UNDP has come up with a Resource Governance Index (RGI). The RGI is a critical governance tool that "assesses transparency and accountability in the oil, gas and mining sector worldwide" (www.centrefornaturalresourcegovernance). In analysing the RGI, Muzulu notes that "the political climate surrounding Zimbabwe's extractive sector is characterized by hostility, mistrust, and lack of transparency" (Zimbabwe Independent, November 16 2012). In our assessment of the present state of affairs in Zimbabwe's extractive industry, we note that the same conclusions reached by [22] still prevail.

Zimbabwe's natural resources are scattered across communal areas hence these valuable natural resources are managed under the national community-based natural resource management initiative, the Communal Areas Management Programme for Indigenous Resources (hereinafter called CAMPFIRE). CAMPFIRE is a "multi-level programme implemented in a complex system … aimed to change the governance of wildlife, seeking to integrate local communities into the wildlife and natural resource management governance system" [17: vii]. Just like accountability, resource governance "is a legal and social obligation on the part of those holding political, bureaucratic or technical positions to carry out certain clearly defined responsibilities or functions" [4: 1].

6 The Efficacy of Resource Governance for Economic Development

In our analysis of resource governance, we are mindful that there are existing policies and institutional frameworks in the country's natural resource management and governance systems. But our concern is; are these existing governance frameworks and institutions strong enough to ensure transparency and good governance in the management of the country's natural resources? In as much as good governance is an important concept in dealing with disaster risk management, it is also critical to apply the same concept of good governance in natural resource management. We are pragmatic in our definition of resource governance. Without much-ado, resource governance is a measurement of critical processes related to the control and management of resources in any setting. Natural resource governance therefore, refers to "the norms, institutions and processes that determine how power and responsibilities over natural resources are exercised, how decisions are taken and how citizens—men, women, indigenous people and local communities—participate in and benefit from the management of natural resources" (International Union for the Conservation of Nature, IUCN website).

In our delineation of issues at stake, we are mindful that resource governance covers many issues, from the public management of natural resources themselves "through to the management of the revenue generated (thereof)" [21]. Resource governance in public management resonates very well with issues of accountability, "an integral part of good governance" [4]. Those who are entrusted with the responsibility for overseeing the usage of our varied resources must always be "held accountable for what they are expected to do to" [4] with these resources. Observably, transparency alone in resource governance cannot attain "sustainable development outcomes ... (it) must be supported by strong regulatory frameworks, competent and strong institutions, flexible, collaborative and adaptive management practices, a solid understanding of the political economy at the local level, sound and sustainable fiscal regimes and a leadership commitment to using the revenue to promote sustainable development outcomes" [11]. Just like accountability, resource governance "is a legal and social obligation on the part of those holding political, bureaucratic or technical positions to carry out certain clearly defined responsibilities or functions" [4].

For effective resource governance, we need flexible, collaborative and adaptive management practices anchored in a solid understanding of the political economy to adapt to changing circumstances and local context that yield great results for the nation. Since governance is not only confined to the state, issues of accountability and resource governance also become "a complex process, going beyond legal and judicial domains. Both civil society and corporate actors demand certain standards in governance and in the process, accountability also acquires a social dimension" [4]. In this context, resource governance becomes a driver of accountability that must not be left to the state alone but must involve all stakeholders that are actively engaged on good governance and disaster management platforms.

Our present reality in many under developed countries show that with effective resource governance, "natural resource wealth can potentially contribute to development outcomes, but many resource-rich countries are plagued with unsustainability, conflicts, dysfunctional institutions, poor governance, corruption and weak economic performance" [11], a tragedy of lack of institutional frameworks and political will that is indicative of lack of good governance structures. We are certain that effective resource governance frameworks of our natural resources will result in good management and increased accountability of the same leading to the demystification of the 'resource curse' that haunting many third world countries. In the same vein, effective resource governance policies and frameworks lead to "sustainable fiscal regimes and a leadership committed to the use of revenues for the benefit of all" (Chêne, ibid.) thereby translating resource wealth into sustainable development outcomes for the nation.

Natural resources can bring considerable wealth to a country thereby "contributing to livelihoods, food security and the green economy, and generating trade and enterprise at local, national and international levels" [11] but many third world countries rich in natural resources and with potential billions of revenue at stake, natural resources tend to be associated with poor governance and corruption geared for the benefit of those in political control. This is all thanks to lack of resource governance frameworks and institutions. Harrison argues for community involvement in the governance of natural resources, she proffers what she calls 'the efficiency argument'. This theory sees people's participation as a means to an end, hence local people's participation tends to "provide local technical and environmental knowledge, including the values and local needs that will better inform the policy and decision-making processes" [17]. Another benefit of community participation is that it provides local understandings about social benefits that are more suitable for the livelihood and local context in resource governance.

7 Conceptualising Disaster Management

Although differences of opinion still exist, there appears to be some agreement that disaster management is an activity of all spheres of government, it relates to an integrated, multi sectoral, multi-disciplinary approach aimed at reducing the risk associated with hazards and vulnerability [23: 96]. According to a definition provided by [30: 10], disaster risk management is a systematic process of using administrative directives, organizations, and operational skills and capacities to implement strategies, policies and improved coping capacities in order to lessen the adverse impacts of hazards and the possibility of disaster. A more comprehensive definition is provided by the South African Disaster Risk Management Policy that understands disaster risk management as a continuous integrated multidisciplinary process of planning and implementing measures aimed at preventing or reducing the risk of disasters hence Disaster Risk Management has to become an integral part of development planning and process in order for it to be successful in mitigating the severity of consequences of disasters emergency preparedness and effective response to disasters, post-disaster recovery and rehabilitation (Act 57 of 2002).

While a variety of definitions of disaster risk management have been suggested, it can be drawn from them that disaster risk management aims to diminish the adversative effects of hazards through collaborative initiatives aimed at for prevention, mitigation and preparedness of disasters. This means that good collaborative governance is the cornerstone of disaster risk management. The ultimate aim is to create resilience in times of disaster. It cannot be disputed that resilience is a vital concept within disaster risk scholarship. In that same vein, [24] argues that the building of resilience should therefore be seen as an integral part of disaster risk reduction activities. Resilience, determines the resistance of relationships within a system and is a measure of the ability of these systems to absorb changes of state variables, driving variables, and parameters, and still persist (Holling 1973, Cited in [14]). As propounded by [30], resilience of a community in respect of potential hazard events is determined by the degree that the community has the necessary resources and is capable of organising itself both prior to and during times of need Cited in [24]. Resources are a key element. There is risk of disaster when a society is exposed to natural hazards but lacks adequate means and ability to protect itself from potential negative impacts (vulnerability) [16]. Thus, the chapter put forward that financing of disaster risk governance through resource governance is essential for sustainable disaster risk management and financing. The following section establishes the nexus between disaster risk management and resource governance.

8 Disaster Risk Management and Resource Governance: The Nexus

For Smith [24], the core function of disaster risk reduction governance is ensuring that the necessary support exists within government to drive the disaster risk reduction agenda. And often disaster governance is characterised as a risk management system, that is collaborative, multi-party, and multi-level [15]. In this nature of governance, the state has the responsibility to ensure the safety and welfare of its citizens, their livelihoods and natural resource endowments [24]. Thus, the state should invest in sustainable funding that is cost-efficient and this can result in sustainable disaster risk management activities. Given that most disasters are incited through natural environmental-prone challenges, governments should finance disaster risk management from natural resources. As discussed in the chapter good resource governance contributes to economic development. This builds the resilience of states towards national disasters. Niekerk [24] holds the view that allocating necessary resources for disaster risk reduction disaster risk governance is one of the fundamental principles and objectives of disaster risk governance. Consequently, building the resilience of states to disasters. Hence, the proposition that every government must create a sustainable financial base of disaster management from resource governance.

Disturbingly, due to financial resources constraints, most countries in the developing countries context are failing to alleviate disasters. The effectiveness of national and local level disaster management activities tends to be hampered by the lack of resources, responsibilities, and political legitimacy/recognition and inadequate stakeholder representation, and/or missing linkages to established networks [15]. This hinders the establishment of long-term engagements in disaster management by most countries. It is a chief role of the state to provide proper and effective institutional and legislative frameworks and capacities for disaster risk management financing through proper resource governance and the most important emphasis in good governance in this approach is the realisation that it requires a multi-sectoral and multi-disciplinary approach [24].

9 Findings

The following sub-section discusses the prevailing state of natural resource governance in Zimbabwe. This section is important in that it provides a framework for how resource governance can be utilised for effective disaster management.

9.1 Institutional and Regulatory Frameworks for Natural Resource Governance

In a country like Zimbabwe, where natural resources are dotted across areas where communities have lived for ages, regulatory frameworks must have an approach that is deliberately geared towards making sure that communities where resources are located also benefit from the extraction of such natural resources since the same communities will be affected by the mining ventures. This leads us to talk about community natural based resource management (CBNRM). CBNRM is "an approach to empowering rural communities with the knowledge, skills and authority to sustainably manage natural resources" [5] that are located in their communities. In our analysis with issues, we believe that deliberate and sustainable natural resource management knowledge and skills must be given to community stakeholders so that communities are empowered to benefit from financial, social and economic processes involved in resource extraction. This style of shared resource governance is a critical aspect of good governance and community beneficiation.

Whither to Zimbabwe's regulatory framework when it comes to the mining industry? In Zimbabwe's natural resource governance, the most important piece of legislation is the Natural Resources Act whose main objective is to control the use of resources, however, this policy is not easily applicable in the communal areas that cover about half of the total land area of Zimbabwe "since it is enforced via legal title to land … (and yet the) land tenure system in the communal areas is based upon traditional usufruct rights, that makes the act ineffective in areas where it is needed the most" (Zimbabwe and the United Nations Framework Convention on Climate Change: Zimbabwe: National Policy (unfccc.int).

Our analysis of Mines and Minerals Act (CAP 21:05), reveals that the Act gives the responsible Minister "excessive power to make unilateral decisions that may have far reaching consequences for the country" [9: 4]. The Mines and Minerals Act has a glaring overreliance on the goodwill of the minister than critical institutions of governance and accountability. A good example of over dependence on the good will of the responsible minister is that the Mines and Minerals Act makes a provision for the creation of a Mining Affairs Board that is appointed by the minister and is also answerable to the same—this flies in the face of good corporate governance since the minister appoints the board and the board is beholden to the same person. We believe that Zimbabwe has adequate expertise capable of monitoring natural resource extraction, but will need to come up with solid resource governance regulatory frameworks that are in respect of the international conventions governing the mining sector industry.

When it comes to the extraction of diamonds, the Zimbabwe government unveiled a diamond policy whose main objective was "to promote the sustainable development of the diamond industry for the benefit of all Zimbabweans" [9]. The policy acknowledges sovereign ownership of diamonds by the Zimbabwean people through the state. The policy tackles a number of critical issues that needs a progressive regulatory framework to enforce it. Of critical concern is that the policy does not provide the criteria for choosing investors. This leaves room for other regulatory frameworks to be enforced that provides room for multiple players thereby opening the window for compromising the legislative framework and this may ultimately lead to corruption. The policy is subservient to the Mines and Minerals Act (CAP 21:05) that was enacted in 1961 and has been amended several times thereby making it outdated and inadequate to deal with contemporary challenges in the mining sector [9].

The Mines and Minerals Act (21:05) accords excessive powers to the responsible minister to the extent that it ultimately disempowers other critical players in the regulatory framework i.e. according to the Act the Board performs "such other functions and duties as may from time to time be required of it by the Minister ... Board shall NOT have power to approve or refuse an application, but instead shall make recommendations thereon to the Minister, and any reference in those sections to an approval or refusal of an application by the Board shall be construed as a recommendation by the Board to the Minister that the application should be granted or refused, as the case may be" (Mines and Minerals Act, 21:05). The delegation of excessive power to one individual renders the regulatory framework obsolete in terms of good corporate governance. We therefore believe that a smart resource governance framework is essential for good corporate governance if the greater good is to be obtained in the management of natural resources in Zimbabwe as is envisaged by a variety of existing legislations and related Acts such as the Mines and Minerals Act 21:05.

We deduce that a good resource governance framework must regulate for the involvement of all relevant public and private stakeholders towards promoting due diligence in terms of scrutinising any related decision-making processes when it comes to the awarding of tenders and contract framing for qualitative resource governance frameworks. Such a process has the potential to eliminate any forms of corruption in the award of mining tenders and contracts. In our interrogation of qualitative resource governance, we define corruption as "the manipulation of frame conditions to attain exclusive benefits to individuals or groups at the cost of social benefits" [3]. In the context of Zimbabwe's natural resource governance, there remains, however, much to be done when it comes to stakeholder involvement, transparency, fair competition and business integrity.

It is undeniable that natural resources are a critical element of sustainable development in any nation or state hence natural resources are better governed under well coined national and communal institutions and regulatory frameworks. Institutions relate to governing bodies/structures while regulatory frameworks relate to relevant laws such as mining laws and contracts that must establish the terms of what a country might gain from extraction of its natural resources. Regulation "determines the opportunities for the sector to function efficiently, and whether this works for the benefit of society at large" [3]. In the rich mining industry, effective regulatory frameworks must "determine the revenues and responsibilities for actors involved in the industry, including the private sector and the government, defining certain requirements regarding exploration, field development activities, technologies applied, the number of operators … and monitoring opportunities" [11].

10 Conclusion

For sustainable natural resource governance to be realised, we are of the opinion that there must be a political will on the part of the government—a political will that will lead to a synchronisation of the varied instruments that are vulnerable to manipulation. We believe that in the realm of good governance, political will results in authentic political governance—a process of critical decision making towards formulating policies and planning. The nature of this process "brings together the state, non-state and private sector actors" [8] and determines the quality of the policy outcomes. This political process must then be accompanied by administrative governance—the process of "policy implementation … at the central and local levels" (Briceño, ibid.). Qualitative resource governance must be premised on the need to transform "government to governance" and to successfully implement good governance frameworks and respond to the complexities of contemporary challenges [27]. Finally, we conclude that a balanced multiple stakeholder approach to resource governance build around critical institutional frameworks promotes transparency and accountability in the extractive industry.

10.1 Resource Governance in the Context of Disaster Risk Management

Hazards can be of natural and manmade origin; or can be complex if any interplay of the natural and manmade factors triggers a disaster [10]. Extreme natural events such as droughts, floods, earthquakes, tropical cyclones, and pandemics can threaten lives, livelihoods, and even entire economies [31]. Ensuring a sound disaster risk management system is a prime function of the State, often characterised by hierarchical involvement of several stakeholders, it is pertinent to interrogate the concept and frameworks of resource governance in Zimbabwe. According to [28], governance is the umbrella under that disaster risk reduction takes place. It is often cast as encompassing the legislative frameworks, institutional and structural engagement. When disasters occur, there are usually three common ways in which disaster risk reduction is financed. According to [19], disaster risk reduction is mostly funded through firstly, national financing of Disaster Risk Reduction, in this level the governance structure is complex: and it comprises of national, provincial and local levels, the second form is through financing outside government that is through international aid financing and lastly national financing from civil society and the private sector.

10.2 Financial Component in Resource Governance

In the occurrence of a disaster, governments are faced with a daunting task of financing the disaster risk management structures. This calls for proper financing mechanisms and governance strategies during pre-disaster planning, in the course of the occurrence of the disaster and post-disaster occurrence. It is against such background that research reveals that in disaster prone countries, financial resource governance must become a fundamental function of the state. Ultimately, the core of financial resource governance revolves around bringing disaster risk reduction to "the heart of economic decision making" and "shift climate and disaster resilience away from being an external environmental agenda to a domestic development priority," (The International Institute for Environment and Development (IIED) Cited in [29]). In the same vein, a call has since been made by the United Nations:

> we call for disaster risk reduction and building of resilience to disasters to be addressed with a renewed sense of urgency in the context of sustainable development and poverty eradication, and, as appropriate, to be integrated into policies, plans, programmes, and budgets at all levels and considered within relevant future frameworks. We invite governments at all levels and relevant sub regional, regional and international organizations to commit to adequate, timely and predictable resources towards a coherent and comprehensive approach for disaster risk reduction in order to enhance resilience of cities and communities to disasters, according to their own circumstances and capacities. (United Nations, 2012: 33 Cited in [19]: 4)

There is valid reasons to suggest that national financing of natural resource management frameworks is essential for the reduction of disaster risks in the extractive resource sector which is increasingly volatile. The importance of integrating DRR into national development plans and priorities in the management of natural resources is seen as the way forward in eradicating conflicts and disasters within the extractive sector [19]. Disturbingly, most governments are found wanting as they do not have effective disaster resource governance mechanisms in place. In most of the countries, the engagement of non-government actors and their financing (including that of the international community) is not clear, showing that more needs to be done to integrate (and leverage) different sources of investment in Disaster Risk Reduction [19]. This is worrying and costly to countries. Without Disaster Risk Finance, disaster response is too expensive [31]. Proper resource governance is the bedrock of disaster risk reduction and conflict transformation in the natural resource sphere. Nevertheless, the lack of Disaster Risk Reduction financing itself, and even the lack of coordination or coherence of that financing, may not be an issue of availability but rather one of prioritisation, an issue that remains to be addressed in many countries with high risk of disaster [19].

The financing of Disaster Risk Reduction is contextual and different frameworks can be adopted by different countries and these are subjective to the state of development of the system of law and governance in each country. According to the [31] with regards to finance governance of disasters, most countries are increasingly asking some difficult questions about how to decide on the details of disaster risk finance strategies such as: Should we set aside funds in a reserve fund, and how large should this reserve fund be? How much reliance should be placed on emergency reallocations of funds away from other parts of our budget to finance disaster and conflict losses? And should we seek to establish a line of credit on that we can immediately draw if a disaster were to occur? There can never be one best way to govern resources in times of disasters. However, states can borrow, modify and implement resource governance models that can be adjusted to fit a number of financing and disaster circumstances. Kellett et al. [19] provides the following framework on different ways in the financial management and governance of Disaster Risk Reduction

- Disaster Risk Reduction must be financed as part of a larger Disaster Risk Management budget that includes preparedness, response and sometimes recovery. This requires management by a single governance structure, such as a national disaster management agency to avoid overlaps and contradictions in the systems management;
- Disaster Risk Reduction as a budget line or special fund must directly target Disaster Risk Reduction as first priority towards promoting sustainable development in any sector;
- Disaster Risk Reduction has to be integrated into development planning and management. This helps in risk reduction when risk is integrated into sector plans and management.

10.3 Recommendations

For effective and good governance mechanisms, institutional and regulatory frameworks systems in resource governance must not only be government based but should also be entrenched around two elements that are community and civil society participation and monitoring frameworks as well as effective information sharing and communication systems that provide qualitative collaborative mechanisms involving the government, the communities and the supportive agencies or service providers. In pursuit of good governance—premised on transparency and accountability; there must be adequate institutional designs that consist of clear, coherent and logically framed legislative frameworks. Such frameworks must be complimented by ethical leadership from those who have been delegated the authority to superintend over such institutions. Our deductions are premised on the fact that at all material times, regulations of the mining industry are critical since mineral resources involve are of national importance to sustainable development because natural resources are a major source of government revenues.

Congruent and effective resource governance frameworks for the mining industry provide a win–win situation for governments and mining firms as both institutions maximize profits from an unadulterated system of resource extraction and revenue collection since nothing is lost through corrupt underhand dealings. Incorruptible regulatory frameworks accompanied by complimentary and strong governance institutions that involve both state and civic society are essential for effective resource governance. The same results in less conflicts within the natural resource industry. Institutional frameworks must allow for the involvement of varied anti-corruption initiatives directed towards promoting good governance initiatives in resource governance. A smart national policy framework must put in place clear ethical codes of conduct to regulate voluntary initiatives since voluntary initiatives have the capacity to provide efficient anti-corruption mechanisms.

Another critical component in resource governance is public access to information. With access to information, those who are entrusted with the responsibility for making critical decisions in resource management are held accountable for what they are expected to do to hence the minimisation of sinister shenanigans. In this context, accountability in resource governance needs to be strengthened not only at national governmental level but across local communities with a bearing on all stakeholders that are involved in the promotion of good governance. This is an important element in promoting transparency and accountability in the extractive industry. We have observed that in the existing Mines and Minerals Act (21:05), there is no provision for stakeholder and public access to information. This missing link makes it very difficult to gather information on important decisions, such as the control of revenue streams and the conditions behind licence contracts. In pursuit of transparency, a central element of good governance, there should be deliberate legal provisions to access information around the selection criteria for mining licences, awarding of contracts, and the details of final contracts themselves. This provision will enhance accountability and promote confidence in the national management systems of natural resources. Finally, we believe that since natural resources are a critical national asset, natural resource governance is even a more critical aspect of governance for the nation's sustainable development goals to be realised through collective frameworks of good governance.

References

1. www.centrefornaturalresourcegovernance. Accessed: 26/07/2021
2. www.indextofpoorcountries. Accessed: 29/07/2021
3. Al-Kasim F, Søreide T, Williams A (2008) Grand corruption in the regulation of oil. U4 Anti-Corruption Resource Centre, Bergen (Norway)
4. Amaratunga D, Haigh R, Hettige S (eds) (2019) Accountability in the context of disaster risk governance. UNDRR. ISBN 978-1-86218-170-0
5. Anderson J, Mehta S (2013) A global assessment of community based natural resources management: addressing the critical challenges of the rural sector. USAID-IRG, Washington, DC

6. Ansell C, Gash A (2008) Collaborative governance in theory and practice. J Public Adm Res Theory 18:543–571. https://doi.org/10.1093/jopart/mum032
7. Bianchi C, Greta N, Rivenbark C (2021) Implementing collaborative governance: models, experiences, and challenges. Public Manag Rev. https://doi.org/10.1080/14719037.2021.187877
8. Briceño S (Dir.) (Dec. 2004) Disaster risk reduction, governance & development (Volume 2, Issue 4). UN/ISDR Africa Educational Series, Cape Town
9. CNRG: website. An analysis of Zimbabwe's diamond policy (April, 2013)
10. Chirenda TG, Tandlich R, Krele V, Luyt CD, Srinivas CS, Iheanetu CU (2018) Legislation, vulnerability and disaster risk management of waterborne diseases in Zimbabwe. Inf Secur 40(1):61–91
11. Chêne M (2017) Natural resource management transparency and governance. Berlin; U4 Helpdesk Answer
12. Denzin NK, Lincoln YS (2005) Introduction: the discipline and practice of qualitative research. SAGE Publishers
13. Emerson K, Nabatchi T, Balogh S (2012) An integrative framework for collaborative governance. J Public Adm Res Theory 22(1):1–29. https://doi.org/10.1093/jopart/mur011
14. Fekete A, Hufschmidt G, Kruse S (2014) Benefits and challenges of resilience and vulnerability for disaster risk management. Int J Disaster Risk Sci 5:3–20
15. Gall M, Cutter SL, Nguyen K (2014) Governance in disaster risk management. Integrated Research on Disaster Risk, Beijing
16. German Federal Ministry for Economic Development (2013) Disaster risk management for all: the inclusion of children, elderly people and persons with disability. BMZ, Berlin
17. Harrison EP (2015) The governance of natural resource management in Zimbabwe: unravelling the relationships between conservation and development. PhD Thesis, The University of Leeds, Leeds
18. International Union for the Conservation of Nature, IUCN website: accessed: 02/08/2021). Zimbabwe and the United Nations Framework Convention on Climate Change: Zimbabwe: National Policy (unfccc.int): Accessed 02/08/2021
19. Kellett J, Caravani A, Pichon F (2014) Financing disaster risk reduction: towards a coherent and comprehensive approach. ODI and the UNDP, London
20. Mason J (2002) Qualitative researching, 2nd edn. SAGE Publications Ltd., London
21. McGrath L, Flemming A (2020) The 2021 resource governance index. Blog @NRGInstitute
22. Muzulu P (2012) Zim industrial sites now ghostly. Zimbabwe Independent (16 Nov 2012)
23. Niekerk VD (2006) Disaster risk management in South Africa: the function and activity towards an integrated approach. Politeia 25(2):95–115
24. Niekerk VD (2011) Introduction to disaster risk reduction. Available Online: https://www.preventionweb.net/files/26081_kp1concepdisasterrisk1.pdf. Accessed 17 August 2021
25. Olowu D, Sako S (eds) (2002) Better governance and public policy, (capacity building and democratic renewal in Africa), Bloomfield. Kumarian Press, Incl
26. Rhodes RAW (1997) Understanding governance: policy networks, governance and accountability. Open University Press, Buckingham
27. Tierney K (2012) Disaster governance. Social, political, and economic dimensions. Ann Rev Environ Resources 37(1):341–363
28. UNDP (2010) Disaster risk reduction, governance & mainstreaming. Available Online: http://www.undp.org/cpr/disred/documents/wedo/ils/ils_esummary.pdf. Accessed on 20 June 2021
29. UNDRR (2021) Decoding public finance for disaster risk reduction and climate investments. Available Online. https://reliefweb.int/report/world/decoding-public-finance-disaster-risk-reduction-and-climate-investments. Accessed on 20 June 2021
30. UNISDR (2009) UNISDR terminology on disaster risk reduction. United Nations International Strategy for Risk Reduction, Geneva
31. World Bank (2016) Disaster risk finance as a tool for development: a summary of findings from the disaster risk finance impact analytics project. The World Bank, Washington

Sifiso Mpofu is a Zimbabwean scholar, and a certified Peace-building and Conflict Resolution and Management practitioner. Sifiso is well-read in Peace-building, Conflict Management, Religion and Peace Studies, as well as Disaster Management. Sifiso is an honorary lecturer at the University of Kwazulu Natal having been a senior lecturer at the University of Zimbabwe before retirement. Sifiso has also lectured at the United Theological College, Holy Trinity College and Christian College of Southern Africa. Sifiso's profile has some international flavour as he has been an examiner with the UKZN, a visiting professor at Emmanuel College of the University of Toronto in Canada, a research fellow with the Council for World Mission and has actively participated in the World Council of Churches and All Africa Conference of Churches academic fora amongst other platforms. In 2000, Sifiso was honoured by the World Alliance of Reformed Churches for his article which analysed the efficacy of financial ethics in addressing global financial challenges. Sifiso has published over twenty book chapters, seventeen articles and edited four books.

Hurdles to Achievement of Climate Justice in Southern Africa and the Search for Possible Solutions

Kudzai Chatiza and Tariro Nyevera

Abstract This chapter presents findings from an analysis of development, policy and academic literature regarding climate justice in Southern Africa. It is based on the question of factors stopping attainment of climate justice. A human rights or social justice lens was applied recognising climate change's differential impacts on individuals, families, communities, subnational territories and countries. The chapter shows that the region is exposed to extreme climate-related events, including floods, droughts, cyclones, unprecedented rainfall and temperature variations. The impacts of these events are exacerbated by poorly performing and inequitable economies saddled with debt. Additionally, racial inequalities, historically uneven development, portions of harsh geography, pre-existing high levels of poor populations without basic services, weak governance systems and regulatory capacity gaps at country and regional levels make climate impacts worse. This results in climate injustice at different scales. Southern Africa's poor and vulnerable residents in specific locations with limited adaptive capacities, inadequate social services and lacking adequate representation are affected the most by water stress, food insecurity and job losses. The chapter concludes by arguing that directly tackling inequalities is critical to climate justice and building resilient communities in Southern Africa. For this to happen climate change policies and strategies need to be mainstreamed across various sectors, actors and from national to local levels. The need to support community, household and individual measures in fighting climate and economic injustice is also elaborated for purposes of building bottom-up resilience where 'external science' and local technical knowledge are integrated based on the leadership of capable state and regional bodies not captured by local and global capital.

Keywords Vulnerability · Climate change · Resilience · Inequality · Poverty · Governance

---

K. Chatiza (✉) · T. Nyevera
Development Governance Institute (DGI), Harare, Zimbabwe
e-mail: kudzai@mweb.co.zw

R. Kasimba and I. Chirisa (eds.), *Disaster Risks and Hazards*, Advances in 21st Century Human Settlements, https://doi.org/10.1007/978-981-96-4577-0_15

1 Introduction

This chapter is based on a literature review on climate justice issues in Southern Africa. The review recognised that climate change is triggering events that impact individuals, families, communities, subnational territories and countries in the region differently. The differential impacts arise from complex political economy factors that are both historical and emergent.

The chapter used a definition of climate change that sees it as 'significant changes in global temperature, precipitation, wind patterns and other measures of climate that occur over several decades or longer' [37]. On the other hand, climate justice was operationalised in relation to the idea that the climate crisis has to be looked at through human rights and social justice lenses. In this construction, issues of fairness and equity are critical. This is in consideration of the need to (i) protect vulnerable people from the negative impacts of climate change, (ii) dampen negative disruptions arising from climate change-induced transformations and (iii) enhance the envisioning and implementing of an equitable post-carbon society [1].

It is acknowledged that attention to climate change and sustainability issues received intense attention in the late 1980s with the Brundtland Commission[1] and since the 2000s Summit held in Southern Africa [21]. Those mostly affected by climate change are the vulnerable people whose resilience and coping strategies are weak. As such, climate injustice piles pressure onto residents of the region already burdened by other multiple injustices connected with colonial and post-colonial displacement, socio-economic, political and spatial marginalisation. These historical and emergent injustices seriously erode the resilience of households and communities in disadvantaged localities within the region. The causes of climate change are mainly greenhouse gas emissions. The variations in socio-economic conditions across Southern Africa are also reflected in the country-related share of greenhouse emissions and between Southern Africa and other regions.

2 Methodology

This study engaged with literature review to explore climate justice/injustice issues in Southern Africa and to offer possible solutions. The literature were gathered from development, policy and academic sources. The methods used to select those sources included: (i) defining the scope of the paper, (ii) coming up with key search words, (iii) evaluating the quality of literature sources and their contribution to this study, and (iv) selecting and utilising the sources that were useful for this paper. Several perspectives were integrated from various empirical studies and international and national development reports. The geographical area of focus for this study was Southern Africa, with case foci on Mozambique, Zimbabwe and Malawi due to the experiences of Tropical Cyclone Idai in 2019. The other countries that this review

[1] Report of the World Commission on Environment and Development: Our Common Future.

focused on include Angola, Botswana, Lesotho, Madagascar Namibia, Saint Helena, South Africa, Swaziland and Zambia.

Thematic analysis was used to identify issues, trends and themes from the country cases. The integration of perspectives and findings from different sources was useful in drawing lessons and formulating possible solutions to the longstanding history of climate injustice in the region.

3 Conceptual Framework

Physical, social, economic and environmental factors are determinants of climate change vulnerability [13]. In framing climate justice in this chapter, three dimensions of climate change are used (see Fig. 1).

These are causes and effects of and responses to climate change. Climate injustice is reflected on all the three dimensions. The dimensions are consistent with the UNFCC's operationalisation of climate variability as "the right to health, the rights of indigenous peoples, local communities, migrants, children, persons with disabilities and people in vulnerable situations and the right to development, and gender equality, empowerment of women and intergenerational equity" [38: 2]. These dimensions are used to analyse the roles of state, private sector, civil society and ordinary citizens as agents that work together and alone in ways that have a bearing on climate change and the events it triggers. For the state, this conceptual framework looks at the presence/ absence and scope of national and local policies, laws and strategies. These are framed with variable input from different non-state agencies. Private sector operations across different socio-economic sectors have a history of and the potential to cause climate

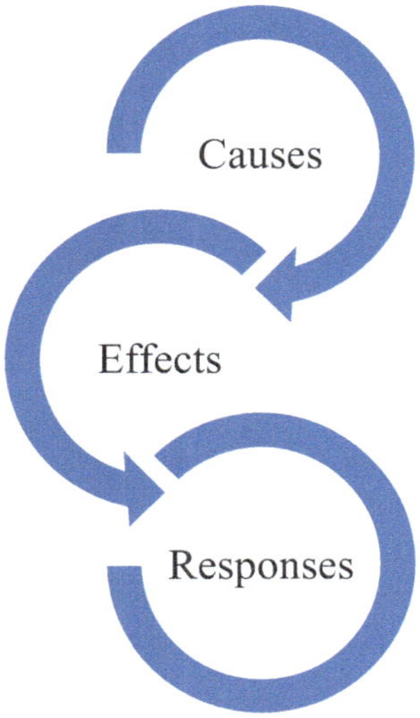

Fig. 1 Conceptual and analytical framework

change effects that endanger natural and socio-economic ecosystems. All actors have the power to positively address the impacts of climate change induced events by tackling spatial and other forms of injustice. In that way they can contribute to sustainability or resilience.

Overall, Southern Africa's contribution to global greenhouse gas emissions is merely 2.19% (World Resource Institute 2029). Yet, socio-economically, the region is considered one of the poorest and most vulnerable to climate change. At global level, therefore, this represents climate injustice [3]. Countries, like Zimbabwe, Mozambique, Zambia and Malawi, bear the greatest impacts [23]. These countries are geographically located in areas that are vulnerable to climate change induced disasters. For instance, Zimbabwe, South Africa, Mozambique and Malawi were hit by three cyclones between 2017 and 2019 while Mozambique additionally suffered from tropical storm Desmond in 2019 and a debt crisis in 2016 [1].

Mozambique was affected by all eight cyclones[2] that have occurred since 2000 as most of them originate in the South and South West Indian Ocean and pass through the Mozambican Channel [1, 5, 6, 43]. Zimbabwe was hit by six while Malawi was affected by four of these cyclones. As such, these appear to be the countries most affected by tropical cyclones in Southern Africa. Clear from analyses following Tropical Cyclone Idai is that a number of shocks due to the climate crisis and ecological degradation sit side-by-side with political and economic governance crises that these three countries and others in the region have been experiencing [1, 6, 18]. Rich countries contribute more to climate change than poor countries yet the least contributors to the problem bear the greatest cost of climate change [1]. This is a dimension of climate injustice that can also be applied at regional and country levels.

There are climate injustice issues that emanate from the effects of climate change at regional, national, sub-national, local, household and individual level. The effects are looked at in the context of vulnerabilities and existing stresses that the population is subjected to. People's vulnerabilities reflect the presence and functionality of roles of state, private sector, civil society and people's agency. Those that suffer the most from climate change impacts are the least able to cope with and are also the least to contribute to climate change. This is as a result of poorly performing and unequitable economies saddled with debt, racial inequalities, uneven development, weak governance systems, poverty and regulatory capacity gaps at country and regional levels. This raises the issue of responses to climate change and how these enhance community resilience. With resilience building, impacts of climate change are minimal. However, where resilience is lacking, climate related disasters negatively affect humanity. The adverse impacts of climate change are manifesting in Africa as evidenced by rainfall variability and extreme events, such as heat waves, floods and droughts [10, 43]. In the last 100 years, temperatures have risen by 0.5 °C [33]. The negative impacts already being experienced by communities in Southern Africa include water stress, food insecurity, decreased yields from rain-fed agriculture and an increase in land that is becoming arid and semi-arid [13] (World Resource Institute 2029). Rainfall patterns are also projected to worsen by about 20% and this

[2] Eline, Hudar, Japhet, Favio, Jokwe, Dineo, Idai and Kenneth.

will trigger increases in the number of consecutive dry days in Namibia, Botswana, Northern Zimbabwe and Southern Zambia [33, 43].

The region's economies are sensitive to the effects of climate change. Extreme weather events have caused cumulative damages amounting to USD6.4 billion in the last decade. Taking the overall cost of the 2016 events, the regional economies would easily surpass USD10 billion. The losses exclude losses in human life, that are difficult quantify.

4 Drivers of Differential Impacts of Climate Change

4.1 Poorly Performing and Inequitable Economies

Southern Africa comprises countries with economies that range from upper and lower middle-income countries to low-income and least developed countries [13]. Economic growth has been fairly stable in the region but in many cases, it has been slowing down (See Table 1).

Economic growth has however not translated into poverty reduction, hence the presence of high-income inequalities [2]. Southern African economies' inadequate performance weigh down efforts at building climate justice (OCHA 2010) [30]. Uneven economic development creates segments of the population with less capacity to reduce risk making them vulnerable to shocks from climate change [13]. Most of Southern Africa has poorly performing economies underlining state fragility due to high external debt, policy inconsistencies and declining investments [9, 10]. Together with weak economic engagement at international level these factors weaken Southern Africa's ability to benefit from investment into green, low carbon growth [17]. Poorly performing economies affect development of critical infrastructure and provision of basic services.

A number of shocks that communities are experiencing in Southern Africa sit side-by-side with economic governance crises [36]. This affects state-led resilience building efforts especially if countries have borrowing limitations due to huge debts and weak international relations [22]. States affected by disasters have little time to recover before other disasters strike largely because the poor performance of their economies result in the lack of key assets from national to individual levels [30]. Poor economic performance also exacerbates food insecurity. This is because the region's economies are heavily dependent on agriculture, that is being impacted heavily by

Table 1 GDP growth in South Africa between 2007 and 2018

Year	2007	2008	2009	2010	2011	2012	2013	2014	2015	2016	2017
GDP rate %	6.8	5.1	0.2	4.5	4.2	4.5	4.0	3.5	2.2	1.5	2.1

Source SADC [30]

climate change. For instance, poor economic performance in Lesotho and Zimbabwe in the latter part of the last decade compounded saw sharp rises in food prices [35].

4.2 Racial Injustices and Geography of Inequality

Differential climate change impacts are being felt in the context of racial injustices. Spatially and socially distinct geographies of inequality were a feature of colonial development. Some of these socio-economic and spatial geographies seen in infrastructure and service disparities have not been transformed by post-independent governments. This explains the appropriateness of adopting the climate justice lens to engage with issues of inequality, racism and colonialism [2]. In Southern Africa, climate change drivers went unchecked when the region experienced state-based or institutionalised racism under colonialism.

Racist governance models exposed most indigenous populations to poverty and vulnerability by pushing them to locations with inadequate services and natural capital constraining coping strategies [9, 16, 22]. Health, gender and social inequality issues specifically in terms of access to public assets and participation as measured by the Gini coefficient were hatched during this period. Inequality is highest in Lesotho at 0.632 followed by South Africa at 0.625, Botswana at 0.605, Namibia at 0.597 and Zambia at 0.575 [2].

South Africa's long history of segregation and discrimination played the biggest part in the inequalities being experienced today [9, 16]. Since independence governments' efforts to improve social services in previously neglected areas [24] have been inadequate to systematically transform socio-economic structures. The typical African enclave economy has thus persisted [9, 22]. Differences in the burden of climate change in Southern Africa are thus linked to inequalities that date back to the colonial era (ibid.). There is consensus on the climate change—inequality nexus [9, 22]. On challenges arising from climate change countries with high inequalities fare poorly [13, 14, 16, 28].

4.3 Physical Geographies

Southern Africa has a varied climate. The range is from arid to humid subtropical conditions. Mozambique, Malawi, Zimbabwe, South Africa and Madagascar are geographically located in areas most prone to cyclones and floods increasing vulnerability to disasters [5, 13, 23]. In addition, Botswana and Namibia are mainly arid landscapes that increases their vulnerability to food security [1]. Rainfall variations are due to ocean currents and prevailing winds. Madagascar receives the highest rainfall, up to 3100 mm while the highest summer temperatures exceeding 40 °C are recorded in the Kalahari Desert of Namibia and Botswana [11]. Different physical or natural features result in the region experiencing climate change related extreme weather events amplified by those features.

Recent cyclones in Southern Africa point to climate change vulnerabilities that related to geography. The Table 2 below shows the major cyclones that have hit the sub-region and their direct impacts. The frequency of the cyclones has increased since 2000. For instance, the region was hit by three cyclones between 2017 and 2019 while Mozambique additionally suffered from tropical storm Desmond in 2019 [1].

Mozambique was affected by all eight cyclones. Zimbabwe was hit by six while Malawi was affected by four. These appear to be the countries most affected by tropical cyclones in Southern Africa. Mozambique experiences the most damage as the energy and intensity of the cyclones will still be very high. Low-lying settlements,

Table 2 Cyclones in Southern Africa and their impacts, 2000–2020

Cyclone	Year	Estimated direct impacts	Countries affected
1. Eline	2000	• 205 deaths • Over 10,000 people were left homeless	Mozambique, Zimbabwe, Madagascar, Botswana and South Africa
2. Hudar	2000	• 3 deaths • Strong winds damaged roofs and downed trees	Mozambique, Malawi and Zimbabwe
3. Japhet	2003	• More than 50,000 people affected • 237,000 ha of crop fields destroyed • Destruction of bridges and infrastructure	Zimbabwe and Mozambique
4. Favio	2007	• 10 deaths • More than 70 people injured	Mozambique and Zimbabwe
5. Jokwe	2008	• More than 10,000 houses destroyed • Widespread energy cuts and crop damages	Mozambique
6. Dineo	2017	• Estimated 20,000 homes were destroyed • Approximately 130,000 people directly affected • Flooding occurred in many parts of region	Zimbabwe, Mozambique, Botswana, South Africa and Malawi
7. Idai	2019	• Up to 3 million people affected and more than 1000 deaths • Nearly a million hectares of cropped land and many settlements destroyed • Billions in infrastructure damage, economic losses, land degradation and flooding • Food and nutrition insecurity especially among children	Mozambique; Malawi and Zimbabwe
8. Kenneth (6 weeks after Idai)	2019		Mozambique, the Comoros and Malawi

Sources African Centre for Biodiversity [1], Chanza et al. [5] , Chatiza [6], Manatsa and Chatiza [18], World Resource Institute [43], OCHA [23], ReliefWeb [27], Brown et al. [4], Reason and Keibel [26]

like Beira City and parts of Tete, Manica and Sofala Provinces in Mozambique, were flooded and inaccessible for weeks while in Zimbabwe Chimanimani District experienced considerable landslides and infrastructure that made some areas inaccessible for months [1, 6, 18]. The Tropical Cyclone Idai disaster was acknowledged as a wakeup call for countries to prioritise strengthening early warning systems [5] (Chanza et al., 2019; World Health Organisation, 2019). All three countries did not receive adequate international support to respond to the disasters.

ACB [1] observes that Mozambique received less than a third of what it needed to cope with and recover from the disaster while Zimbabwe got less than a fifth and Malawi less than half. In essence, climate justice considerations were not met in the most recent climate change induced disaster. Further, in Zimbabwe. Chatiza [6] observed that institutional development support was weak and responses to the emergency by international agencies circumvented local and national governments. In many respects this undermined and did not prioritise state capacity development [1] (ibid.). A just and decolonised socio-ecological system is needed as ACB [1] observes. This can only be built on a broadened disaster management approach where the rights and competences of local institutions are prioritised.

4.4 High Population in Poverty

A key contextual issue for climate justice is poverty. Approximately 45% of the population in Southern Africa live on less than USD1 per day (USAID [39]; see also Institute of Security Studies, ISS 2020[3]). The poorest people often reside in coastal, low-lying, informal and desert settlements. These areas particularly receive inadequate services and state capacities to boost resilience are weak [25, 34]. The Fig. 2 below shows poverty trends that have implications for addressing climate justice deficits. While the percentage of people living in extreme poverty is projected to slow, the absolute numbers are set to rise. The critical concern is that the rise is likely to be in the same previously deprived or underserved areas where most lack basic services, like water, sanitation and health. The structural reasons for poverty in Southern Africa are traceable to colonial, post-colonial separate development, extractivism and weak states (ISS 2020).

Climate change impedes poverty reduction at individual and household levels [1]. People become trapped in the vicious cycle of poverty that those who may have managed to emerge from poverty will slide back when climate-induced disasters hit. Most of the poor in Africa depend on natural resources and rain-fed agriculture. They are thus least able to cope with the shocks of climate change-induced droughts, floods, soil erosion and other natural disasters [35].

[3] https://issafrica.org/iss-today/extreme-poverty-set-to-rise-across-southern-africa.

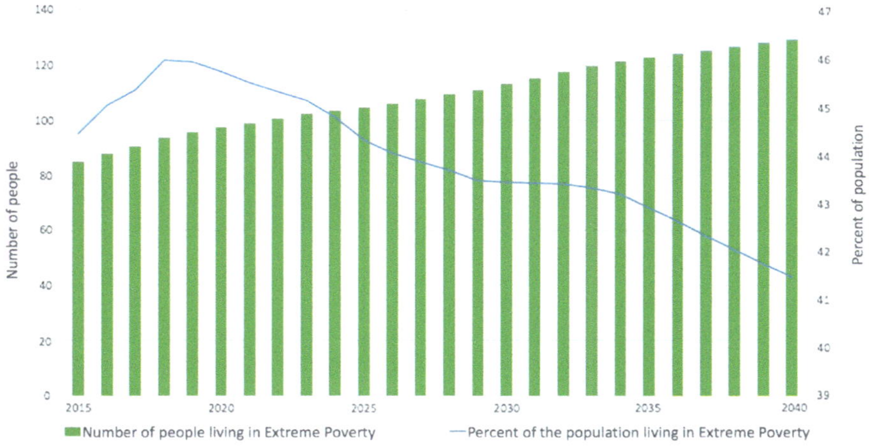

Fig. 2 Extreme poverty forecast, Southern Africa 2015–2040. *Source* ISS (2020) (see footnote 3)

4.5 *Weak Basic Services*

Climate change is contributing to public health threats since it affects the fundamental requirements for health [that are] safe drinking water, clean air, sufficient food and secure shelter [causing] many direct and indirect adverse health impacts. The provision of basic services in poor and informal settlements in almost all countries in Southern Africa is limited [1, 35, 40]. Additionally, those living in areas that lack basic services are disproportionately exposed to climate risks [13]. With informal settlements for instance, WASH services are limited in contexts of poor environmental conditions and dilapidated housing [40].

Examples of settlements lacking basic services include the Townships in South Africa, informal peri-urban settlements in Zimbabwe and low-income settlements located in floodplains, such as Maxaquene in Mozambique [40]. As such, heat waves, floods, droughts and other extreme weather events that are being experienced in Southern Africa [44] directly affect the availability, quality and safety of key health enablers, like shelter, water and sanitation including the social and physical infrastructure delivering them. Inland flash and coastal floods have been a growing and frequent threat [11] with public health implications. They result in a rising disease burden [44] and deaths [6].

Central and local governments sometimes do not prioritise infrastructure development, operation and maintenance in informal and low-income settlements [11]. Such neglect is higher in informal and poor settlements. Most of these settlements lack water, sewer, roads and waste management [40]. Local authorities are often preoccupied with clearing these informal sectors without clearly articulated policies on people's health and well-being. In times of emergencies, it becomes challenging to provide adequate survival packages [11].

4.6 Weak Governance Systems

Governance performance is indicative of a country's readiness to provide social protection services to its citizens in the wake of climate change. This is because governance involves:

> '…the provision of the political, social and economic public goods and services that every citizen has the right to expect from their state and that a state has the responsibility to deliver to its citizens" (Mo Ibrahim Foundation [19]: 17).

Having noted the impacts of climate change, it is evident that climate change is emerging more as a security threat and not just an environmental challenge. Under conditions of poor governance, the insecurities stemming from geographical, political, social and environmental insecurity stress communities increasing their vulnerability. This calls for efficient and inclusive governance frameworks that prioritise resilience of critical sectors and vulnerable populations [42]. Unfortunately, democratic governance and political commitment to climate change remains low in the region.

The political fragility of some countries weakens the coordination development objectives between central and local government. For instance, informal developments on wetlands, uncontrolled urban sprawl into flood-prone areas depict poor law enforcement in protecting the natural resource base and in observing human rights in terms of improving human security [13]. This shows weak spatial governance frameworks. Governance performance is not uniform across the region. Botswana and South Africa are the top-scoring countries, but have been on a concerning downward path since 2015 [19]. Angola and Swaziland/Eswatini[4] are at the bottom of the ranking, though they are on a steady path of improvement since 2010.

Having noted that the most vulnerable communities are the hardest hit by climate change it therefore follows that low governance scores increase the risk that citizens of such jurisdictions are exposed to public health and climate change induced disasters [1]. This forces communities to rely on individual resilience competencies, that are often inadequate. Without effective governance alignment with global and continental objectives (such as Agenda 2030, SENDAI framework, Agenda 2063) that address social protection services becomes inadequate [20, 42]. Good governance helps countries recover from the effects of climate change within the country ownership framework that Agenda 2030 promotes. As such, weak governance has ramifications for the state capacity to address climate change challenges.

[4] At the time of first drafting the chapter Eswatini had not descended into radical chaos that hit it from late June 2021, https://www.africanews.com/2021/07/06/missed-the-ongoing-crisis-in-eswatini-here-s-what-you-need-to-know/.

4.7 Climate Change Information Dissemination

Climate information particularly disseminated through early warning systems (EWS) raises the critical awareness on climate hazards. Yet, experiences in Southern Africa show that key information is often lacking particularly for local communities who need it to respond accordingly [18]. A study in South Africa revealed that agriculture related information was not adequately disseminated to farmers [15]. Cherotich et al. [8]: 28 point out that "Choice of the dissemination channels can influence access and use of climate information and service disseminated to enable the vulnerable groups...build adequate response capacities".

In Zimbabwe, weather forecasts are mainly broadcast in English. As such, some social groups miss the information. In other countries scientific knowledge on climate change has not been integrated with local knowledge systems [8]. Tropical Cyclone Idai experiences indicated that community members were warned by departments responsible for meteorology, but this turned out to be incomplete and not on time leading to inadequate responses by residents [18]. Barriers to flow of information include poverty, illiteracy, lack of infrastructure and socio-economic factors. These are not factored into information sharing strategies [15]. A multi-disciplinary climate information distribution strategy addressing variable information needs helps improve local adaptive capacity and reduces impacts of climate hazards.

4.8 Climate Change Institutions

Regional consolidation of climate change instruments in Southern Africa is in its infancy. Protocols, policies and initiatives across sectors at national and regional levels are fragmented. A holistic framework prioritising climate change adaptation in the region is still lagging [5]. Opportunities to adopt appropriate measures exist at country level. ISD [12] and Kamara et al. (2018), note cases of participation by the private sector and CSOs. Yet a critical mass of actions remains outstanding. SADC's work on disaster reduction is guided by frameworks, such as the Sendai Framework, Africa Regional Strategy for Disaster Risk Reduction, Plan of Implementation of the Africa Regional Disaster Risk Reduction Strategy, Regional Early Warning Centre and Hyogo Framework for Action. Some of these have influenced regional social protection-related instruments.

Examples of progress are seen in SADC's Declaration and Treaty, the Charter of the Fundamental Social Rights, the Code on Social Security, the Protocol on Gender and Development, and the Protocol on Health and Protocol on Education [35]. SADC's work on climate change is guided by the SADC Water Sector Climate Change Adaptation Strategy and the SADC Policy Paper on Climate Change. The latter guides countries to institute the political, economic, administrative and infrastructure systems necessary for the region to deal with the uncertainty caused by climate change [32]. The focus is on water governance, infrastructure development and water management.

SADC structures and processes are connected to AU institutions through the African Ministerial Conference on Environment (AMCEN) established in 1985. This is a platform for Ministers responsible for environment, that convenes on issues relevant to climate change [32]. It provides advocacy on environmental protection in Africa [31]. The 12th AMCEN recognized the need to integrate Africa's existing and new climate change programs under a consolidated framework [41].

At national level, all countries have climate change policies and strategies and programs that address climate change issues [12, 31]. Countries also have dedicated agencies working on climate change. The agencies are usually the Ministries responsible for the environment and resource-related sectors. There are some gaps in integrating national climate change in local development planning [17]. UNECA [36] confirms the implementation lethargy citing lack of political will. At the same time global climate change denialists that continue to misinform and undermine multilateral responses also exist at regional, national and local levels. Evidence suggests that coordinating institutions are critical in integrating climate change policies across sectors [12] and territories. Chatiza [7] notes that technical, financial and logistical capacities of key institutions fall short of meeting current needs and are more reactive than proactive.

4.9 Conclusion, Policy Options and Recommendations

Southern Africa is vulnerable to the impacts of climate change and this is already manifesting in the form of rainfall variability and extreme events, such as heat waves, floods, cyclones and droughts. The differential impacts of climate change induced events expose the region's historical and emergent disparities that are economic, social, geographical and political. This makes an application of a climate injustice lens critical in Southern Africa. The costs of climate change mostly affect the vulnerable populations who have weak coping mechanisms. Issues of poverty, inequality, lack of basic services, geographic location in hazard-prone areas and lack of social protection are key drivers of climate change vulnerability. Social protection strategies at all levels are therefore critical in the fight against climate change related hazards. This requires acknowledging and directly fighting inequalities.

The chapter recommends mainstreaming of climate change policies and strategies across various sectors at national and local level. Strengthening disaster preparedness and responsiveness at all levels is important in mitigating the impacts of climate disasters that are set to increase in frequency and gravity. Institutional coordination and capitation are also critical. Stronger institutions will work towards building the resilience and adaptive capacity of vulnerable and previously disadvantaged communities.

Support for community, household and individual measures in fighting climate change is important. In addition, integrating indigenous knowledge systems and scientific knowledge increases adaptive capacity. Social safety nets for the poor and vulnerable groups are critical in coping with climate change impacts guided by competent and accountable state and regional institutions. Unfortunately, these currently trip on fragmented and fragmenting layers of policies, laws and local to international bodies that are under-resourced and weak. State capture by local and global capital, extractive-driven development models and debt also weigh down efforts at state-led building of climate justice.

References

1. African Centre for Biodiversity (ACB) (2020) Neo-colonial economies and ecologies, smallholder farmers and multiple shocks: the case of cyclones Idai and Kenneth in Mozambique and Zimbabwe, discussion paper, December 2020
2. Asongu SA, Odhiambo NM (2020) Inequality and gender inclusion: Minimum ICT policy thresholds for promoting female employment in Sub-Saharan-Africa. Telecommun Policy 44(4):101900
3. Awojobi ON, Tetteh J (2017) The impacts of climate change in Africa: a review of the scientific literature. J Int Acad Res Multi 5(11):39–52
4. Brown D, Chanakira RR, Chatiza K, Dhliwayo M, Dodman D, Masiiwa M, Muchadenyika D, Mugabe P, Zvigadza S (2012) Climate change impacts, vulnerability and adaptation in Zimbabwe. Climate change working paper 3. International Institute for Environment and Development (IIED) London, pp 1–40
5. Chanza N, Siyongwana PQ, Williams-Bruinders L, Gundu-Jakarasi V, Mudavanhu C, Sithole VB, Manyani A (2020) Closing the gaps in disaster management and response: drawing on local experiences with tropical cyclone Idai in Chimanimani, Zimbabwe. Int J Disaster Risk Sci 1–12
6. Chatiza K (2019) Tropical cyclone Idai in Zimbabwe: AN analysis of policy implications for post-disaster institutiona development to strengthen disaster risk management, Oxfam
7. Chatiza K (2020) Cyclone Idai in Zimbabwe: an analysis of policy implications for post-disaster institutional development to strengthen disaster risk management. Oxfarm
8. Cherotich VK, Saidu B, Bebe BO (2012) Access to climate change information and support services by the vulnerable groups in semi-arid Kenya for adaptive capacity development. Afr Crop Sci J 20(2):169–180
9. Dasgupta S, Emmerling J, Shayegh S (2020) Inequality and growth impacts from climate change—insights from South Africa. Working paper 20–10
10. Davis-Reddy CL, Vincent K (2017) Climate risk and vulnerability: a handbook for Southern Africa, 2nd edn. CSIR, Pretoria, South Africa
11. Godfrey S, Tunhuma FA (September 2020) The climate crisis: climate change impacts, trends and vulnerabilities of children in Sub Sahara Africa. United Nations Children's Fund Eastern and Southern Africa Regional Office, Nairobi
12. Institute for Sustainable Development (ISD) (2020) Building climate-resilient and equitable cities during COVID-19. https://www.iisd.org/sustainable-recovery/building-climate-resilient-and-equitable-cities-during-covid-19/
13. International Organization for Migration (IOM) (2017) Spaces of vulnerability and areas prone to natural disaster and crisis in six SADC countries: disaster risks and disaster risk management capacity in Botswana, Malawi, Mozambique, South Africa. IOM Publications, Zambia and Zimbabwe
14. Islam N, Winkel J (2017) Climate change and social inequality. https://www.un.org/esa/desa/papers/2017/wp152_2017.pdf

15. Jiyane GV, Fairer-Wessels F (2012) Dissemination of information on climate change: a case study of women mussel harvesters at KwaNgwanase in KwaZulu-Natal. Mousaion 30(1):19–38
16. Machaka LL (2020) Linking racism and climate justice during Covid-19 era. https://www.cidse.org/2020/08/04/linking-racism-and-climate-justice-during-covid-19-era/
17. Madziwa F, Betzold C (2014) 20 years of African CSO involvement in climate change negotiations: priorities, strategies and actions. Heinrich Böll Stiftung and Inter Press Services (IPS), Africa
18. Manatsa D, Chatiza K (2020) Building resilience to natural disasters in populated mountain ecosystems: the case of tropical cyclone Idai in Chimanimani. TSURO Trust
19. Mo Ibrahim Foundation (2019) Agendas 2063 and 2030: is Africa on track? African Governance Report
20. Mo Ibrahim Foundation (2020) 2020 Ibrahim index of African governance
21. Nhamo G (2017) New global sustainable development agenda: a focus on Africa. Sustain Dev 25(3):227–241
22. Nyiwul L (2020) Climate change adaptation and inequality in Africa: case of water, energy and food insecurity. J Clean Prod 278:123393
23. OCHA (2019) Southern Africa Cyclones Idai and Kenneth Snapshot. https://reliefweb.int/sites/reliefweb.int/files/resources/SA_Cyclone_and_Flooding_Snapshot_10072019.pdf
24. Open Society Initiative for Southern Africa (2011) Inequality in Southern Africa: options for redress. OSISA, South Africa
25. Rawlins J, Kalaba FK (2020) Adaptation to climate change: opportunities and challenges from Zambia. In: African handbook of climate change adaptation, pp 1–20
26. Reason CJC, Keibel A (2004) Tropical cyclone Eline and its unusual penetration and impacts over the southern African mainland. Weather Forecast 19(5):789–805
27. ReliefWeb (2019) Joint call for action to address the impacts of climate change and a deepening humanitarian crisis in Southern Africa. https://reliefweb.int/report/zimbabwe/joint-call-action-address-impacts-climate-change-and-deepeninghumanitarian-crisis
28. Ruiters G (2001) Environmental racism and justice in South Africa's transition. Politikon: South Afr J Polit Stud 28(1):95–103
29. SADC (2010) Southern Africa sub-regional framework on climate change programmes report. In: Draft working document. SADC–UNEP. Available online: https://www.sadc.int/files/4813/5293/3518/Southern_Africa_Framework_of_Subregional_Climate_Change_Programmes.pdf
30. SADC (2020) SADC facts and figures. Available online: https://www.sadc.int/about-sadc/overview/sadc-facts-figures/
31. SADC (2011) Climate change adaptation in SADC A strategy for the water sector. https://www.sadc.int/files/2213/5293/3544/SADC_Climate_Change_Adaptation_for_the_Water_Sector_booklet.pdf
32. SADC (2010) Southern Africa sub-regional framework on climate change programmes report. Draft working document. SADC–UNEP. https://www.sadc.int/files/4813/5293/3518/Southern_Africa_Framework_of_Subregional_Climate_Change_Programmes.pdf
33. Shepard D (2019) Global warming: severe consequences for Africa. UNDP. Available online: https://www.un.org/africarenewal/magazine/december-2018-march-2019/global-warming-severe-consequences-africa
34. UNDP (2020) Strengthening climate resilience of agricultural livelihoods in agro-ecological regions I and II in Zambia. Available online: https://www.adaptation-undp.org/projects/GCF-Zambia#:~:text=The%20%22Strengthening%20climate%20resilience%20of,climate%20resilient%20agricultural%20production%20and
35. UNECA (2017) Disaster risk reduction in Southern Africa. United Nations Economic Commission for Africa, Addis Ababa
36. UNECA (2020) Climate change and development in Africa post COVID19: some critical reflections, discussion paper, African Climate Change Policy Centre
37. UNFCC (2012) Fact sheet: climate change science—the status of climate change science today. Available online: https://unfccc.int/files/press/backgrounders/application/pdf/press_factsh_science.pdf

38. UNFCCC (2015) Paris agreement to the United Nations framework convention on climate change, 12 Dec 2015, T.I.A.S. No. 16-1104
39. USAID (2015) Climate change information fact sheet SOUTHERN AFRICA. https://www.climatelinks.org/sites/default/files/asset/document/Southern%20Africa%20Climate%20Info%20Fact%20Sheet_FINAL.pdf
40. United Nations (2020) Emergency appeal. United Nations, South Africa
41. Westermann O, Förch W, Thornton P, Körner J, Cramer L, Campbell B (2018) Scaling up agricultural interventions: case studies of climate-smart agriculture. Agric Syst 165:283–293
42. World Economic Forum (2020) 4 ways our COVID-19 recovery can tackle climate change and global inequality. Available online: https://www.weforum.org/agenda/2020/09/4-priorities-for-climate-action-and-social-equity-in-the-COVID-19-recovery
43. World Resource Institute (2019) Climate disasters are getting worse. Here's how developing countries are insuring against them. Available online: https://www.wri.org/blog/2019/09/climate-disasters-are-getting-worse-heres-how-developing-countries-are-insuring-against
44. Ziervogel G (2018) Climate adaptation and water scarcity in Southern Africa. Curr Hist 181–186

Nuancing Urban Informality in Light of Policy to Contain COVID 19 in Zimbabwe

Mervis Chiware and Bernard Nkala

Abstract This chapter analyzed urban informality in light of policies aimed at curbing Covid 19 given that Zimbabwe is a low-income country where income is earned informally and daily for food purchasing. Using the qualitative approach, the study picked the Mbare Magaba complex in Harare and Ngozi Mine industrial complex in Bulawayo and reviewed the effectiveness of disaster risk management policy programmes implemented by the local and national government in building disaster and risk resilience amongst informal traders. Desk reviews were conducted to establish any policy gaps. It emerged that there is a lack of integrated approaches in policies employed by the government in building livelihood resilience amongst urban informal businesses in response to Covid 19. Findings demonstrated a lack of urban informality engagement from policy formulation through to implementation. Although the government continued to review policy measures to contain covid 19, informal businesses are viewed as accelerating risk drivers hence policy measures do not factor in urban informality perspectives on why they preferred to tolerate or accept covid 19 risks. The study proposes the need for stakeholder engagement in building consensus on an understanding of covid 19 risk effects and the importance of disaster resilience. Social distancing and hand hygiene measures are feasible when supported by government handouts to families thus locally designed interventions are effective in containing Covid-19.

Keywords Urban informality · Covid 19 · Resilience

M. Chiware
University of Zimbabwe, Department of Governance and Public Management, Harare, Zimbabwe

B. Nkala (✉)
Bindura University of Science Education, Department of Economics, Harare, Zimbabwe
e-mail: bnkalahsb@gmail.com

R. Kasimba and I. Chirisa (eds.), *Disaster Risks and Hazards*, Advances in 21st Century Human Settlements, https://doi.org/10.1007/978-981-96-4577-0_16

1 Introduction (Background, Overview and Objectives)

The COVID-19 pandemic is one of the most serious challenges the world has faced in recent times and some of the measures that are currently being used to counteract the pandemic may impact our future lives in non-trivial ways. While the pandemic has been simply devastating, its impact has been particularly severe on the informal sector. Against this backdrop, this chapter discusses urban informality and COVID-19 regulations. According to Zimbabwe Broadcasting Corporation news (2 August 2021), the informal sector employs more than eighty percent of the workers. The informal sector in Zimbabwe has been growing as one of the largest contributors to the National Gross Domestic Product, between 1991 to 2015 and its contribution averaged 61%. Zimbabwe's informal sector provides a long-standing source of household's livelihood especially for urban dwellers working to supplement their income through informal exchanges given the high reported unemployment rates [1]. Urban informality is now a shared policy challenge in the face of covid 19, and emerging markets amongst low-income economies. The rising urban informality is partially attributed to lack of development and poor governance systems that are unable to build risk and disaster resilience among informal communities. With the rise of COVID-19, Zimbabwe adopted several policy measures to combat its spread. According to Newsday (27 March, 2020) President Emmerson Mnangagwa imposed a nationwide lockdown for 21 days from Monday to help curb the spread of the fast-spreading coronavirus. The enforcement of national lockdown in March 2020 introduced a new normal order of survival to both business and informal activities. Section 4 of SI 110 of 2020 announced an indefinite extension of the lockdown with a two weeks review. The urban informal sector was cast with lots of doubts and uncertainty as to how they were going to survive amid government's call for an indefinite extension of the lockdown. The COVID-19 announced policy measures included restrictions on movement as public transport was banned except for Zimbabwe Passengers Company, total closure of the informal businesses except for shops designed under essential services. According to the Covid lockdown measures all citizens were required to stay at home except of course in respect of essential movements relating to seeking health services, to purchase and procure food and medicines and other critical services supplies. Only state and health workers and those who maintain key infrastructure such as power, water and sanitation among others were to be exempted from the lockdown. The continued enforcement of restrictions into town disrupted the informal activities as customers could not easily travel. Research is limited globally and in particular to Zimbabwe as to guiding the informal sector on how to adapt to COVID-19 principles and guidelines. People in the informal markets, small and micro-entrepreneurs, women in precarious employment conditions, historically excluded groups, such as indigenous and afro-descendants, must be at the centre of the policy response. The chapter will be guided by the following research questions. Which features of the informal sector intensify the spread of COVID-19? How should policies to mitigate Covid 19 negative impacts be crafted to include the urban informal sector?

Overcrowded Mbare Market in Zimbabwe | *Ngozi Mine Informal settlement in Bulawayo*

Source: My Guide Zimbabwe myguidezimbabwe.com

2 Methodology

Using the qualitative research approach, the study reviewed the effectiveness of disaster risk management policy programmes implemented by the local and national governments in building disaster and risk resilience amongst informal traders. A case study approach was preferred to draw observations, experiences, and information on survival patterns of urban informal traders and settlers amid COVID-19 and related impacts. Mbare Magaba complex in Harare and Ngozi Mine informal complex in Bulawayo were used as proxies of urban informal sectors. These informal sites were purposively selected as they are situated within major cities Harare and Bulawayo where the cost of living is high, and informality constitutes a bigger proportion of business. Systematic site vists were then conducted to verify some conlcusions made on existing knowledge at informal settlements. Rich conclusions were drawn from the systematic site visits where behaviour was observed and in some insances traders and settlers were interviewed to confirm some issues before conclusions were made. The two informal settlements pose as high potential risk zones and can offer rich information and lessons drawn from informal urban livelihoods patterns. In addition, there is already existing literature on informal business making it convenient to review its relevancy in COVID-19 response context.

In-depth desk reviews were conducted to establish disaster policy systems gaps, assessing their effectiveness in building resilience within COVID-19 pandemic amongst the urban informal settlements. The thematic review of the literature was guided by the rational choice theory and Risk Cost–Benefit analysis and evaluation principles in risk assessment. The theories are relevant in understanding policy options, concepts, self-interest, and local authorities in risk management COVID-19 phases. The Risk Cost–Benefit Analysis (RCBA) in risk management supports the decision-making process where effects of the policy are projected before being encountered and if such projected inferences are sound, informed decisions are undertaken but if not sound, decision-makers may decide to incur risks and costs that are greater than expected benefits. The risks, cost and benefit analysis provide sound theoretical underpinnings when reviewing implemented COVID-19 policy measures on urban informal sector identifying perceived risks, costs, and benefits. On the other hand, the rationale choice theory proffers arguments that individuals are central in analysis and understanding collective outcomes in designing policy. Therefore, the assumption is relevant in defining urban informal traders and settlers as capable of making rational calculations and choices to achieve survival outcomes in line with their community resilience. Urban informality presents its rationale of maximising survival interests amongst the informal sector, regardless of the risks and dangers associated with breaching COVID-19 containment policy measures. The use of published national data in form of pronouncements and statistical updates on COVID-19 related impacts guided the study in coming up with updated focused discussions and reviews. This was made easier given that COVID-19 information is now shared on national forums and made public and easy to attain including the informal sector. In the case of personal experiences, the study could not utilise direct individual responses for confidentiality reasons instead all information was reported as representative of urban informal case sites studied.

3 Theoretical Framework

3.1 Systems Thinking

The prevailing policy measures on COVID 19 tend to be biased towards the reductionist perspective thus becoming limited in understanding nuances of urban social classes and missing the complexity of the informal sector livelihoods [2]. Systems thinking becomes one of the alternative theories that can help in reviewing COVID 19 policy measures within the urban informal sector. According to Da Luke and Stamatakis systems thinking can help policymakers understand and influence the spread of infection and its multifaceted consequences across the community since society is itself a complex adaptive system. The World Health Organization (WHO) has in a recent landmark publication strongly advocated the employment of a 'Systems Thinking' approach in formulating plans for resolving complex social and health issues including COVID 19. COVID-19 has been described as a "wicked problem", a complex issue involving multiple interlinked systems with no single cause and no clear solution [3]. Systems thinking can provide a framework to look beyond the chain of COVID-19 infection and better understand the multiple implications of decisions and actions in face of such a complex situation involving many interconnected factors. The solution to the current pandemic must go beyond reductionist strategies, which are strategies that seek to understand complex systems through understanding their parts. According to Solomonian et al., to truly address this crisis, we must critically examine its causes and their relationship to our worldview. Causal loop diagrams (CLDs) are tools to depict the causal connections between components of a system and illustrate how changes in one component cascade in changes in others and back to itself, via feedback loops, potentially affecting the status of the entire system. COVID-19 pandemic resulted in unprecedented impacts on all domains of life—economy, entertainment, transport, education and it highlighted another critical domain in the Systems Thinking mindset—the domain of 'unintended consequences'. The research intends to discuss the negative consequences of COVID-19 policy measures on the urban informal sector and proffer recommendations on how inclusivity can be achieved in the stages of policy formulation. Consequences are the negative impacts of social distancing, curfew, market closures, etc. on people's livelihoods, mental health, domestic violence, and other non-COVID-19 medical ailments. Imad Hassan et al. posit that systems thinking facilitates an in-depth understanding of system dynamics and its tools enable users to re-design their systems, thereby radically creating the results they truly desire. In line with this, the chapter will discuss the importance of COVID-19 policy re-design to get compliance from informal traders. There are 5 pivotal steps in Systems Thinking: (1) root cause analysis, (2) selection of and focusing on strong-leverage areas, (3) system design or redesign coupled with (4) measures to nullify the impact of any unintended consequences resulting from these interventions and (5) continuous learning and improvement from the whole exercise.

4 Literature Review

4.1 The COVID 19 Pandemic Widespread and Urban Informality

The huge impact of COVID-19 is predicted to be felt by the informal sector post covid 19. Scholars have predicted high rates of unemployment that will drive scores of populations into the informal sector and informal cross-border trade with high prospects of relocating those working. The growing prevalence of urban informality in Zimbabwe also suggests the potentially high risk of exclusion of the informal sector in both public and private sector-driven responses to the COVID-19 pandemic [4].

The impact of COVID-19 in several developing countries has been devastating and heavy especially on members in the informal sector. The informal economy is susceptible to economic, and fiscal shocks. In countries with widespread urban informality, GDP per capita is about one-quarter that of economies with low informality [5]. In addition, there are lower per capita incomes, the prevalence of intense poverty, weaker growth in output, investment, and productivity. Urban informalities are often allied with underdevelopment, production processes typically labour-intensive production, with less educated and more poorly paid workers.

The widespread urban informality occupants suffer inferior education and health, wide gender inequality and sanitation facilities, making it very difficult to contain the spread of the COVID-19 pandemic. In Sub-Saharan Africa, informal employment is high and in Zimbabwe accounts for more than 80% of total employment and in other informal economies, the output is about 62% of official GDP.

The informal sector depends on self-funding, posing the inhabitants vulnerable to disruptions to cashflows stimulated by extenuation and other control practices. Urban informality makes observance of COVID-19 lockdowns and particularly social distancing challenging, consequently negatively affecting national governments' ability to curb the spread of the infections. This is supported by Loayza and Pennings who argue that the design and arrangement of urban informality tend to undermine policy efforts to contain the spread of COVID-19 The informal workforce is in crowded conditions where social distancing is difficult to implement, occupied in minor income activities without strong income savings and lack of access to formal social support, thus the majority may struggle to abide by government lockdown orders. Informal workers' prominent presence in the services sector means that they are more expected to lose jobs or suffer severe income losses during lockdowns [5]. Some informal sector features cause its members to suffer more severe economic losses and the features have a greater effect of intensifying the spread of COVID-19 amongst informal workers [6] and worsen its adverse health.

The political leaders in most developing countries have been generally silent about how they plan to manage COVID-19 within the growing challenge of widespread informality. The absence of social reform or government support programs targeted towards the urban informal sector in low-income economies, present an opportunity for the further spread of COVID 19. Greenstone and Nigam established that confirmed COVID-19 cases rapidly rose in informal settings starting end of March 2020, amid a lower level of testing. According to their projection, physical distancing for a period of three to four months could save about 1.7 million lives, which is equal to about $8 trillion employing the value of a statistical life. The analysis supported the widely announced measure of upholding physical-distancing measures in many regions in the inception stages of the COVID-19 pandemic.

Therefore, the COVID-19 pandemic becomes a wake-up call for urban city authorities to reconsider their engagement framework with the informal traders and settlements. The pandemic highlights how governance, health and equity are intertwined, and demonstrates the fact that effective urban governance cannot be achieved without collaboration with informal sector residents and real governance actors. Turning a blind eye on public services and accountability in informal settlements and will be counterproductive both in the control of the COVID 19 pandemic and the prevention of future epidemics given urban informality.

5 Global Policies to Contain COVID 19

COVID-19 induced panic among communities globally. Globally as of 6 August 2021, there have been 200,840,180 confirmed cases of COVID-19, including 4,265,903 deaths, reported to World Health (https://Covid19.who.int) According to the World Health Organization the global strategy to reduce COVID 19 should include mobilizing all sectors and communities to ensure that every sector of government and society takes ownership of and participates in the response and in preventing cases through hand hygiene, respiratory etiquette and individual-level physical distancing. The combination of policies such as social distancing, lockdowns, and quarantines, imply a slowdown or even a complete stop in production and consumption activities for an uncertain period, crashing markets and potentially leading to the closure of businesses and ultimately sending millions of workers home. In addition, Hevia and Neumeyer argue that a social distancing environment has the following results (a) Many firms go out of business. This is especially true of firms in intensive social contact industries (travel and entertainment) and small and medium enterprises (SMEs) with little working capital and limited credit lines. Restarting these businesses may be a long and costly process. (b) Firms depleting their capital will lay off workers. We know from previous recessions that after spikes in unemployment, matching workers and vacancies in the recovery is a slow process. (c) Layoffs reduce aggregate demand. (d) Households and firms facing increased uncertainty save more on the safest assets and ditch risky ones.

The most important strategy for the population to undertake is to frequently wash their hands and use portable hand sanitiser and avoid contact with their face and mouth after interacting with a possibly contaminated environment. To reduce the risk of transmission in the community, individuals should be advised to wash hands diligently, practice respiratory hygiene (i.e., cover their cough), and avoid crowds and close contact with ill individuals, if possible. A large informal sector is associated with poorer access for many to public health and sanitation facilities, making it harder to contain the spread of the pandemic.

Early screening, diagnosis, isolation, and treatment are necessary to prevent further spread. Preventive strategies are focused on the isolation of patients and careful infection control, including appropriate measures to be adopted during the diagnosis and the provision of clinical care to an infected patient. Guiner et al. define quarantine of persons as the restriction of activities of or the separation of persons who are not ill but who may be exposed to an infectious agent or disease, to monitor their symptoms and ensure the early detection of cases. Quarantine is different from isolation, which is the separation of ill or infected persons from others to prevent the spread of infection or contamination. Starting from a relatively lower level, confirmed COVID-19 cases have been rising rapidly in low-income countries with extensive informality since the end of March 2020, despite a lower level of testing.

6 Results and Discussion of Findings

6.1 COVID 19 Policy Challenges Associated with Urban Informality

Generally, governments' capacity to reform, ability to provide social safety nets during recessions and implement effective policy responses to pandemics is weak and mainly affected by limited resources and administrative structures to influence positive change especially on informal groups. COVID 19 policy measures and strategies and any attempt to control the pandemic is theoretically feasible if only it is compatible with the blanket rules of social distancing [7]. The application of containment measures applied across regions has been applied without an in-depth consideration of the urban settlement nuances within major cities. The tenacity of the marginalised groups of informal groupings who embody resilience within socio-economic, and geographical exclusion is a bright side that is yet to be utilized by policymakers and researchers working on dynamics of urban informality [2]. The prevailing inequalities in urban residents in low-income countries predispose them to heightened risks of COVID 19 transmission.

Existing literature zeros on stereotyping urban informality into classes (the poor and the rich), those who work and access services through unregulated systems. This line of thinking not only raise the morality aspects of essentialism of things but can lead to mistaken diagnosis and management of COVID 19 that is accelerated through human interaction regardless of class. Modelling studies on the management of COVID 19 indicated a strong association between environmental factors and the transmissibility of the virus [7]. The policy agenda on COVID 19 has limited attention on the potential negative impacts of blanket public health interventions in managing the crisis.

6.2 The Impacts of COVID-19 Policy Measures on Urban Informality in Zimbabwe

In Zimbabwe, the COVID-19 pandemics accelerated the prevalence of hostility in the urban informal sector. The COVID-19 crisis in Harare and Bulawayo has aggravated urban development challenges for low-income groupings, because of its disproportionate impacts on informal economic activity. As highlighted above COVID-19 implemented policy measures in Zimbabwe included tightening border control, the call for quarantine of members suspected of COVID-19, lockdowns, self-isolation, promoting working from home, ban and restricted travel -suspension of all public transport operations excluding ZUPCO, closure of marketplaces and formal and informal gatherings. The above-implemented policy measures had impacts on informal sector traders and settlements. Impact of COVID-19 in Zimbabwe is reviewed in relation to Mbare markets in Harare and Ngozi mine settlements in Bulawayo.

6.3 *Enforced Lockdown and Informal Sector Survival*

The government's lockdown targeted restricting the movement of the public [8] to help enforce social distancing at informal markets like Mbare in Harare. The enforced lockdown measure to date has restricted millions of households dependent on informal economic activities to sustain their families. In analysis, the implemented lockdown in 2020 under the COVID-19 measures impacted negatively on informal sector activities thus posing a great threat to the survival of residents. Most informal traders and settlers at Mbare in Harare and Ngozi Mine in Bulawayo live with very limited savings or capacity to save, surviving through trading at open places and largely depend on volumes of walk-in customers. Such urban residents survive on informal activities to sustain their low incomes hence can barely resist continued trading at uncontrolled, undesignated points therefore disregarding the call for social distancing. The lockdown measure though aimed at reducing risks of exposure of contracting and spreading COVID 19, traders at Mbare lost income as potential customers dropped. In the long run, the informal sector failed to sustain income inflows as they solely depend on daily sales. The announced lockdown resulted in the closure of Mbare markets and implied that informal traders were to conduct their business from home. The loss of income from the closure of Mbare markets affected informal traders' income inflows thus triggered dangers of households failing to afford basics and worse hygiene kits and personal protective equipment (PPEs).

The government fashioned lockdown measure policy increased the vulnerability of urban informal residents as they often had to defile the measures in search of survival thus exposing them to high risk of contracting COVID 19. Pronouncement of lockdown somehow lacked inclusivity as it assumed the livelihoods at informal settlements is like that of informal business environment. Ngozi mine residents survive on open space trading and mining scrap metal and wares from Bulawayo city dumping site. The Ngozi mine grew to become the major informal settlement in Bulawayo now associated with overcrowding, shacks, and lack of infrastructure to support survival. Therefore, the announced lockdown meant residents spent the greater part of their time at crowded shacks with limited PPEs.

One of the major characteristics of informal settlements is that they provide labour to the economy. The lockdown affected the survival of many informal households as some companies reduced salaries and wages calculated on daily attendance to work. Most household heads are employed by small scale informal sector business owners at Mbare especially as advertisers in open spaces. When the lockdown was announced, informal business owners cut salaries of engaged members by more than half citing lack of savings to sustain payment of salaries during the lockdown period. Literature has not yet established the actual numbers losing their jobs due to covid, but it is estimated that the economy was to shrink by between 15 and 20% during

2019 and 2020. This is a massive contraction pointing to job losses and serious social consequences for a country already burdened with escalating urban informality.

Informal traders and settlers from Mbare and Ngozi Mine perceived lockdowns as the exertion of state authority, an extension of national and local politics hence preferred playing hide and seek game to avoid law enforcing agents while neglecting dangers of contracting COVID-19. The government recorded running battles with informal traders dealing in open spaces outside closed markets in Mbare and surrounding areas during the time of lockdown. In Bulawayo Ngozi mine, a joint operation between police and the army took turns to patrol at Ngozi mine mainly to discourage crowding of residents and open trading. Resnick [9] argue that Zimbabwe has a history of cracking down on informal traders. The government has been reported as cracking on informal traders, vendors in the cities and these constitute the urban poor who have no means of surviving. Law enforcement agents have been reported raiding vegetable markets and confiscating produce sold at illegal points thereby destroying the potential survival of the informal traders in the urban [10]. The majority of informal traders do not have insurance to cushion them against financial mishaps for example COVID 19, hence are not able to survive without trading daily.

COVID-19 has seen an increase in calls for not just temporary economic relief to individuals and families impacted by the economic consequences of the pandemic, but for full adoption of universal basic income. The implementation of lockdowns has not supported reforming the informal settings, instead of the policy measures to curtailing the pandemic, it further divided the social classification and widened inequalities consequently increasing the impacts of COVID 19 amongst the poor urban communities. The pronouncement of lockdown measures lacked the social inclusion element of the informal sector thus the nature of informal urban livelihood forced the residents to defy the COVID-19 policy measures in search for survival.

7 Masking and Social Distancing on Urban Informality

As the COVID-19 cases dropped, the government relaxed lockdown measures and allowed informal markets to operate but under strict observance of masking and social distancing. Social distancing is one of the primary response measures promoted by the government in line with WHO COVID-19 preventing guidelines but lacks relevancy as it assumes the informal urban traders have adequate working space, access to basic infrastructures like water and social safety nets to survive. Reality points that social distancing is not enforceable at overcrowded Mbare markets and informal settlements like the Ngozi mine in Bulawayo due to absence of alternative space for trading. Informal traders continued to defy the call for social distancing due to the need for survival. The COVID-19 impacts are fashioning a disruptive new normal for informal traders through enforcement of social distancing guidelines which for urban poor households is burdensome and impossible. Promoting masking by residents become easy for members who can buy adequate personal protective equipment not informal traders who struggle for bread and butter. Effective masking at Ngozi mine is difficult as families can not afford to purchase PPEs. Most families headed by informal traders have an average of six members who all rely on the family head to provide finances including PPEs. Such households at Ngozi mine have to forgo the risk of COVID 19 and concentrate on saving the family from the risk of loss of income and daily survival. The daily income raised from informal selling is rather channelled towards putting food on the table than purchasing PPEs thus exposing all members to the risk of contracting and spreading COVID-19. Critics have also observed that expecting compliance to masking within informal circles works well if supported by government handouts in form of PPEs to cushion those households not able to provide enough protection equipment for members. The call for social distancing at Mbare markets caused delays in trading hours thus resulted in a reduction in daily sales as customers were restricted entrance to markets allowing only a sizeable number inside at any given time. Trading within reduced hours reduced potential sources of income and therefore impacted negatively on the survival of traders who rely on turnover in sales.

Mbare Musika Traders during Covid 19 Era, 2021

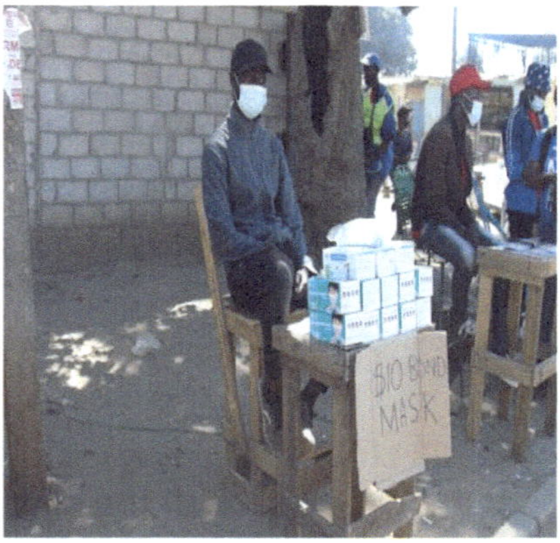

Urban Open trading at Mbare in Harare 2021

The concern about informal settlements and traders in implementing government advice on social distancing and upholding high hygiene measures in the absence of running water, sanitation and continued high population densities presents complexities and is perceived only as rhetoric.

8 Policy on Travel Restrictions on the Urban Informal Sector and Livelihoods

The travel restrictions were implemented mainly to refrain the public from daily commuting in (peri-)urban settlements or areas into the city. The ban and restricted travel suspended all public transport operations excluding ZUPCO as these were viewed as hot spots in the transmission of COVID 19. Suspending public transport has been criticised as having created regressive socio-economic livelihoods standards for urban informal dwellers. Travel restrictions affected informal traders in the transportation of raw materials and finished products to markets other than Harare and Bulawayo. Informal traders at Mbare industries rely on orders from South Africa as they find it cheaper and without implications of paying taxes to the government. Equally at Ngozi mine, traders in scrap metal need to use public transport for transporting their wares to West-commonage industries in Bulawayo for selling. The ban on public travel grossly affected the sustainability of informal business operations as traders ended up using expensive transport to reach their market. This scenario was supported by a theory that established that travel restriction translates to a loss of potential revenue for informal business traders and the causal employers who survive on a daily rate. In Zimbabwe, there is a strong reliance on informal sector activities. In the medium-term problems of nutrition and low-income prospects are likely to affect informal traders as resulting from current lockdown measures [9]. Interfering with transport impacts on the survival prospects of urban traders as businesses dependent on consistency and reliability in supplying customers same commodity over time.

9 Basic Health, Hygiene and Hand Sanitising as COVID-19 Policy Measures in the Informal Sector

Risks within Mbare and Ngozi Mine settlements are multifaceted, encompassing cross-cutting issues of health, basic sanitary and infrastructural challenges. Urban informal settlements of Ngozi Mine in Bulawayo lack proper sanitary facilities and running water essential to uphold high levels of hygiene pushed under covid 19 safely and prevention measures. Many traders at Mbare and Ngozi mine informal settlements lack access to basic social amenities due to underfunding of development thereby ripping traders and residents their resilience to survive without contracting COVID 19.

Government and local authorities should be committed to guaranteeing regular water supply to all urban areas including refuse collection, however, this does not extend to Ngozi Mine informal settlements in Bulawayo. Most informal settlers survive on selling wares at nearby cities townships mainly dirty open spaces where there is no running water and basic sanitary and health facilities. This affects the informal sector's resilience in the COVID 19 environment.

The informal sector can not afford the expectation to conduct mandatory testing of its informal employees hence in line with rational choice theory, decide to maximise their self-interest of sustaining their households through informal trading in the face of COVID 19. In Mbare mandatory testing of employees can not be sustained as most of the businesses can not afford to purchase vaccines from the government as per the instruction. Only formal companies can come open and embrace the call for mandatory testing. At Mbare markets, traders lack basic health measures like recording the temperature of both traders and customers as they view the exercise as time-consuming and at times fail to procure necessary equipment like thermometers. Many traders at Mbare markets operate from open spaces and some operate from the downstream of the Mukuvisi river running adjacent to the markets maximising on utilising dirty water diluted with sewer for cleaning their materials. The absence of a reliable source of water from Harare City Council forces informal traders to utilise sewer infested water thereby exposing themselves to contracting diseases that weakens their immune systems therby incresing the possibility of contracting COVID-19. When these traders fall sick, it affects family survival as the household members rely out of pocket to seek medical treatment. Meeting health expenses for informal households becomes a nightmare without savings.

Although policy authorities expect informal traders to comply with government restrictive measures in fear of covid 19, the risks associated with contracting covid weighed were viewed as lower than those of losing income from informal trading.

10 Conclusion and Recommendations

The challenge established in Zimbabwe's urban informal settlements and traders of Mbare and Ngozi mine is consistent with the theory that its very nature, informality is hard to measure and even harder to address with policies [11]. Novel integrated comprehensive policy responses are expected from local authorities, to avoid a great deviation in economic prospects that the cities envisage attaining.

COVID-19 has risen amid these urban risks and presented the potential to diminish communities' resilience. The national and local governments are presented with the reality to reform policy and redesign measures that promote the safety of urban informal traders from COVID-19 and impacts while building effective resilience. Like in other African countries, the COVID-19 pandemics in Zimbabwe exposed severe national government deficiencies in social security systems and the ability to support urban vulnerable groups in the face of covid.

As the government continued to review policy measures to contain COVID-19 in Zimbabwe, informal businesses are still viewed as accelerating risk drivers hence policy measures do not factor in urban informality perspectives on why they preferred to tolerate or accept COVID-19 risks. The following are recommended policy measures for adoption in dealing with urban informality issues in the face of COVID 19.

(i) Building urban resilience and promoting sustainability in implementing control measures would be key action in COVD 19. Informal traders need to be reorganized to only operate from decongested marketplaces to avoid overcrowding and be able to implement social distancing [12]. Local authorities must set up additional new markets to prevent informal traders overcrowding while doing business thereby compromising social distancing in the traditional open market. Decongestion measures need to foremost ensure the provision of running water at all marketplaces and proper waste disposal strategies including at urban informal settlements.

(ii) Government should adopt an integrated urban informality engagement approach. It is crucial for urban planning to adopt an integrated urban informality engagement approach from policy formulation through to implementation. Urban policy needs to be fashioned incorporating interrelated aspects or urban resilience, central in endorsing social and economic classification amongst informal communities while achieving public safety and urban development. The informal sector provides crucial support to formal sector dealings through the provision of housing, jobs and social services thereby pointing to existing interdependencies. Urban policy measures on covid 19 need to fully back up informal service provision while balancing calls for enforcing maximum safety essential in safeguarding urban informal traders and settlers. Several scholars argue on the need for inclusive social protection systems [9].

(iii) Informal sector Stakeholder engagement is very essential. There must be stakeholder engagement in building consensus on an understanding of covid 19 risk effects and the importance of disaster resilience. Informal community action should participate in defining their covid 19 risks and this would assist in developing contextually relevant and effective strategies for enforcing social distancing and subsequent compliance amongst the informal traders and residents. Involvement of urban informal leadership in discussing of covid 19 impacts and risks would reveal possible information gaps that could easily be addressed through the use of community political voice. The informal sector can define what works best for them and in the process be guided on complying with expected covid 19 health containment measures. Community engagement amongst informal urban dwellers in policy reviews is likely to lead to effective use of public facilities, provision of appropriate care, quarantine, and support at the community level while advancing the essential informal livelihood undertakings. The urban informal sector, if consulted are capable of providing reliable real-time data on covid 19 impacts, disseminate accurate information counteracting misleading information on covid 19 in informal settlements.

Resnick [9] provides suggestions on the need for local government to engage market leaders about how best do business during the announced stoppage and the reopening times. Resnick recommends the need to relax bylaws that restrict the public selling from outside their homes, to support continued subsistence income inflows thereby discouraging the travel of people to marketplaces. Other recommended policy measure includes opening markets every other day and permitting trading around the clock to reduce congestion at markets [9].

(iv) Social Distancing and hand hygiene are only achievable with government handouts. Social distancing and hand hygiene measures are feasible when supported by government handouts to families thus locally designed interventions are effective in containing COVID-19. Protective equipment and sanitisers must be provided to informal traders to promote continuity of business in a safe environment. Urban informal sector groups must be able to develop local disaster management response structures [13] and advocacy platforms to deal with COVID 19 impacts and possible risks.

(v) Government to partner with informal sector leader in COVID 19 vaccine advocacy. With the growing call for accepting COVID 19 vaccines, urban informal settlements need to be planned for and educated on the need to be vaccinated. Collaboration and partnerships would serve as the best vehicles to achieve head immunity at informal markets as all traders understand and accept to be vaccinated. Thereafter trading in crowded congested markets may become easy having minimised risks of transmitting COVID 19.

References

1. Chagonda T (2020, 12 April) Zimbabwe's shattered economy poses a serious challenge to fighting COVID-19. Conversation. Accessed from https://theconversation.com/zimbabwes-shatteredeconomy-poses-a-serious-challenge-to-fighting-covid-19-135066
2. Banks N, Lombard M, Mitlin D (2020) Urban informality as a site of critical analysis. J Dev Stud 56(2):223–238
3. Lawrence M (2020, April 09) The "Wicked Problem" of the Covid-19 pandemic. Retrieved 12 Aug 2020, from https://www.iass-potsdam.de/en/blog/2020/04/wicked-problem-covid-19-pandemic
4. Wilkinson A et al (2020) Local response in health emergencies: key considerations for addressing the COVID-19 pandemic in informal urban settlements. Inst Environ Develop Stud Univ Sussex (IIED) 32(2):503–522. https://doi.org/10.1177/0956247820922843)
5. Balde R, Boly M, Avenyo E (2020) Labour market effects of COVID-19 in Sub-Saharan Africa: an informality lens from Burkina Faso, Mali and Senegal. UNU-MERIT working paper 2020-022, University in Maastricht, Netherlands
6. Government of Zimbabwe (2020) Public health (COVID-19 prevention, containment and treatment) (National Lockdown) (Amendment) Order, 2020 (No. 5). Statutory Instrument 99 of 2020. [CAP. 15:17]. Government Printer, Harare. https://www.smeaz.org.zw/media/com_acymailing/upload/si_2020_099_public_health_order_5.pdf
7. Onditi F, Obimbo M, Muchina SK, Nyadera I (2020) Modeling a pandemic (COVID-19) management strategy for urban slums using social geometry framework. Eur J Dev Res. https://doi.org/10.1057/s41287-020-00317-5
8. International Institute for Environment and Development (2020) Working with informality for more resilient, equitable responses to COVID-19. Article, 10 July Accessed from https://www.iied.org/working-informality-for-more-resilient-equitable-responses-covid-19
9. Resnick D (2020) COVID-19 lockdowns threaten Africa's vital informal urban food trade. 31 March, IFPRI Blog. [Blog] Retrieved from https://www.ifpri.org/blog/covid-19-lockdowns-threatenafricas-vital-informal-urban-food-trade

10. Mhlanga D, Ndhlovu E (2020) Socio-economic implications of the COVID-19 pandemic on smallholder livelihoods in Zimbabwe. Preprints 2020:2020040219. https://doi.org/10.20944/preprints202004.0219.v1
11. Ohnsorge F, Yu S (eds) (2021) The long shadow of informality: challenges and policies. Advance Edition. Creative Commons Attribution CC BY 3.0 IGO, License
12. ZCTU (2020) Response to the impact of covid-19 (coronavirus) pandemic on workers and the Zimbabwean economy, March 31
13. Scoones I (2020) Surviving COVID-19: fragility, resilience and inequality in Zimbabwe. African Arguments, Debating Ideas. 27 March [Blog] Retrieved from https://africanarguments.org/2020/03/27/surviving-covid-19-fragility-resilience-and-inequality-inzimbabwe/

The Future of Managing Disasters and Risks in Zimbabwe and Beyond

Gift Mhlanga and Innocent Chirisa

Abstract The chapter explores the available options for the future management of disasters and risks in disaster-prone countries, with particular focus on Zimbabwe. The country has implemented various frameworks for disaster risk management, including the National Disaster Risk Management Policy, the Zimbabwe National Drought Plan, the Zimbabwe Drought Risk Management Strategy and Action Plan, and the draft Comprehensive Agricultural Policy Framework. Therefore, the main objective of the chapter is to examine and ascertain whether such policies are sufficient in the reduction of disasters in the country. The chapter is based on a desktop research study that employs textual analysis of information gathered from Google Scholar research engine, together with other websites with official information such as governmental policies and statutes. The research confers that while hazards may not be prevented, risk can be managed with a focused attention to limiting exposure and reducing vulnerability. Most developing countries, particularly Zimbabwe, only implement targeted Disaster Risk Reduction (DRR) practices. However, such a traditional approach is not adequate in managing disasters and risks. A change of perspective in DRR involves the implementation of both integrated and targeted DRR approaches. It is concluded that risk-informed interventions have greater chances of success in mitigating hazards associated with disasters. By and large, the chapter recommends that the use of a needs and risk assessment for preparedness, response and recovery stages of a disaster. In this regard, developing early warning systems through advanced technology, capacity building, and livelihood and economic assets protection, are the initiatives that may be adopted. Ultimately, prioritising initiatives that are catalytic and developing a longer-term strategy for institutional change on DRR, backed by national financing mechanisms are the basis of a sustainable DRR management strategy.

G. Mhlanga
Freelance Researcher, Chipinge, Zimbabwe

I. Chirisa (✉)
Office of the Vice Chancellor, Zimbabwe Ezekiel Guti University, Bindura, Zimbabwe
e-mail: innocent.chirisa@gmail.com

Department of Urban and Regional Planning, University of the Free State, Bloemfontein, South Africa

R. Kasimba and I. Chirisa (eds.), *Disaster Risks and Hazards*, Advances in 21st Century Human Settlements, https://doi.org/10.1007/978-981-96-4577-0_17

Keywords Disaster risk reduction · Drought mitigation · Early warning systems · Resilience

1 Introduction

Zimbabwe is a disaster-prone country owing to its global location and other human-induced, hazard-triggering issues. As such, both natural and man-made disasters are common in the South-African country. Corresponding to global trends, disasters in Zimbabwe evolving to be more frequent, with a heavier toll [17]. This indicates the need to adopt an effective disaster risk management framework with a goal to reduce the morbidity and mortality of human beings in disaster-prone areas. Noteworthy is that the most common disasters in Zimbabwe are droughts and floods, and these are exacerbated by climate change [14]. Since the disasters are propelled by climate change, this connotes dynamism in disaster risk as hazard patterns are expected to change, leading to more intense and frequent extreme weather events [1]. As such disaster risk management needs to conform to such changes so as to apprehend the increasing disaster risks and hazards. Inversely, a static disaster risk management framework is ineffective in mapping the prevention, mitigation, preparedness, response, recovery, and reconstruction phases of disaster management. In this vein, the Government of Zimbabwe and UN [14: 03] defines disaster management as "organization, planning and application of measures, preparing for, responding to, and recovering from disasters." According to Manyena [16], Zimbabwe is among the fifty-five African countries that are signatories for the Sendai Framework for Disaster Risk Management (SFDRM). The goal of the Framework is to: "To attain the expected outcome, the following goal must be pursued: Prevent new and reduce existing disaster risk through the implementation of integrated and inclusive economic, structural, legal, social, health, cultural, educational, environmental, technological, political and institutional measures that prevent and reduce hazard exposure and vulnerability to disaster, increase preparedness for response and recovery, and thus strengthen resilience" (UNISDR 2015: 07). This implies that Zimbabwe moots to contribute to such a global milestone through adopting tro strategies and policies. In this regard, the main objective of the study is to examine and ascertain whether such strategies and policies are sufficient in the reduction of disasters in the country.

2 Background and Overview

Zimbabwe is prone to be affected by both protracted and sudden disasters. These include both man-made and natural disasters. Cyclone induced disasters have sudden devastative effects. Mavhura [17] points out that Cyclone Eline that affected Zimbabwe in 2000 induced floods in northern Lowveld of Zimbabwe that left 120 people dead, over 250,000 people affected, and approximately US$7.5 million in economic losses. On the other hand, Mhlanga et al. [19] reveal that Cyclone Dineo that landed in Zimbabwe in 2017 caused flooding and left a trail of destruction in various villages that included killing seven people and more than 4000 homesteads, houses and Government structures demolished by the floods. It was established that Tropical Cyclone Idai that hit Zimbabwe in 2019 affected 270,000 people in Zimbabwe: 51,000 were displaced, more than 340 died and many others went missing (UNICEF, 2019). The Zimbabwe Ministry of Local Government [20] convey that floods are mostly common in the following areas: Middle Sabi, Muzarabani, Tsholotsho, Kamativi-Confluence of Gwayi and Shangani, Malipati-Mwenezi and Bubi, Tuli-Shashe.

Some parts of Zimbabwe are also prone to geological hazards, particularly earthquakes and landslides. It is established that whereas the rest of Zimbabwe lies a continental shield, the south-eastern part and the north-western parts of the country have Karroo sediments [20]. Seismic activities in these parts of the country become common due to the fact that the country is located within the far southern tip of the Great East African Rift Valley system [17]. Subsequently, the Save-Limpopo basin and the mid-Zambezi basin are prone to experience both earthquakes and landslides. As such, the Chipinge district experiences mild earthquakes from time to time, qualifying the area to be earthquake-hazardous. Additionally, the Meteorological Service Department reported that areas around Kariba, particularly Binga, felt a 4.5 earthquake on the 09th of August 2021. On the other hand, landslides are common in these areas as they are secondary effects after heavy rains, storms, and cyclones. Oxfam [24: 16] submits that "After nearly two days of heavy rains, brought on by Tropical Cyclone Idai, several landslides engulfed homes in the [Chimanimani] town on March 15 [2019]". This portrays that such landslides are hazardous enough to disrupt the welfare of people and their livelihoods.

Zimbabwe also suffers from biological disasters due to increasing exposure to gastro-intestinal tract infections (GTIs), notably cholera, diarrhoea and typhoid. Mavhura [17] points out that the country has been experiencing cholera outbreaks since 1975 and these usually occurred on a 5- to 10-year cycle. However, the year 1998 marked a change in the cholera outbreak cycles and they have since been reported on a yearly basis. The Government of Zimbabwe (2013) reports that the 2008/2009 cholera outbreak was the worst ever recorded outbreak that claimed 4000 lives and over 100,000 cases reported. Subsequently, UNOCHA [28] reveals that in 2018, Zimbabwe recorded the second largest cholera outbreak with more than 10,730 cases and 69 deaths reported. Typhoid is another GTI that has become a perennial disease in Zimbabwean urban areas. Since 2017, the country has been battling a typhoid fever outbreak, that has affected Harare and Gweru cities with more than 6100 cases reported and 17 deaths since the outbreak began [28]. On the other hand, in 2020, an outbreak of diarrhoeal disease was reported in the area of Luveve in Bulawayo with 1962 cases and 12 deaths, along with a typhoid outbreak in Harare, with 695 cases and 10 deaths [29]. The outbreak of these GTIs in Zimbabwean urban areas has been compounded by poor planning in water, sanitation and hygienic (WASH) issues, with over 7.3 million people being affected by the current WASH challenges in Zimbabwe [29]. Therefore, as the country is still grappling with these challenges, the GTIs have become a regular feature on the Zimbabwean urban landscape.

Besides the hydro-meteorological hazards and biological hazards, the other hazards that affect Zimbabwe are technological hazards, particularly road traffic accidents (RTAs). The RTAs are a hazard that have become a cause for concern for the country. Traffic accidents are likely to continue rising due to the increase in the number of vehicles including those in transit to neighbouring countries, deteriorating road infrastructure, non-deterrent fines for road traffic offences, lack of age limit for public transport drivers, absence of law that forces passengers in public and private vehicles to wear seat belts, high admissible alcohol content for drivers and lack of road markings, signs and traffic controlling lights (Government of Zimbabwe 2012). While some of the issues are being addressed by the government through the Zimbabwe National Roads Agency (ZINARA), the condition of some road networks have triggered national disasters. Oxfam [24] concur that whereas heavy rains reduce the accessibility of some areas due to landslides, poor road maintenance is the denominator for RTAs. As such, Statutory Instrument (SI) 47 of 2021 is a notice of declaring a state of emergency by the President of the country in terms of section 27(2) of the Civil Protection Act [Chapter 10:06]. Section 3 of the SI states that "Following the incessant rain that has of late fallen countrywide resulting in excessive damage to the road infrastructure network, a state of disaster exists in all rural and urban areas in Zimbabwe with effect from promulgation of this notice…" This declaration allowed the Minister of Transport and Infrastructural Development to be in charge of the maintenance of both primary and secondary road networks in the country as a solution to reduce RTAs.

The fact that most national policies and strategies are implemented as intervention measures to reduce drought effects proves that drought is the most significant hazard in Zimbabwe. The Government of Zimbabwe and UN [14] assessment and the National Drought Plan for Zimbabwe [12, 13] portray that the agricultural sector is guided by various policies and strategies. Some of the regulatory and guiding instruments are the following: The National Policy on Drought Management (NPDM); The Water Policy; The Irrigation Policy; The Zimbabwe Drought Risk Management Strategy and Action Plan (2017–2025); The Food and Nutrition Security Policy (2011); and the Draft Comprehensive Agricultural Policy Framework (2012–2032). However, despite all the legal instruments for the mitigation of drought effects, the country continues to suffer from the hazard. The World Bank (2021: 01) puts forward that in the 2019/2020 rainy season, "Zimbabwe experienced the worst drought in a decade…and a cyclone [Idai] causing damages of more than US$600 million." This proves that the country is disaster-prone despite all the policies to reduce the risks associated with the hazards.

Most developing countries suffer from disaster impact due to poor disaster risk management. The Asian Development Bank [1] carried out an assessment of resilience development in the following countries of South-east Asia: Cambodia, Indonesia, Myanmar, the Lao People's Democratic Republic, the Philippines, Thailand, and Viet Nam. From the study, it was discovered that "In the last 3 decades, disasters triggered by natural hazards have resulted in more than 400,000 deaths,

397 million affected population, and $133 billion direct physical losses" in these developing countries [1: 02]. As such disasters derail sustainable development, and this validates the implementation of effective DRM strategies since such strategies pave way for resilient development (Twigg, 2015). One recommendation from the study is that "...the role of public investment planners should not be limited to being passive users of disaster risk information, but instead should be at the center in defining their risk information-related needs and supporting the development of such information in close partnership with hydro-meteorological and geological institutes and relevant national agencies that address DRM and climate change" [1: 01]. As such, it is pointed out that disaster risk information is necessary in disaster risk reduction planning, and for the information to have more impact, there is need for integrated effort among stakeholders. On the other hand, the [23], in its Action Plan for Disaster Risk Reduction, points out that in assessing disaster and extreme event risk and creating public awareness on disaster risk reduction, there is need for establishment of central coordination mechanism and repository of disaster risk information. This confirms the significance of disaster information-gathering and propagation in disaster risk reduction.

Information dissemination is a significant factor that may influence disaster risk management. This is portrayed by a study in the arid regions of Tunisia carried out by Dhehibi et al. [9]. One of the objectives of the study was to assess the efficacy of numerous agricultural technology diffusion means introduced in the livestock–barley system in semi-arid Tunisia [9]. In other words, the research was meant to ascertain whether technological dissemination of information was effective in the implementation of strategies meant to reduce the effects of drought. In this context, Azumah et al. [2] opine that mass media and information and communication technologies (ICT) are effective means of passing information to smallholder farmers and thus raise awareness and enrich livelihood-improving knowledge. However, the study results indicated that technology-led extension approaches, particularly using text messages, were considered as complementary and not an alternative to the conventional face to face extension approaches [9]. This corresponds with the Government of Zimbabwe and UN [14: 16] observation that "In terms of the information relayed through cell phones, the major obstacle is that most of the time the information is in English and Shona language that pose a challenge to certain communities such as the Shangani community in Chiredzi." This infers that technology may pose as a communication barrier to the stakeholders who are not well-versed in such ways of communication and if the language used is not familiar. On the other hand, Nangombe [22] highlights that most smallholder farmers in Zimbabwe complain about the lack of access to weather trends and climate data due to poor awareness campaigns by the MSD to educate the farmers where weather and climate information may be accessed and how the information may be interpreted and used. In general, this is a challenge to most less developed countries that have no wide coverage of internet connectivity and such means of communication have not yet diffused to these places, necessitating face-to-face agricultural extension communication as the most effective way of communication.

The Sendai Framework for Disaster Risk Reduction (2015–2030) is an internationally-acclaimed blueprint for disaster management. Wilkinson et al. [32] opine that achieving positive outcomes under the Sendai Framework is attainable, but developing a national strategy for disaster risk reduction (DRR) is a pre-requisite for reducing the impact of disasters. In other words, as signatories for the Sendai Framework, countries need to implement their own DRR strategies aligned to the Framework well before 2030 since disaster risk reduction cannot be achieved through projects but long-term programmes. The UNDP-Zimbabwe [26] points out that building resilience requires an integrated approach and a long-term commitment to improving three critical capacities: absorptive capacity, adaptive capacity, and transformative capacity. In this vein, if countries need to contribute to the outcome and goal of the Sendai Framework, the long-term aspect of the strategies to be adopted need to be put into consideration if a significant reduction in disaster risks, and thus sustainable development is to be realised. Delve et al. [7: 01], emphasizing the need for resilience building in the Horn of Africa, put forward that "By viewing sustainable development through a lens of disaster risk reduction, communities worldwide can adopt a more proactive approach that supports long-term development planning and emergency interventions when needed". This confirms that DRR strategies are long-term measures that may only be achieved if long-term changes are adopted. Among the long-term changes that may facilitate a reduction in disasters, Jeggle [15: 05] opines that institutional factors are a necessity, pointing out that the "…greatest demands for the future are the expanded understanding, official leadership and strategic professional engagement to establish and sustain a full time commitment to identify, assess and then manage [both present] and future disaster risks."

3 Conceptual Framework

Disaster risk management involves the implementation of practices that are meant to foster sustainable development through capacity improvement on measures to counter disaster risks. Since disasters disrupt the status quo, it means it affects every community aspect of development, implying that all people within a particular community become victims. Therefore, effective disaster risk management need to be inclusive and holistic. Such a systems-approach to disaster risk management may be explained through the structural–functional theory. The theory holds that society may only be explained adequately through realising that it is a complex system with various interdependent parts that work together to promote stability. Stability in this sense emphasises the capacity of a system, community or society potentially exposed to hazards to adapt by resisting or changing in order to reach and maintain an acceptable level of functioning and structure [17]. Therefore, within the structural–functional paradigm, disaster risk management encompasses the effort and resources channelled towards maintaining stability through building resilience, and this is the responsibility of all stakeholders [10]. This is in line with the proclamation by the

UNISDR [27: 04] that DRR refers to the "practice of reducing disaster risks through systematic efforts to analyse and manage the causal factors of disasters, including through reduced exposure to hazards, lessened vulnerability of people and property, wise management of land and the environment, and improved preparedness for adverse events." As such, the conceptual framework serves as the benchmark for analysing DRR strategies in Zimbabwe.

4 Methodology

The study employs a desktop research methodology. This a research approach which uses secondary data. As such, the secondary data sources which were used in the study include the following: journal articles, governmental reports and articles; as well as organizational reports. These data sources were mined from internet research engines, particularly Google Scholar. The selection of the articles involved an internet search of the key words such as "disaster risk reduction" and "Sendai Framework", and "resilience". The secondary data was then thematically analysed according to the objectives of the study to come up with the research findings.

5 Results

The Civil Protection Act [Chapter 10:06] is the principal Act and regulatory framework that guides and regulates all disaster risk reduction (DRR) activities in the country [11]. Commenting on the "Protection" aspect of the Act, Mavhura [17] observes that the Act is inclined towards ex-post-DRR approaches rather than ex-ante approaches to disasters, reducing the risk reduction management to response, recovery, and rehabilitation phases of disaster cycles only. In this respect, the Civil Protection Act becomes ineffective and archaic in managing disasters as laid out in the Sendai Framework for Disaster Risk Management, an international blueprint upon that Zimbabwe is a signatory member. The other pointers to the ineffectiveness of the Civil Protection Act are the following: high centralization of authority; heavy militarization of disaster management; and repudiation of local communities in managing disasters. These flaws compounded the Tropical Cyclone Idai-induced disaster in Chimanimani. Oxfam [24] reveals that it took five days for the national response team to access the Chimanimani area due to limited resources. In this regard, UNOCHA [28] points out that only one Mi8 helicopter was deployed for immediate rescue services in Chimanimani and this was derailed by poor visibility. On the other hand, Chanza et al. [4: 663] reveals the heavy-handedness of the military in relief interventions:

> "Survivor 3 expressed dissatisfaction about the treatment native villagers sometimes got: "Soldiers whipped us while we were waiting in queues to get relief items. Their arrangement should have separated the disaster-affected people from the rest of the villagers."

Therefore, questions are raised if this commandeering style could work effectively at community level where villagers are expected to take a leading role in building their own resilience against disasters [17]. Nonetheless, the top-down approach to disaster management, as signalled in the Civil Protection Act, was proven to be ineffective in the aftermath of the cyclone since local people, local-based interventions, were the first responders to the disaster [4]. By and large, such flaws contributed to the proposition of a DRR Bill. The Government of Zimbabwe [13] posits that since 1995, effort had been driven towards a review of DRR legislation, and these efforts led to a Bill that was meant to be deliberated upon in parliament in 2004. Unfortunately, seventeen years down the line, the Bill is yet to be conformed into law.

It is highlighted that among the priorities of the Bill were the following: Repeal of the current Civil Protection Act and to be replaced by the Emergency Preparedness and Disaster Management Act; establishment of an Emergency Preparedness and Disaster Management Authority whose major functions include developing a risk reduction strategy in order to minimize vulnerability to both natural and man-made or technological hazards; establishment of an integrated early warning system for emergencies and disasters; and capacitating of local authorities to manage emergencies and disasters at the local level [13]. The Government of Zimbabwe and UN [14] posits that the DRM Bill had been revised on three occasions and this resulted in the 2011 DRM Bill and this conformed to the Hyogo Framework for DRR. However, Oxfam [24] argues that although most of the proposals in the Bill promoted resilience building, they were tabled prior to the enactment of the 2013 Constitution, that creates a devolved three-tier government system, and thus the decentralization proposals need to be revised accordingly. Besides the 2013 Constitution hurdle, the Bill also predates the Sendai Framework for Disaster Risk Reduction (2015–2030). As such, in line with Pillar 2 of the Sendai Framework—Strengthening governance and institutions to manage disaster risk—the DRM system in Zimbabwe is predominantly focused on civil protection and emergency management, as opposed to a holistic approach to DRM [14: 23].

Critiquing the Civil Protection Act, Mavhura [17: 613] observes that "...terms such as resilience, vulnerability, mitigation, contingency plans and DRR are not appearing in all the 44 sections and numerous subsections of the Act." Despite the absence of a comprehensive disaster risk management legislation and framework, the Zimbabwe is guided by various legislative instruments that are unwittingly geared towards the achievement of the first Sendai Framework priority and this is "Understanding disaster risk". The priority highlights that effective disaster risk reduction policies and strategies are those mapped and implemented basing on risk assessments taken prior to the implementation of the policies. In other words, an understanding of what constitutes disaster risks paves the way for measures that may be implemented to curb the challenges. As such, the country has various institutions put in place to assess disaster risks. In this regard, the Government of Zimbabwe and UN [14] reveals that some risk identification and mapping initiatives have been undertaken by Government with support from partners. Examples of such initiatives include the Zimbabwe Resilience Building Fund (ZRBF) in 2015 and this was developed with the assistance from the World Food Programme (WFP) and the United Nations

Development Programme (UNDP). Moreover, in 2015 the Ministry of Agriculture, Mechanisation and Irrigation Development, with the support from the Food and Agricultural Organization (FAO) developed a national agricultural hazard risk profile that identifies the major hazards that affect the country. In addition, the MSD, through a network of fifty synoptic weather stations (32 of them being automatic weather stations) it controls, is able to map risks related to climate and disseminate the information to concerned stakeholders [14]. These institutions, as already established, and are functional, are repositories of disaster risk data that may be used to inform disaster risk policies in the country. However, the disintegrated approach to disaster management dampens such initiatives.

Conversely, Zimbabwe has taken some strides towards the third pillar of the Sendai Framework—Investing in resilience. This priority is highlighted in the Zimbabwe National Development Strategy (NDS) 1 (January 2021–December 2025). It is pointed out that the "NDS1 Period will prioritise increasing investment in Disaster Risk Management Systems including early warning systems and contingency planning to enhance country preparedness" (Ministry of Finance and Economic Development 2020: 51). In this vein, paragraph 853 of the NDS1 conveys that the following strategies will be implemented in line with DRR management: capacitating the Civil Protection Department; promoting linkages between the Meteorological Services Department (MSD) and the University of Zimbabwe for utilisation of the High Performance Computer; establishing and implementing a citizen feedback mechanism; and strengthening provision of psycho-social support mechanisms in the aftermath of disasters for affected communities, to reduce post disaster trauma. As such, these strategies do not address all the components of the disaster risk management cycle that include mitigation, preparedness, response, and recovery phases, but are mostly inclined towards the last two phases. Moreover, the NDS1 affirms the authority vested in the Civil Protection Act as the principal legislative instrument to manage disasters. Nonetheless, the [14] report indicates that since the MSD started using the High Performance Computer Centre, the accuracy of weather forecasts issued by the department has improved. Moreover, Nangombe [22] highlights several institutions that have the mandate for early warning in the agricultural sector so as to reduce the impact of droughts. The institutions include the following: the Meteorological Office, National Early Warning Unit (NEWU), Famine Early Warning systems Network (FEWSNET), Drought Monitoring Centre (DMC) and the Food and Nutrition Council in collaboration with ZIMVAC. On the other hand, the Government of Zimbabwe and UN [14] reveal that a web-based Disaster Information Management System (www.dcpzim.org) has been developed with support from IOM to serve as the information management (IM) platform for the Government. Additionally, although there is no legal basis to enforce full participation, the Postal and Telecommunications Authority of Zimbabwe (POTRAZ) reached an understanding with mobile network operators to dispatch early warning or early action messages to people at risk.

However, despite the elaborate early warning systems, the country suffers both protracted and sudden disasters. Nangombe [22] argues that the early warning system for drought in Zimbabwe at the national level is not very effective due to lack of coordination. This corresponds with the Government of Zimbabwe and UN [14: 14] findings that "Awareness on DRM is less common in line ministries that are not directly involved in drought, flood or health emergency response, thus indicating a real need for further awareness raising and training on DRM concepts". These problems may be attributed to a lack of a comprehensive, trans-disciplinary and holistic DRR framework to guide disaster management in the country.

6 Discussion

The fact that the Civil Protection Department falls under the Ministry of Local Government and Public Works portrays that disaster risk management in Zimbabwe does not take a holistic approach, and thus, is not effective. As such, the Department is answerable to the respective Minister. However, the other duties of the Minister include, for example, overseeing all local governments in the country, both rural and urban local authorities. Such a daunting task may sometimes overwhelm the capacity of the Minister's office such that other duties may be overlooked or flouted. In this case, issues to do with disaster management may be regarded as far-fetched from the day-to-day issues the Minister's office may be grappling with. However, Ministries such as the Ministry of Agriculture, and the Ministry of Environment and Climate Change, have portfolios that may relate more closely with proactive disaster risk management. Noteworthy is the fact that, for the Civil Protection Unit to fall under the Ministry of Local Government is a colonial legacy whereby during the colonial era, the management of disaster risks was referred to as civil defence and was meant to serve the interests of the white minority at national level [17]. Therefore, such a relic in disaster management cannot be effective in dealing with dynamic disaster risk issues.

As highlighted above, the socio-economy of the country is being guided by the NDS1 and this development blueprint does not give any reference to the Sendai Framework. This implies that the country does not realise the positive relationship between disaster risk reduction and sustainable development. The only reference to disaster management in the NDS1 is through highlighting the authority of the Civil Protection Department as promulgated by the Civil Protection Act. This legislation instrument, and the present basis of disaster risk management, is not enough to stir the country free of heavy tolls from disasters as it is lacking in many aspects of proactive disaster risk management. As such, the country is superficial in its commitment of reducing disaster risks (as evidenced by the country being a signatory) yet ignoring the stipulation of the Framework to achieve the enhancement of disaster preparedness, it is important to: "Prepare or review and periodically update disaster preparedness and contingency policies, plans and programmes with the involvement of the relevant institutions, considering climate change scenarios and their impact on disaster risk, and facilitating, as appropriate, the participation of all sectors and relevant stakeholders" (Paragraph 33 (a) of the Sendai Framework). In this regard, Oxfam [24] puts forward that in the case of Tropical Cyclone Idai, lack of institutional development was seen in the inadequate readiness of the Meteorological Services Department (MSD), the army, police, and in the education and health sectors in initiating effective preparation and response measures.

The disaster management legislation and accompanying policies and strategies fail to recognise local participation in disaster risk management in Zimbabwe. Most of the policies feed into national disaster reduction strategies, relegating the local aspect of disaster resilience. However, all the four pillars of the Sendai Framework, a disaster management framework that can usher nations into sustainable development through DRR, acknowledge the significance of the local hand in such endeavours. For example, it is pointed out that in achieving an understanding of disaster risks (the first Pillar), there is need to ensure the use of traditional, indigenous and local knowledge and practices so as to complement scientific knowledge in disaster risk assessment and the development and implementation of DRR policies. This is a necessary component of disaster risk management since the local communities would bear the responsibility of disaster risk management. On the other hand, as a means of investing in DRR for resilience (the third Pillar), the Framework highlights that there is need to promote the mainstreaming of disaster risk assessment, mapping and management into rural development planning and management of, inter alia, mountains, rivers, coastal flood plain areas, drylands, wetlands and all other areas prone to droughts and flooding…" (Paragraph 30(g) of the Sendai Framework). Such an incorporation of local people in rural areas is significant in the Zimbabwean context. Oxfam [24] observes that there is no integration of rural area spatial development planning and disaster risk management yet better spatial planning and settlement development is critical to resilience. As such, there is often a conflict of interest between traditional leaders and local authorities [4]. Because of such lack of coordination and inclusivity, the vulnerability of people to natural disasters such as flooding is compounded through increased exposure as was the case in Chimanimani on the wake of Tropical Cyclone Idai. Therefore, if local communities are not included in disaster risk management initiatives, the impact of disasters would increase. In this regard, WHO [31] iterates that many lives can be saved in the first hours after an emergency through effective local response, before external help arrives and this was evidenced during the landfall of Tropical Cyclone Idai in Chimanimani [4].

The Zimbabwean risk reduction system is characterised by a plethora of institutions, particularly early warning systems, without effective coordination. In this context, cases of moral hazard and principal-agent problems occur whereby particular institutions may pursue other duties, unwittingly delegating the early warning mandate to other government departments. For example, in the case of droughts, it was highlighted that the MSD, AGRITEX, ZINWA, National Early Warning Unit (NEWU), Famine Early Warning systems Network (FEWSNET), World food Programme (WFP), Drought Monitoring Centre (DMC) and the Food and Nutrition Council in collaboration with ZIMVAC monitor drought hazards and maintain early warning systems. As many institutions focus on one eminent issue, duplication of resources may lead to inefficiency. Moreover, the Government of Zimbabwe and UN [14: 17] remarks that "One major area that needs attention in Zimbabwe is improving early action because early warning is available but it is not translated into any meaningful preventive measures." This means that although the needed infrastructure and resources are available, the information gathered is not put to better use as evidenced by recurrent disasters. For example, the victims of Tropical Cyclone Idai reveal that the early warning systems were not elaborative enough to inspire them to migrate to safer areas before the cyclone struck [4]. As such, the early warning systems become ineffective owing to a gap in institutional frameworks to coordinate disaster risk management.

On the other hand, the various policies and strategies meant to mitigate disasters, particularly in the agricultural and environmental sectors, portray that disaster risk management in Zimbabwe is sectoral. As the country does not have an updated, all-hazards disaster management framework, particular sectors formulate their own disaster management blueprints meant for particular disasters. This is exemplified by the National Drought Plan for Zimbabwe that was drafted specifically for reducing the impact of droughts in Zimbabwe. However, the intervention measures proposed may also be used for other mitigation measures such as the development of irrigation schemes, that negates the need to have both the Water Policy and the Irrigation Policy, as it is, at present. On the other hand, the Rapid Impact and Needs Assessment (RINA) carried out by the government of Zimbabwe and other partners as part of the recovery process after the Tropical Cyclone Idai-induced disaster proposed recovery and resilience needs in such a sectoral basis [33]. Such sectoral disaster risk management is costly as it means spreading resources so as to curb one particular disaster risk. Oxfam [24: 11] comments that:

> In Zimbabwe, the agencies relevant to a holistic and high-performing emergency management framework do not always work in an integrated manner. They are not effectively connected to their sub-national centres for effective data gathering, information dissemination and communication.

Such lack of integration and a holistic approach to disaster risk management implies that disaster coordination will be curtailed as each sector will pursue its own blueprint in trying to mitigate the same disaster challenge.

7 Conclusion and Recommendations

Zimbabwe does not have an updated disaster risk management framework that is aligned to the Sendai Framework, an internal blueprint for disaster risk reduction upon that the country is a signatory member. As it is, the Sendai Framework was drafted with the aim to reduce disaster risks. As such, aligning Zimbabwe's disaster risk management legislature to the Framework will signify the country's sincerity in its commitment to reduce disaster risks. Since regulatory institutions are meant for signalling what actions are legal and otherwise, the passing of the Disaster Risk Management Bill into law will whip all the involved stakeholders into legal order, and thus propel the quest for DRR to greater success. Nonetheless, the Bill needs to be aligned to the Sendai Framework so as to make it a tool for future disaster risk management. As such the following factors need to be incorporated into the Bill: local community participation; coordinated early warning systems; and an all-hazards and holistic approach to disaster risk management.

References

1. Asian Development Bank (2018) Understanding disaster risk for advancing resilient development. Knowledge Note, Metro Manila
2. Azumah SB, Donkoh SA, Awuni JA (2018) The perceived effectiveness of agriculturaltechnology transfer methods: evidence from rice farmers in Northern Ghana. Cogent Food Agric 4:1–11
3. Bhaiseni B (2017) Poverty as a threat to environmental sustainability: implications for socialwork practice in Zimbabwe. Afr J Soc Work 7(2):25–29
4. Chanza N, Siyongwana PQ, Manyani A (2020) Closing the gaps in disaster management and response: drawing on local experiences with tropical cyclone Idai in Chimanimani, Zimbabwe. Int J Disaster Risk Sci 11:655–666
5. Coetzee C, van Niekerk D (2012) Tracking the evolution of the disaster management cycle: a general system theory approach. Ja 'mba': J Disaster Risk Stud 4(1):1–9
6. Crossman ND (2018) Drought resilience, adaptation and management policy (DRAMP) framework
7. Delve R, Guyo G, Hulufo H, Lembara S (2013) Building resilience: a guide to facilitating community managed disaster risk reduction in the horn of Africa. CRS and IIRR publishing
8. Department of Civil Protection (2012) Zimbabwe National contingency plan: 2012–2013. Government Printers, Harare
9. Dhehibi B, Rudiger U, Moyo HP, Dhraief MZ (2020) Agricultural technology transfer preferences of smallholder farmers in Tunisia's Arid Region. Sustainability 12(421):1–18
10. Dynes RR (1970) Organized behavior in disaster. Heath Lexington Books, Lexington, Mass
11. Government of Zimbabwe (1989) Civil protection act chapter 10:06. Government of Zimbabwe, Harare
12. Government of Zimbabwe (undated (a)) National drought plan for Zimbabwe. Government Printers, Harare
13. Government of Zimbabwe (undated (b)) Disaster risk reduction efforts in Zimbabwe. Government Printers, Harare
14. Government of Zimbabwe and UN (2017) Capacity assessment of the disaster risk management system in Zimbabwe. Capacity for Disaster Reduction Initiative, CADRI. https://www.GovernmentofZimbabweandUN.net/sites/default/files/Zimbabwe-Report-May-2017.pdf
15. Jeggle T (2013) Strategy report on disaster risk management, reduction and response in Mongolia. For The UNISDR/North-East Asia Office, Inchon, Republic of Korea, and The Office of the United Nations Resident Coordinator in Mongolia
16. Manyena B (2016) After Sendai: is Africa bouncing back or bouncing forward from disasters? Int J Disaster Risk Sci 7(1):41–53
17. Mavhura E (2016) Disaster legislation: a critical review of the civil protection act of Zimbabwe. Nat Hazards 80(1):605–621
18. Mavhura E, Manyena B, Collins AE (2017) An approach for measuring social vulnerability in context: the case of flood hazards in Muzarabani district, Zimbabwe. Available at:https://doi.org/10.1016/j.geoforum.2017.09.008
19. Mhlanga C, Muzingili T, Mpambela M (2019) Natural disasters in Zimbabwe: the primer for social work intervention. Afr J Soc Work 9(1):46–55
20. Ministry of Local Government (2009) Disaster management: a resource book for educational institutions in Zimbabwe. In: Megan A (ed). Civil Protection Organisation of Zimbabwe, Harare

21. Muzingili T (2016) Greening the profession: a missing perspective in contemporary socialwork practice in Zimbabwe. Afr J Soc Work 6(2):29–39
22. Nangombe SS (2015) Drought conditions and management strategies in Zimbabwe. In: Proceedings of the regional workshops on capacity development to support national drought management policies for Eastern and Southern Africa and the Near East and North Africa Regions. pp 84–89
23. National Disaster Management Committee, Republic of the Union of Myanmar (2017) Myanmar action plan for disaster risk reduction. Nay Pyi Taw
24. Oxfam International (2019) Tropical cyclone Idai in Zimbabwe: an analysis of policy implications for post-disaster institutional development to strengthen disaster risk management. Oxfam Briefing Paper—November 2019, Oxfam House, Oxford
25. UNDP (2016). Mapping of selected hazards affecting rural livelihoods in Zimbabwe. A district and ward analysis
26. UNDP Zimbabwe (2015) Building resilience In Zimbabwe: towards a resilience strategic framework. Government Printers, Harare
27. UNISDR (2009) Terminology on disaster risk reduction. United Nations Office for Disaster Risk Reduction, Geneva. https://www.unisdr.org/files/7817_UNISDRTerminologyEnglish.pdf
28. UNOCHA (2019) Zimbabwe humanitarian appeal revision: February 2019–April 2020. Government Printers, Harare
29. UNOCHA (2020) Zimbabwe situation report, 7 Aug 2020
30. WHO (2009) Cholera in Zimbabwe: epidemiological bulleting No. 27, Week 24 (7–13 June 2009), Harare
31. WHO (2019) Health emergency and disaster risk management framework. World Health Organization, Geneva
32. Wilkinson E, Twigg J, Weingärtner L, Peter K (2017) Delivering disaster risk reduction by 2030: pathways to progress. Overseas Development Institute, London
33. World Bank, Government of Zimbabwe and GFDRR (2019) Zimbabwe Rapid Impact andNeeds Assessment (RINA) report, May 2019
34. Zimbabwean Government Gazette Extraordinary (2021) Statutory instrument 47 of 2021: civil protection (declaration of state of disaster: Rural and Urban Areas of Zimbabwe) (road infrastructure network) notice, 2021. Government Printers, Harare

The manufacturer's authorised representative in the EU is Springer Nature Customer Service Centre GmbH, Europaplatz 3, 69115 Heidelberg, Germany. If you have any concerns regarding our products, please contact ProductSafety@springernature.com

Printed and bound by CPI Group (UK) Ltd, Croydon, CR0 4YY
01/10/2026
02225936-0007